OCR A level
Biology A

Second Edition

2

Sue Hocking
Frank Sochacki
Mark Winterbottom

ALWAYS LEARNING

PEARSON

Published by Pearson Education Limited, 80 Strand, London, WC2R 0RL.

www.pearsonschoolsandfecolleges.co.uk

Text © Pearson Education Limited 2015
Edited by Mary White, Jo Egré and Priscilla Goldby
Designed by Elizabeth Arnoux for Pearson Education Limited
Typeset by Tech-Set Ltd, Gateshead
Original illustrations © Pearson Education Limited 2015
Illustrated by Tech-Set Ltd, Gateshead and Peter Bull Art Studio
Cover design by Juice Creative
Picture research by Caitlin Swain and Ewout Buckens
Cover photo © **Science Photo Library**: Living Art Enterprises, LLC

The rights of Sue Hocking, Frank Sochacki and Mark Winterbottom to be identified as authors of this work have been asserted by them in accordance with the Copyright, Designs and Patents Act 1988.

First edition published 2008
This edition published 2015

21
10 9 8 7

British Library Cataloguing in Publication Data
A catalogue record for this book is available from the British Library

ISBN 978 1 447 99080 2

Websites
Pearson Education Limited is not responsible for the content of any external internet sites. It is essential for tutors to preview each website before using it in class so as to ensure that the URL is still accurate, relevant and appropriate. We suggest that tutors bookmark useful websites and consider enabling students to access them through the school/college intranet.

Printed in Italy at Lego S.p.A

This resource is endorsed by OCR for use with specification OCR Level 3 Advanced GCE in Biology A (H420).

In order to gain OCR endorsement this resource has undergone an independent quality check. OCR has not paid for the production of this resource, nor does OCR receive any royalties from its sale. For more information about the endorsement process please visit the OCR website www.ocr.org.uk

Acknowledgements

Pearson would like to thank Peter Kennedy for his contribution to the previous edition.

The publisher would like to thank the following for their kind permission to reproduce their photographs:

(Key: b-bottom; c-centre; l-left; r-right; t-top)

123RF.com: 123rf.com 243b, 270tl, Ewelina Kowalska 188tl, Mariusz Prusaczyk 279r, Ruud Morijn 296tr, tom tietz 10l; **Alamy Images:** Edwin Remsberg 189r, FLPA 189t, Linda Richards 207tl, Nigel Cattlin 82tl, 244, 295t, Nigel Cattlin 82tl, 244, 295t, Nigel Cattlin 82tl, 244, 295t, Simon Rawles 89r, Streeter Photography 222tl; **Corbis:** Pierre Vauthey / Sygma 279l, Robert Dowling 194r; **Digital Vision:** 40; **DK Images:** Dave King 115; **Fotolia.com:** adogslifephoto 209, alinamd 186r, blueringmedia 58, byrdyak 207bl, chokniti 274tr, Chris Brignell 195r, Circumnavigation 298r, dreamnikon 190b, Ewais 275, giedriius 186l, granitepeaker 298l, Grzegorz Piaskowski 295b, hal_pand_108 271t, kaphotokevm1 299, ktsdesign 44-45, LianeM 195l, micro_photo 293, Olha Rohulya 207tr, palm30 270br, pankajstock123 66l, pavel vashenkov 303, rufar 186-M, 205b, Tylinek 294, ueuaphoto 193; **Frank Sochacki:** 282, 283l; **Getty Images:** DEA Picture Library 18, Konrad Wothe 21, VINCE BUCCI / Staff 223r; **Pearson Education Ltd:** Malcolm Harris 194l, Trevor Clifford 256; **PhotoDisc:** S. Meltzer. Photolink. Photodisc 233br; **Science Photo Library Ltd:** Alfred Pasieka 22-23, Astrid & Hanns-Frieder Michler 154, Bill Barksdale / Agstockusa 274br, Biophoto Associate 187l, Biophoto Associates 20br, 66r, 103tr, 104l, 178, British Antarctic Survey 301, Chris Knapton 233t, Clouds Hill Imaging Ltd 160-161, Cnri 25, 68l, 76, Darryl Leja, Nhgri 211, David Parker 221l, David Scharf 248l, Dr Kari Lounatmaa 114tr, Dr Keith Wheeler 121r, Dr Kenneth R. Miller 117l, Dr P. Marazzi 171, Dr. John Brackenbury 15bl, Dr.Jeremy Burgess 277, Eye Of Science 137tr, 169, 242, Fletcher & Baylis 147t, Frans Lanting, Mint Images 288-289, Geoff Kidd 179, George Bernard 243-Ml, Hubert Raguet/Look At Sciences 130, J.C. Revy, Ism 81l, 81r, J.C. Revy, Ism 81l, 81r, James King-Holmes 237, Jim West 297, John Durham 112-113, Jon Stokes 128b, Karl H. Switak 15tr, Laguna Design 157, Lawrence Berkeley National Laboratory 122t, M.I. Walker 20cr, 86tr, Manfred Kage 24cr, 91, Marek Mis 114br, Mark Smith 137br, Martin Krzywinski 214-215, Martin Shields 78-79, 221r, 222bl, Martyn F. Chillmaid 88l, 283r, Maximilian Stock Ltd 248r, Microfield Scientific Ltd 148t, Microscape / Science Photo Library 104tr, National Library Of Medicine 180l, Omikron 82bl, Patrick Landmann 147b, Power And Syred 176-177, Professors P. Motta & T. Naguro 140tr, Ramon Andrade 3dciencia 167tr, Ria Novosti 88r, Richard Bizley 134-135, Russell Kightley 148b, Science Photo Library 205t, Sinclair Stammers 122b, Siu 24cl, UK Crown Copyright Courtesy Of Fera 224, W K Fletcher 10r; **Shutterstock.com:** baranq 8-9, BMJ 271bl, Cheryl E. Davis 270tr, Cynthia Farmer 270bl, Dario Sabljak. 102bl, Dr. Morley Read 255, Elena Larina 200tl, Jason Benz Bennee 265, javarman 300t, Kirill Kurashov. Shutterstock 233bl, Lakomanrus 243tr, Lestodd 252, lzf 89l, Maisna 296l, Maxim Golubchikov 251b, Maya2008 62-63, Olaf Speier 251t, Olinchuk 240-241, Paul Aniszewski 268-269, Piotr Marcinski 199l, Rhimage 274l, Ryan M. Bolton 300b, Sari ONeal 188bl, Tony Campbell 271br, zcw 243-MR, Zurijeta 199r

All other images © Pearson Education

We are grateful to the following for permission to reproduce copyright material:

Article on p.18 from Scientists solve the mystery of whether dinosaurs were hot or cold blooded - and reveal they were somewhere in between *Mail online*, 12/06/2014 (Mark Prigg), http://www.dailymail.co.uk/sciencetech/article-2656673/Scientists-solve-mystery-dinosaurs-hot-cold-blooded-reveal-between.html, MailOnline; Article on p.40 from Care 2, http://www.care2.com/greenliving/male-fish-turning-female-due-to-pollution.html Jake Richardson. 09/08/2010, Care2.com with permission; Article on p.58 adapted from Could a cure for MS and diabetes be on the way? *The Express* (Willey, Jo), http://www.express.co.uk/news/uk/506643/MS-Breakthrough-Cure-Diabetes; Extract on p.74 from © 2013 Steele et al. This is an open-access article distributed under the terms of the Creative Commons Attribution License Use of HbA1c in the Identification of Patients with Hyperglycaemia Caused by a Glucokinase Mutation: Observational Case Control Studies DOI: 10.1371/journal.pone.0065326 June 14, 2013 Anna M. Steele Kirsty J. Wensley, Sian Ellard, Rinki Murphy, Maggie Shepherd, Kevin Colclough, Andrew T. Hattersley, Beverley M. Shields; Article on p.108 adapted from Building tastier fruits and veggies (no GMOs required), *Scientific American*, 311:56-61; Investigation on p.118 adapted from Separating photosynthetic pigments using thin layer chromatography (TLC) adapted from SAPS protocol, adapted from the Science and Plants for Schools practical, available free from www.saps.org.uk; Article on p.130 from Tom Ireland: Algal biofuel – in bloom or dead in the water?, *The Biologist*, Vol 61, p.20–23 Feb/Mar 2014; Article on p.210 from 'Me, myself, us. The human microbiome – looking at humans as ecosystems that contain many collaborating species could change the practice of medicine', *The Economist*, by NEDERLANDS ECONOMISCH INSTITUUT. Reproduced with permission of SPRINGER NEW YORK LLC in the format Book via Copyright Clearance Center; Extract on p.236 from *50 genetics ideas you really need to know*, Quercus Publishing Inc. (Henderson, M 2008) p.52, © Mark Henderson 2009. Reproduced by permission of Quercus Editions Limited; Extract on p.239 adapted from http://www.sciencemag.org/content/302/5644/415.short, Science 17 October 2003: Vol. 302 no. 5644 pp. 415-419 DOI: 10.1126/science.1088547. Copyright © 2003, American Association for the Advancement of Science; Article on p.264 from How human cloning could cure diabetes, *The Telegraph*, 29/04/2014 (Sarah Knapton), http://www.telegraph.co.uk/news/science/science-news/10794029/How-human-cloning-could-cure-diabetes.html, © Telegraph Media Group Limited 2014; Article on p.284 from *New Scientist* (www.newscientist.com/article/dn25795-going-vegetarian-halves-co2-emissions-from-your-food.html#.VNZv8EfXerU), ©2014 Reed Business Information - UK; All rights reserved. Distributed by Tribune Content Agency; Article on p.304 adapted from *National Geographic*, www.nationalgeographic.com/news/2015/01/150127-antarctica-translucent-fish-microbes-ice.

Contents

Module 6
Genetics and ecosystems

How to use this book

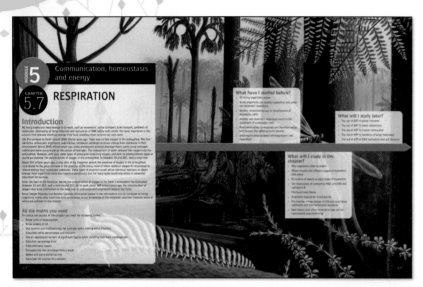

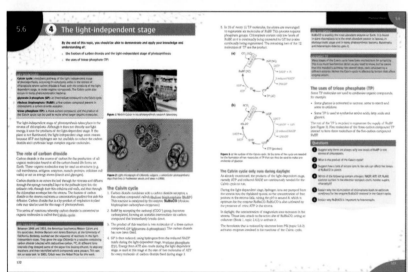

Welcome to your OCR A Level Biology A student book. In this book you will find a number of features designed to support your learning.

Chapter openers

Each chapter starts by setting the context for that chapter's learning.

- Links to other areas of Biology are shown, including previous knowledge that is built on in the chapter and future learning that you will cover later in your course.

- The **All the maths you need** checklist helps you to know what maths skills will be required.

Main content

The main part of the chapter covers all of the points from the specification you need to learn. The text is supported by diagrams and photos that will help you understand the concepts.

Within each topic, you will find the following features:

- **Learning objectives** at the beginning of each topic highlight what you need to know and understand.

- **Key terms** are shown in bold and defined within the relevant topic for easy reference.

- **Worked examples** show you how to work through questions, and how your calculations should be set out.

- **Investigations** provide a summary of practical experiments that explore key concepts.

- **Learning tips** help you focus your learning and avoid common errors.

- **Did you know?** boxes feature interesting facts to help you remember the key concepts.

At the end of each topic, you will find **questions** that cover what you have just learned. You can use these questions to help you check whether you have understood what you have just read, and to identify anything that you need to look at again. Answers to all questions in this student book are available at: http://www.pearsonschoolsandfecolleges.co.uk/Secondary/Science/16Biology/OCR-A-level-Science-2015/FreeResources/FreeResources.aspx

Thinking Bigger

At the end of each chapter there is an opportunity to read and work with real-life research and writing about science. These sections will help you to expand your knowledge and develop your own research and writing techniques. The questions and tasks will help you to apply your knowledge to new contexts and to bring together different aspects of your learning from across the whole course. The timeline at the bottom of the spread highlights which other chapters of your book the material relates to.

These spreads will give you opportunities to:

- read real-life material that's relevant to your course
- analyse how scientists write
- think critically and consider relevant issues
- develop your own writing
- understand how different aspects of your learning piece together.

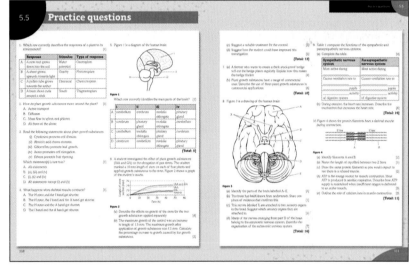

Practice questions

At the end of each chapter, there are practice questions to test how fully you have understood the learning.

Answers to all questions in this student book are available at: http://www.pearsonschoolsandfecolleges.co.uk/Secondary/Science/16Biology/OCR-A-level-Science-2015/FreeResources/FreeResources.aspx

Maths Skills

At the end of the book there is a **Maths Skills** section that focuses on key mathematical concepts to provide greater depth of explanation and enhance your understanding through worked examples.

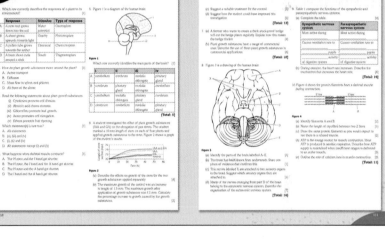

Preparing for your exams

The book concludes with a section that offers some practical advice about preparing for your exams, including sample questions and answers that allow you to see where common mistakes are made and how you can improve your responses.

Getting the most from your ActiveBook

Your ActiveBook is the perfect way to personalise your learning as you progress through your OCR A Level Biology course. You can:

- access your content online, anytime, anywhere
- use the inbuilt highlighting and annotation tools to personalise the content and make it really relevant to you
- search the content quickly.

Highlight tool

Use this to pick out key terms or topics so you are ready and prepared for revision.

Annotations tool

Use this to add your own notes, for example, links to your wider reading, such as websites or other files. Or make a note to remind yourself about work that you need to do.

CHAPTER

5.1

COMMUNICATION AND HOMEOSTASIS

Introduction

All organisms are able to respond to stimuli. Stimuli are changes in the environment that elicit a response. These could be:

- changes to the external environment, such as changes in temperature, sounds, or the appearance of a predator
- changes inside the organism, such as a change in temperature, pH or water potential.

A response can be brought about by communication between cells within the body. This is known as cell signalling. Communication may be chemical and/or electrical, and any suitable response will improve the chances of survival.

Internal conditions may change as a result of changing external conditions or as a result of cell activity. When internal conditions change, the new conditions could affect the activity of enzymes and, therefore, the metabolism in the body. Homeostasis is the maintenance of a constant internal environment despite any external changes. Homeostasis ensures that conditions inside the body remain at the optimum for enzyme activity. Communication between cells is fundamental to homeostasis and organisms use both chemical and electrical systems to monitor and respond to any deviation from the body's steady state.

All the maths you need

To unlock the puzzles of this chapter you need the following maths:

- Recognise and make use of appropriate units in calculations
- Recognise and use expressions in decimal and standard form
- Use ratios, fractions and percentages
- Estimate results
- Use an appropriate number of significant figures
- Construct and interpret frequency tables and diagrams, bar charts and histograms
- Understand and use the symbols: =, <, <<, >>, >, ~
- Calculate the circumferences, surface areas and volumes of regular shapes

What have I studied before?

- The meaning of homeostasis
- The use of hormones to control growth and metabolism in plants and animals
- The use of nerves to send messages around an animal's body quickly
- How mammals can regulate their body temperature
- The various physiological mechanisms in the skin to help regulate temperature
- The structure of the plasma membrane and its role as a selectively permeable barrier
- The role of the plasma membrane in cell signalling
- The structure of proteins
- The action of enzymes and factors that affect enzyme activity
- The role of the blood circulatory system in transport

What will I study later?

- How nerve cells carry messages
- How nerve cells communicate with each other
- How hormones are released
- How hormones act upon their target cells

What will I study in this chapter?

- The need for communication systems
- The need for a constant internal state (homeostasis)
- Maintenance of a constant internal state by negative feedback
- Temperature regulation in ectotherms and endotherms

By the end of this topic, you should be able to demonstrate and apply your knowledge and understanding of:

* the need for communication systems in multicellular organisms
* the communication between cells by cell signalling

KEY DEFINITION

cell signalling: the way in which cells communicate with each other.

Survival and activity

You may recall from your AS level work that cell metabolism relies on enzymes and that enzymes need a specific set of conditions in which to work efficiently. All living things need to maintain a certain limited set of conditions inside their cells. These include:

* a suitable temperature
* a suitable pH
* an aqueous environment that keeps the substrates and products in solution
* freedom from toxins and excess inhibitors.

Without these conditions the cells will become inactive and die. In multicellular organisms, cells are specialised and rely upon one another; therefore they must be able to communicate in order to coordinate their activities.

The threat from changing environments

Changing external environments

All living organisms have an external environment that consists of the air, water or soil around them. This external environment changes, which may place stress on the living organism. For instance, a cooler environment will cause greater heat loss. If the organism is to remain active and survive, the changes in the environment must be monitored and the organism must change its behaviour or physiology to reduce the stress. The environmental change is a **stimulus** and the way in which the organism changes its behaviour or physiology is its **response**.

The environment may change slowly as the seasons pass. These changes elicit a gradual response. For example, the arctic fox (*Alopex lagopus*) has a much thicker white coat in winter and a thinner grey/brown coat in summer. The change in the coat provides greater insulation and camouflage in winter, ensuring the animal can survive. Yet in summer, the animal does not overheat.

(a)

(b)

Figure 1 Arctic fox with (a) winter coat and (b) summer coat.

DID YOU KNOW?

The winter coat of the arctic fox is efficient enough to keep the animal warm while lying asleep. Therefore the fox may actually have a problem overheating when it runs around looking for prey.

However, the environment may change much more quickly. The appearance of a predator or moving from a burrow into the sunlight are rapid changes. Again, the change (a stimulus) must be monitored and the organism must respond to the change.

Changing internal environments

Most multicellular organisms have a range of tissues and organs. Many of the cells and tissues are not exposed to the external environment – they are protected by epithelial tissues and organs such as skin or bark. In many animals the internal cells and tissues are bathed in tissue fluid. This is the environment of the cells.

As cells undergo their various metabolic activities, they use up substrates and create new products. Some of these compounds may be unwanted or even toxic. These substances move out of the cells into the tissue fluid. Therefore, the activities of the cells alter their own environment.

For example, one waste product is carbon dioxide. If this is allowed to build up in the tissue fluid outside the cells, it will alter the pH of the tissue fluid and could disrupt the action of enzymes and other proteins. The accumulation of excess waste or toxins in this internal environment must act as a stimulus to cause removal of these waste products so that the cells can survive. In this example, the reduced pH of the blood stimulates greater breathing activity that expels the carbon dioxide from the body (see topic 5.5.9).

This build-up of waste products in the tissue fluid may also act directly on the cells, which respond by reducing their activities so that less waste is produced. However, this response may not be good for the whole organism.

Maintaining the internal environment

The composition of the tissue fluid is maintained by the blood. Blood flows throughout the body and transports substances to and from the cells. Any wastes or toxins accumulating in the tissue fluid are likely to enter the blood and be carried away. In order to prevent their accumulation in the blood they must be removed from the body by excretion (see topic 5.2.1).

It is important that the concentrations of waste products and other substances in the blood are monitored closely. This ensures that the body does not excrete too much of any useful substance but removes enough of the waste products to maintain good health. It also ensures that the cells in the body are supplied with the substrates they need.

Coordinating the activities of different organs

A multicellular organism is more efficient than a single-celled organism, because its cells are differentiated. This means that its cells are specialised to perform particular functions. Groups of cells specialised to perform a particular function form tissues and organs. The cells that monitor the blood may be in a different part of the body well away from the source of the waste product. They may also be some distance from the tissue or organ specialised to remove the waste from the body. Therefore, a good communication system is required to ensure that these different parts of the body work together effectively.

A good communication system will:

- cover the whole body
- enable cells to communicate with each other
- enable specific communication
- enable rapid communication
- enable both short-term and long-term responses.

Cell signalling

Cells communicate with each other by the process of **cell signalling**. This is a process in which one cell will release a chemical that is detected by another cell. The second cell will respond to the signal released by the first cell.

The two major systems of communication that work by cell signalling are the:

- **neuronal system:** an interconnected network of neurones that signal to each other across synapse junctions. The neurones can conduct a signal very quickly and enable rapid responses to stimuli that may be changing quickly.
- **hormonal system:** a system that uses the blood to transport its signals. Cells in an endocrine organ release the signal (a hormone) directly into the blood. The hormone is transported throughout the body, but is only recognised by specific target cells. The hormonal system enables longer-term responses to be coordinated.

You may recall from your AS level work that cell signalling involves molecules that have a specific shape which is complementary to that of the cell surface receptor. This is essential to enable signals to be specific.

> **LEARNING TIP**
>
> Remember that the shape of the signalling molecule is complementary to the shape of the receptor molecule. Do not say 'similar shape' or 'complementary to receptor'.

Questions

1. (a) State three examples of a stimulus and a corresponding response in animals.
   (b) For each of your examples, suggest whether it uses the neuronal system or the hormonal system for communication, and state why.

2. Describe two examples of a stimulus and response in plants.

3. List the organs that are associated with excretion.

4. List the organs that are associated with maintaining the internal environment of a mammal, stating what role each organ plays.

5. Explain why a good communication system must:
   (a) cover the whole body
   (b) enable specific communication
   (c) enable rapid communication, and
   (d) enable both short- and long-term responses.

By the end of this topic, you should be able to demonstrate and apply your knowledge and understanding of:

∗ the principles of homeostasis

KEY DEFINITIONS

effector: a cell, tissue or organ that brings about a response.
homeostasis: maintaining a constant internal environment despite changes in external and internal factors.
negative feedback: the mechanism that reverses a change, bringing the system back to the optimum.
positive feedback: the mechanism that increases a change, taking the system further away from the optimum.
sensory receptors: cells/sensory nerve endings that respond to a stimulus in the internal or external environment of an organism and can create action potentials.

Homeostasis

Homeostasis is used in many living organisms to maintain conditions inside the body, despite changes in external and internal factors. Aspects maintained by homeostasis may include:

- body temperature
- blood glucose concentration
- blood salt concentration
- water potential of the blood
- blood pressure
- carbon dioxide concentration.

The mechanism of homeostasis

Any response to changes in the environment requires a complex mechanism, which may involve a series of tissues and organs that are coordinated through cell signalling. The standard response pathway is:

stimulus → receptor → communication pathway (cell signalling) → effector → response

A number of specialised structures are required for this pathway to work:

- **Sensory receptors** such as temperature receptors. These receptors may be on the surface of the body, such as temperature receptors in the skin. They monitor changes in the external environment. Other receptors are internal to monitor conditions inside the body, for example, temperature receptors in the brain. When one of these receptors detects a change it will be stimulated to send a message to an effector.

- A communication system such as the neuronal system or the hormonal system. This acts by signalling between cells. It is used to transmit a message from the receptor cells to the

effector cells via a coordination centre which is usually in the brain. The messages from the receptor to the coordination centre are known as the input. The messages sent to the effectors are known as the output.

- **Effector** cells such as liver cells or muscle cells. These cells will bring about a response.

Feedback

When the effectors respond to the output from the coordination centre, they bring about a response that will change the conditions inside the body. Such changes will be detected by the receptors. This will have an effect upon the response pathway. In effect, the input will change. This effect is known as feedback.

Negative feedback

In order to maintain a constant internal environment, any change away from optimum conditions must be reversed. In this way, conditions inside the body will be returned to the optimum. This mechanism that brings the conditions back towards the optimum is known as **negative feedback** (see Figure 1).

When conditions change, the receptors detect this stimulus and send an input to the coordination centre. The coordination centre sends an output to the effectors and the effectors respond to this output. When the effectors bring about a change that reverses the initial change in conditions, the system moves closer to the optimum and the stimulus is reduced. The receptors detect the reduction in stimulus and reduce the input to the coordination centre. The output from the coordination centre to the effectors is also reduced, so the effectors reduce their activity. As the system gets closer to the optimum, the response is reduced.

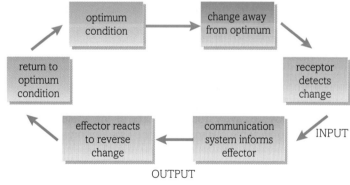

Figure 1 Negative feedback.

For example, if the internal temperature rises too high, the response is to do something that brings the body back towards its optimum temperature (see Figure 2). As a result the stimulus is reduced.

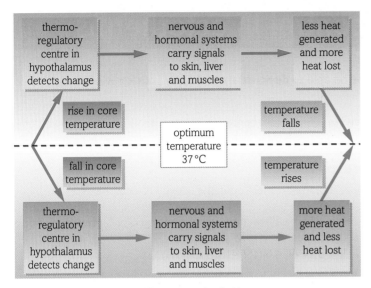

Figure 2 Temperature control by negative feedback.

For negative feedback to work, a number of processes must occur:

1. A change to the internal environment must be detected.

2. The change must be signalled to other cells.

3. There must be an effective response that reverses the change in conditions.

Maintaining a constant internal environment

A negative feedback system can maintain a reasonably constant set of conditions. However, the conditions will never remain perfectly constant: there will be some variation around the optimum condition. When a stimulus occurs it may take time to respond and the response may cause a slight 'overshoot'. However, as long as this variation is not too great, the conditions will remain acceptable. A thermostatically controlled heated room will never get too cold or too hot. Similarly, when negative feedback is applied to living systems, the conditions inside a living organism will remain within a relatively narrow range. The conditions will remain 'warm' enough to allow enzymes to continue functioning efficiently, but 'cool' enough to avoid damage to the body's many other proteins.

Positive feedback

Positive feedback is less common than negative feedback. When positive feedback occurs, the response is to increase the original change. This destabilises the system and is usually harmful.

For example, below a certain core body temperature enzymes become less active and the **exergonic** reactions that release heat are slower and release less heat. This allows the body to cool further and slows the enzyme-controlled reactions even more. This causes the body temperature to spiral downwards.

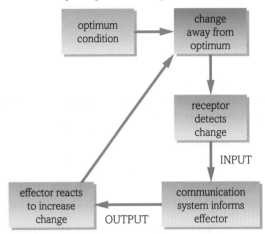

Figure 3 Positive feedback.

There are, however, some occasions when positive feedback can be beneficial. Positive feedback is used to stimulate an increase in a change.

An example is seen at the end of pregnancy to bring about dilation of the cervix. As the cervix begins to stretch this causes the posterior pituitary gland to secrete the hormone oxytocin. Oxytocin increases the uterine contractions which stretch the cervix more, which causes secretion of more oxytocin. Once the cervix is fully dilated, the baby can be born. The birth ends the production of oxytocin.

The activity of neurones also relies on positive feedback.

Questions

1. Define the term homeostasis.

2. The first paragraph in this topic lists six conditions that are maintained in homeostasis. For each condition explain why maintaining an optimum is essential.

3. What is the role of (a) the sensory receptors and (b) the effectors?

4. What is meant by input and output?

5. Explain the difference between negative feedback and positive feedback.

6. Describe and explain the effect of positive feedback on an animal if its body temperature rises too high.

By the end of this topic, you should be able to demonstrate and apply your knowledge and understanding of:

* the behavioural responses involved in temperature control in ectotherms

Controlling body temperature

Changes in body temperature can have a dramatic effect upon the activity of cell processes. As temperature rises, molecules have more kinetic energy. They move about more quickly and collide more frequently. This means that essential chemical reactions occur more quickly. However, in cooler conditions the opposite is true and chemical reactions slow down.

The structure of proteins can also be affected by changes – especially increases – in body temperature. Many proteins have a metabolic function; for example, enzymes increase the rate of biological reactions. Enzymes are globular proteins and have a very specific tertiary structure, giving them a specific three-dimensional shape. In the case of enzymes, the shape of the active site is complementary to the shape of the substrate and any change in shape will affect their ability to function normally. If temperature is allowed to increase too much, enzymes change shape and their function is lost.

Some enzymes are very sensitive to temperature change. If the body temperature drops by 10 °C, the rate of enzyme-controlled reactions falls by half. Many reactions in cells release heat, which can help to maintain the temperature, but if the temperature drops and reactions slow down, less heat is released. This allows the body to cool further. This is a form of positive feedback. As the body cools, the organism is less and less able to function normally. However, if the temperature rises just a few degrees above the optimum, enzymes may denature and cease to function.

The core temperature is the important factor, as all the vital organs are found in the centre of the body. Peripheral parts of the body may be allowed to increase or decrease in temperature to some extent without affecting the survival of the organism.

Endotherm or ectotherm?

Endotherms control their body temperature within very strict limits. They use a variety of mechanisms to control body temperature and are largely independent of external temperatures (see topic 5.1.4).

Ectotherms are not able to control their body temperature as effectively as endotherms. They rely on external sources of heat and their body temperature fluctuates with the external temperature. However, using various behavioural mechanisms, some ectotherms are able to control their body temperature in all but the most extreme conditions.

Temperature control in ectotherms

Ectotherms do not use internal energy sources to maintain their body temperature when cold. However, once they are active their muscle contractions will generate some heat from increased respiration. Temperature regulation relies upon behavioural responses that can alter the amount of heat exchanged with the environment.

If ectotherms are not warm enough, they try to absorb more heat from the environment. They may:

* move into a sunny area
* lie on a warm surface
* expose a larger surface area to the sun.

If ectotherms are too hot they try to avoid gaining more heat and try to increase heat loss to the environment. They may:

* move out of the sun
* move underground
* reduce the body surface exposed to the sun.

KEY DEFINITIONS

ectotherm: an organism that relies on external sources of heat to maintain body temperature.
endotherm: an organism that uses heat from metabolic reactions to maintain body temperature.

DID YOU KNOW?

Many ectotherms can successfully maintain their body temperature at over 30 °C. It is therefore not appropriate to call them 'cold-blooded'.

Some examples

Example	Behavioural adaptation	Benefit
Snake	Basks in the sun. In the UK, adders can often be found lying on an exposed path beside vegetation.	Absorbs heat directly from the sun.
Locust	In the early morning, locusts sit side-on to the sun exposing a large surface area, but at midday they face the sun head-on exposing a smaller surface area. They may also climb to the top of a plant at midday to get away from the soil surface.	In the cool morning they can absorb more heat, but at midday when the sun is hotter they absorb less heat. The soil surface gets hot and radiates heat; if the locust moves away from the soil it gains less heat from the soil.
	Increases both the rate of breathing and the depth of breathing movements when it is hot.	More water evaporates from the tracheal system, cooling the body.
Lizard	Many lizards use burrows or crevices between rocks. They will hide in the burrow during the hottest part of the day and the coolest part of the night.	An underground burrow tends to have a more stable temperature than the air. In the hottest part of the day it will be cooler in the burrow, but at night the burrow may be warmer than the air outside.
Horned lizard	Can change its shape by expanding or contracting its ribcage.	Expanding the ribcage increases the surface area exposed to the sun, so more heat can be absorbed.

Table 1 Behavioural adaptations of ectotherms to maintain body temperature.

Advantages and disadvantages of ectothermy

Advantages

Ectotherms rely on external sources of heat to keep warm. They do not use up energy to keep warm. Therefore:

- Less of their food is used in respiration.
- More of the energy and nutrients gained from food can be converted to growth.
- They need to find less food.
- They can survive for long periods without food.

Disadvantages

They are less active in cooler temperatures. This means that they are at risk from predators while they are cold and unable to escape and they cannot take advantage of food that is available while they are cold.

Figure 1 A locust will climb up a twig to get away from the hot soil.

Figure 2 A horned lizard basking in the sun.

DID YOU KNOW?

Bees are ectotherms. However, the temperature of a bee swarm or colony is successfully maintained at 35 °C. The bees keep open passages through the swarm to allow movement of air. Some worker bees move about to generate heat and others flap their wings at the entrances to the air passages causing the air to circulate.

Questions

1 Explain why the body temperature must be controlled.

2 Suggest why it is easier to catch a fly early in the morning than at midday.

3 Early in the morning dragonflies can often be seen flapping their wings but not taking off or flying. Suggest why they may flap their wings in this way.

4 Explain how basking on a hot rock in the sun can help a lizard to control its body temperature.

5 In the search to find ways of producing sufficient food for the human population, many scientists believe that insects may be the best source of protein. Explain why ectotherms can produce protein more efficiently than endotherms.

By the end of this topic, you should be able to demonstrate and apply your knowledge and understanding of:

* the physiological and behavioural responses involved in temperature control in endotherms

KEY DEFINITION

hypothalamus: the part of the brain that coordinates homeostatic responses.

Endotherms

Endotherms do not rely on external sources of heat – they can use physiological adaptations and behavioural means to control their body temperature.

Temperature regulation mechanisms

Temperature regulation relies on effectors in the skin and muscles. The skin is the organ in contact with the external environment. Therefore, many of the physiological adaptations to control body temperature involve the skin. The changes that take place in the skin alter the amount of heat being lost to the environment.

Many chemical reactions in the body are exergonic – they release energy in the form of heat. Endotherms can increase respiration (an exergonic reaction) in the muscles and liver simply to release heat – they are using some of their energy intake to stay warm. They also have other useful physiological mechanisms, such as directing blood towards or away from the skin to alter the amount of heat lost to the environment.

Tables 1 and 2 show the range of physiological and behavioural responses used by endothermic organisms to maintain their constant body temperature.

Behaviour if too hot	Behaviour if too cold
Hide away from sun in the shade or in a burrow.	Lie in the sun.
Orientate body to reduce surface area exposed to sun.	Orientate body towards sun to increase surface area exposed.
Remain inactive and spread limbs out to enable greater heat loss.	Move about to generate heat in the muscles or, in extreme conditions, roll into a ball shape to reduce surface area and heat loss.
Wet skin to use evaporation to help cool the body. Cats lick themselves and elephants spray water over their bodies.	Remain dry.

Table 2 Behavioural adaptations used by endotherms to maintain body temperature.

LEARNING TIP

The rate of respiration is maintained at a constant, as temperature is constant. Remember that more respiration means more glucose being fed into the respiration pathway to produce more heat.

Organ	Response if body too hot	Response if body too cold
Skin	Sweat glands secrete fluid onto the skin surface; as this evaporates it uses heat from the blood as the latent heat of vaporisation.	Less sweat is secreted, so less evaporation means less heat is lost.
	Hairs and feathers lie flat to reduce insulation and allow greater heat loss.	Hairs and feathers stand erect to trap air, which insulates the body.
	Vasodilation of arterioles and precapillary sphincters directs blood to the skin surface so more heat can be radiated away from the body (see Figure 1).	Vasoconstriction of arterioles and precapillary sphincters leading to skin surface (see Figure 1). Blood is diverted away from the surface of the skin and less heat is lost.
Gaseous exchange system	Some animals pant, increasing evaporation of water from the surface of the lungs and airways. Evaporation uses heat from the blood as the latent heat of vaporisation.	Less panting, so less heat is lost.
Liver	Less respiration takes place, so less heat is released.	Increased respiration in the liver cells means that more energy from food is converted to heat.
Skeletal muscles	Fewer contractions mean that less heat is released.	Spontaneous muscle contractions (shivering) release heat.
Blood vessels	Dilation to direct blood to the extremities so that more heat can be lost.	Constriction to limit blood flow to the extremities, so that blood is not cooled too much – this can lead to frostbite in extreme conditions.

Table 1 Physiological mechanisms used by endotherms to maintain body temperature.

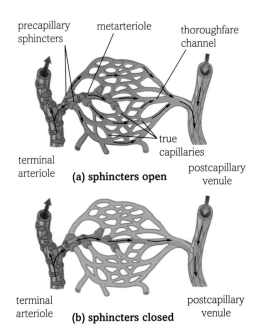

precapillary sphincters   metarteriole   thoroughfare channel

terminal arteriole   **(a) sphincters open**   postcapillary venule

true capillaries

terminal arteriole   **(b) sphincters closed**   postcapillary venule

Figure 1 (a) Vasodilation and (b) vasoconstriction.

Advantages and disadvantages of endothermy

Advantages

There are many advantages. Endotherms can:

- maintain a fairly constant body temperature whatever the temperature externally
- remain active even when external temperatures are low, which means they can take advantage of prey that may be available or escape from potential predators
- inhabit colder parts of the planet.

Disadvantages

However, there are disadvantages. Endotherms:

- use a significant part of their energy intake to maintain body temperature in the cold
- need more food
- use for growth a lower proportion of the energy and nutrients gained from food
- may overheat in hot weather.

DID YOU KNOW?

A shrew has to eat almost its own body mass of food each day to avoid starving to death.

Control of temperature regulation

The maintenance of a core body temperature is important. If the core temperature changes this alters the temperature of the blood. Temperature receptors in the **hypothalamus** of the brain

detect this change. The hypothalamus then sends out impulses to cause different responses that will reverse the change. Some responses need to be quick in order to prevent further change in body temperature – the neuronal system transmits the output from the hypothalamus in order to make these responses rapid. Other responses may need to be longer term; the hormonal system transmits the output to cause these responses.

If the core temperature is too low, the hypothalamus will bring about:

- changes in the skin to reduce heat loss
- release of heat through extra muscular contraction
- increased metabolism in order to release more heat from exergonic reactions.

If the core temperature rises above the optimum, the hypothalamus will bring about the opposite changes. This is an example of negative feedback (see Figure 2 in topic 5.1.2).

The role of peripheral temperature receptors

The thermoregulatory centre in the hypothalamus monitors blood temperature and detects changes in the core body temperature. However, an early warning that the body temperature may change could help the hypothalamus to respond more quickly and reduce variation in the core body temperature. If the extremities start to cool down or warm up this may eventually affect the core body temperature. Peripheral temperature receptors in the skin monitor the temperature in the extremities. This information is fed to the thermoregulatory centre in the hypothalamus. If the thermoregulatory centre signals to the brain that the external environment is very cold or very hot, the brain can initiate behavioural mechanisms for maintaining the body temperature, such as moving into the shade.

Questions

1. Explain how vasodilation and vasoconstriction in the skin help to control the body temperature.

2. Elephants have large ears. Explain the role of these ears in temperature regulation.

3. Penguins in Australia are about 25 cm in height, while those found in Antarctica can be over a metre tall. The size difference is an adaptation to the environment. Explain why these birds are so different in size.

4. Explain why a shrew needs to eat its own body weight per day, while an elephant can survive by eating less than one per cent of its body weight per day.

5. Suggest why the arctic fox has little or no fur on its lower legs in winter.

THINKING BIGGER

WHAT IS A MESOTHERM?

The view of dinosaurs in the Victorian period and the early part of the twentieth century was as large lizard-like animals that were slow and lumbering. In 1968 Robert T. Bakker, a young palaeontologist, wrote a brief article entitled 'The Superiority of Dinosaurs', in which he suggested that dinosaurs were 'fast, agile, energetic creatures'. Ever since then scientists have argued about whether dinosaurs were endotherms or ectotherms.

SCIENTISTS SOLVE THE MYSTERY OF WHETHER DINOSAURS WERE HOT OR COLD BLOODED

The hot question of whether dinosaurs were warm-blooded like birds and mammals or cold-blooded like reptiles, fish and amphibians finally has a good answer.

Dinosaurs, for eons Earth's dominant land animals until being wiped out by an asteroid 65 million years ago, were in fact somewhere in between.

Scientists evaluated the metabolism of numerous dinosaurs using a formula based on their body mass as revealed by the bulk of their thigh bones and their growth rates as shown by growth rings in fossil bones akin to those in trees.

The study, published in the journal *Science*, assessed 21 species of dinosaurs including super predators *Tyrannosaurus* and *Allosaurus*, long-necked *Apatosaurus*, duckbilled *Tenontosaurus* and bird-like *Troodon* as well as a range of mammals, birds, bony fish, sharks, lizards, snakes and crocodiles.

'Our results showed that dinosaurs had growth and metabolic rates that were actually not characteristic of warm-blooded or even cold-blooded organisms,' said University of Arizona evolutionary biologist and ecologist Brian Enquist.

'They did not act like mammals or birds nor did they act like reptiles or fish.

'Instead, they had growth rates and metabolisms intermediate to warm-blooded and cold-blooded organisms of today. In short, they had physiologies that are not common in today's world.'

There has been a long-standing debate about whether dinosaurs were slow, lumbering cold-blooded animals – as scientists first proposed in the 19th century – or had a uniquely advanced, more warm-blooded physiology.

As scientists unearthed remains of more and more fast-looking dinosaurs like *Velociraptor*, some championed the idea dinosaurs were as active and warm-blooded as mammals and birds.

The realisation that birds arose from small feathered dinosaurs seemed to support that view.

University of New Mexico biologist John Grady said the idea that creatures must be either warm-blooded or cold-blooded is too simplistic when looking over the vast expanse of time.

'Like dinosaurs, some animals alive today like the great white shark, leatherback sea turtle and tuna do not fit easily into either category', Grady added.

'A better answer would be 'in the middle.' By examining animal growth and rates of energy use, we were able to reconstruct a metabolic continuum, and place dinosaurs along that continuum. Somewhat surprisingly, dinosaurs fell right in the middle,' Grady said.

Meet the mesotherms

The researchers called creatures with this medium-powered metabolism mesotherms, as contrasted to ectotherms (cold-blooded animals with low metabolic rates that do not produce much heat and bask in the sun to warm up) and endotherms (warm-blooded animals that use heat from metabolic reactions to maintain a high, stable body temperature).

Grady said an intermediate metabolism may have allowed dinosaurs to get much bigger than any mammal ever could.

Warm-blooded animals need to eat a lot so they are frequently hunting or munching on plants. 'It is doubtful that a lion the size of *T. rex* could eat enough to survive,' Grady said.

Source
- Prigg, M. (2014) Scientists solve the mystery of whether dinosaurs were hot or cold blooded. *Daily Mail* (http://www.dailymail.co.uk/sciencetech/article-2656673/Scientists-solve-mystery-dinosaurs-hot-cold-blooded-reveal-between.html)

Where else will I encounter these themes?

| Book 1 | 5.1 | YOU ARE HERE | 5.2 | 5.3 | 5.4 | 5.5 | 5.6 |

Let's start by considering the nature of the writing in the article.

1. The writer uses the terms 'warm-blooded' and 'cold-blooded'. Explain why scientists no longer use this language.
2. The term 'mesotherm' is used to describe the dinosaurs. Explain what it means.

Now we will look at the biology in, or connected to, this article. Don't worry if you are not ready to give answers to these questions yet. You may like to return to the questions once you have covered other topics later in the book. Use the timeline at the bottom of the page to help you to put this work in context with what you have already learned and what is ahead in your course.

DID YOU KNOW?

Very large organisms can maintain their body temperature much more easily than smaller organisms because of their small surface area to volume ratio. Thus, large dinosaurs may have been able to maintain a high body temperature despite being ectothermic. This phenomenon is known as 'gigantothermy'.

3. Animals obtain energy for living processes from their food. What process inside cells releases energy from food? (Note: this is covered in detail later in this book.)
4. What organelle is involved in releasing energy from food?
5. Cells release energy in the form of ATP. What additional element do cells need in order to produce a lot of ATP?
6. How is the circulatory system of mammals adapted to supply sufficient substrates to enable their cells to produce enough heat to stay warm?
7. List the physiological mechanisms that mammals use to stay warm that are not found in reptiles.
8. What were the main advantages to dinosaurs of being able to keep their body temperature higher than their surroundings?
9. Use the theory of evolution by natural selection to explain how this feature may have evolved.

Refer back to chapter 2.1 in your AS level course if you need to.

Think about how substrate delivery is made more efficient.

Activity

Devise a flow diagram to explain how scientists work using the debate over dinosaur body temperature as an example. Your diagram should explain why new ideas are not always accepted at first but may become accepted once more research and evidence is accumulated.

Your flow diagram should explain why a scientist:

- makes observations
- suggests a hypothesis
- attempts to explain the hypothesis with scientific knowledge
- sets out to test the hypothesis
- conducts an investigation to find evidence
- reports the findings of the investigation in peer journals
- invites other scientists to comment or repeat the investigation.

1. All responses follow a pattern. Which row correctly identifies this pattern? [1]

   A. stimulus – sensory neurone – relay neurone – motor neurone – response

   B. stimulus – sensory receptor – coordination – effector – response

   C. sensory neurone – central nervous system – motor neurone – effector – response

   D. sensory receptor – sensory neurone – motor neurone – effector – response

2. The following are examples of feedback.

   (i) When core temperature gets too high, the animal will behave in a way that brings its core temperature down.

   (ii) When the water potential of the blood rises, the animal produces more urine.

   (iii) If the core temperature drops too low, enzyme activity decreases so metabolism releases less heat.

   (iv) When a blood vessel is damaged, platelets start to cling to the injured site and release chemicals that attract more platelets.

   (v) A low pH in the blood causes increased ventilation, which removes carbon dioxide from the blood.

   Which row correctly identifies examples of positive and negative feedback? [1]

Row	Positive feedback	Negative feedback
A	iii only	i, ii, iv and v
B	v only	i, ii, iii and iv
C	iii and iv	i, ii and v
D	iii and v	i, ii and iv

3. The following statements describe an animal's responses to temperature change:

   (i) increased sweating

   (ii) hides in a burrow

   (iii) hair becomes erect

   (iv) vasodilation in the skin

   (v) basks in the sun.

   Which row correctly identifies the responses shown by ectotherms and endotherms? [1]

Row	Ectotherms only	Endotherms only	Both
A		i, iii and iv	ii and v
B	v	i, iii and iv	ii
C	ii and v	i, iii and iv	
D	i, iii and iv	ii and v	

4. Sensory receptors are known as transducers. What is the role of a sensory receptor? [1]

   A. To inform the brain of changes in the environment

   B. To alter the local environment after a change

   C. To release hormones

   D. To convert stimuli into nervous impulses

5. What is the correct definition of negative feedback? [1]

   A. A response that reverses a change

   B. A response that enhances a change

   C. A response that removes the stimulus

   D. A response that reduces a stimulus

   [Total: 5]

6. (a) Define the term 'homeostasis'. [3]

   (b) List three organs involved in homeostasis and state their role in homeostasis. [6]

   (c) Explain why positive feedback is not normally a feature of homeostatic mechanisms. [3]

   [Total: 12]

7. (a) State precisely where the centre for temperature regulation is situated in a mammal. [1]

(b) Describe the role in homeostasis of the centre for temperature regulation. [3]

(c) List three responses to a rise in body temperature seen in an endotherm such as a human. [3]

(d) You are trekking at high altitude in the Himalayas when your guide warns that a storm is approaching and it will be accompanied by a sudden drop in temperature. You do not have time to walk to the nearest village. Suggest what you should do to prepare, and justify your decision. [4]

[Total: 11]

8. Figure 1 shows the core temperature of two camels that were kept together during the day. Camel A had recently been allowed access to water, but camel B had not had water.

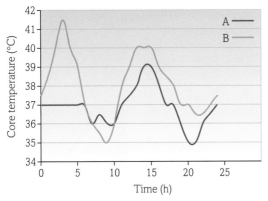

Figure 1

(a) Suggest why the graph records core temperature rather than skin temperature. [2]

(b) Describe the temperature changes seen in camel A throughout the 24 hour period. [3]

(c) Suggest at what time external temperatures were at their highest. [1]

(d) Explain why the core temperature of camel B fluctuates more than that of camel A. [3]

[Total: 9]

9. (a) The table below shows some responses to changes in core body temperature in ectotherms.

Explain how each response helps to regulate body temperature. [4]

Response	Explanation
A lizard will go underground or hide in a crevice when temperatures are high.	
Adders are venomous snakes found in the UK. They can often be found in sunny patches on a footpath.	
The horned lizard can expand its rib cage when lying in the sun.	
Early in the morning a dragonfly can be seen sitting on a plant stem flapping its wings.	

(b) Describe the role of skin temperature receptors in mammals. [3]

[Total: 7]

10. Negative feedback is used to regulate body temperature. Use your knowledge of negative feedback to describe the possible responses of a small mammal such as the jerboa (Figure 2), which lives in deserts where the temperature can fluctuate between freezing and 40 °C. [8]

Figure 2 A jerboa.

[Total: 8]

Communication, homeostasis and energy

EXCRETION AS AN EXAMPLE OF HOMEOSTATIC CONTROL

Introduction

Excretion is the removal of metabolic waste from the body. Certain waste products, such as carbon dioxide and urea, are released from cells as a result of normal metabolism. If these metabolic wastes were allowed to accumulate, they would alter the conditions within the body. Therefore, removal of these waste molecules is essential and makes a major contribution to homeostasis.

Examples of excretion include the removal of:

- carbon dioxide from the blood, by the lungs
- urea from the blood, by the kidneys
- substances in the bile produced by the liver.

The organs involved in excretion also have other roles in homeostasis:

- The liver metabolises toxins that have been consumed, such as alcohol.
- The liver metabolises molecules, such as hormones and drugs, that have entered the bloodstream.
- The kidneys also play an important role in regulating the water potential of the blood.

All the maths you need

To unlock the puzzles of this chapter you need the following maths:

- Recognise and make use of appropriate units in calculations
- Recognise and use expressions in decimal and standard form
- Use ratios, fractions and percentages
- Estimate results
- Use an appropriate number of significant figures
- Construct and interpret frequency tables and diagrams, bar charts and histograms
- Understand and use these symbols: $=$, $<$, $\ll$, $\gg$, $>$, $\sim$
- Translate information between graphical, numerical and algebraic forms

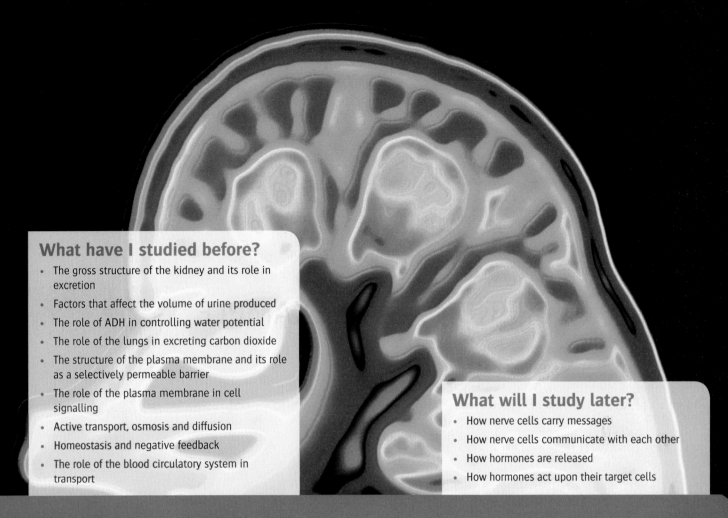

What have I studied before?

- The gross structure of the kidney and its role in excretion
- Factors that affect the volume of urine produced
- The role of ADH in controlling water potential
- The role of the lungs in excreting carbon dioxide
- The structure of the plasma membrane and its role as a selectively permeable barrier
- The role of the plasma membrane in cell signalling
- Active transport, osmosis and diffusion
- Homeostasis and negative feedback
- The role of the blood circulatory system in transport

What will I study later?

- How nerve cells carry messages
- How nerve cells communicate with each other
- How hormones are released
- How hormones act upon their target cells

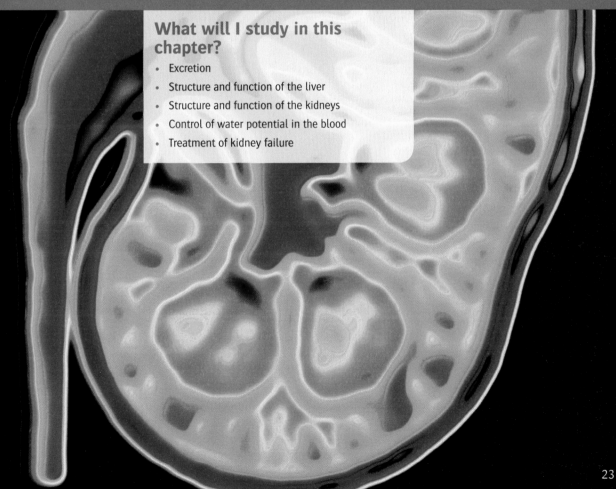

What will I study in this chapter?

- Excretion
- Structure and function of the liver
- Structure and function of the kidneys
- Control of water potential in the blood
- Treatment of kidney failure

Excretion

By the end of this topic, you should be able to demonstrate and apply your knowledge and understanding of:

* the term excretion and its importance in maintaining metabolism and homeostasis

> **KEY DEFINITIONS**
> **excretion:** the removal of metabolic waste from the body.
> **metabolic waste:** a substance that is produced in excess by the metabolic processes in the cells; it may become toxic.

Excretion is the removal of **metabolic waste** from the body. This means the removal from the body of the unwanted products of cell metabolism.

What products must be excreted?

Many substances need to be excreted. Almost all products that are formed in excess by the chemical processes occurring in the cells must be removed from the body, so that they do not build up and inhibit enzyme activity or become toxic. The main excretory products are:

* carbon dioxide from respiration

* nitrogen-containing compounds, such as urea (i.e. nitrogenous waste)

* other compounds, such as the bile pigments found in faeces.

> **LEARNING TIP**
> Do not confuse excretion with egestion. Egestion is the elimination of faeces from the body. Faeces are the undigested remains of food and are not metabolic products.

The excretory organs

The lungs

Every living cell in the body produces carbon dioxide as a result of respiration. Carbon dioxide is passed from the cells of respiring tissues into the bloodstream, where it is transported (mostly in the form of hydrogencarbonate ions) to the lungs. In the lungs the carbon dioxide diffuses into the alveoli to be excreted as you breathe out (see Book 1, topic 3.2.8).

The liver

The liver is directly involved in excretion. It has many metabolic roles and some of the substances produced will be passed into the bile for excretion with the faeces, for example, the pigment bilirubin (see topics 5.2.2 and 5.2.3).

The liver is also involved in converting excess amino acids to urea. Amino acids are broken down by the process of deamination. The nitrogen-containing part of the molecule is then combined with carbon dioxide to make urea.

The kidneys

The urea is passed into the bloodstream to be transported to the kidneys. Urea is transported in solution – dissolved in the plasma. In the kidneys the urea is removed from the blood to become a part of the urine (see topics 5.2.4–5.2.6). Urine is stored in the bladder before being excreted from the body via the urethra.

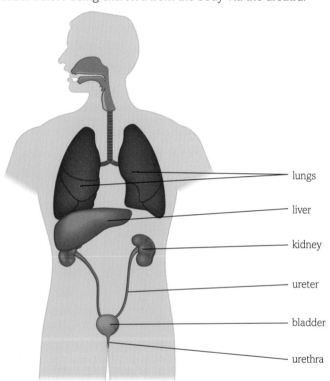

Figure 1 The main organs involved in excretion.

The skin

The skin is also involved in excretion, but excretion is not the primary function of the skin. Sweat contains a range of substances including salts, urea, water, uric acid and ammonia. Urea, uric acid and ammonia are all excretory products. The loss of water and salts may be an important part of homeostasis – maintaining the body temperature and the water potential of the blood (see topics 5.1.2 and 5.1.4).

The importance of excretion

Allowing the products of metabolism to build up could be fatal. Some metabolic products such as carbon dioxide and ammonia are toxic. They interfere with cell processes by altering the pH, so that normal metabolism is prevented. Other metabolic products may act as inhibitors and reduce the activity of essential enzymes.

Carbon dioxide

Most carbon dioxide is transported in the blood as hydrogencarbonate ions. However, from your AS level work you may recall that forming hydrogencarbonate ions also forms hydrogen ions:

$$CO_2 + H_2O \rightarrow H_2CO_3 \text{ (carbonic acid)}$$

The carbonic acid dissociates to release hydrogen ions:

$$H_2CO_3 \rightarrow H^+ + HCO_3^-$$

This occurs inside the red blood cells, under the influence of the enzyme carbonic anhydrase, but can also occur in the blood plasma.

The hydrogen ions affect the pH of the cytoplasm in the red blood cells. The hydrogen ions interact with bonds within haemoglobin, changing its three-dimensional shape. This reduces the affinity of haemoglobin for oxygen, affecting oxygen transport. The hydrogen ions can then combine with haemoglobin, forming haemoglobinic acid. The carbon dioxide that is not converted to hydrogencarbonate ions can combine directly with haemoglobin, producing carbaminohaemoglobin. Both haemoglobinic acid and carbaminohaemoglobin are unable to combine with oxygen as normal – reducing oxygen transport further.

In the blood plasma, excess hydrogen ions can reduce the pH of the plasma. Maintaining the pH of the blood plasma is essential, because changes could alter the structure of the many proteins in the blood that help to transport a wide range of substances around the body. Proteins in the blood act as buffers to resist the change in pH.

If the change in pH is small then the extra hydrogen ions are detected by the respiratory centre in the medulla oblongata of the brain. This causes an increase in the breathing rate to help remove the excess carbon dioxide.

However, if the blood pH drops below 7.35 it may cause headaches, drowsiness, restlessness, tremor and confusion. There may also be a rapid heart rate and changes in blood pressure. This is respiratory acidosis. It can be caused by diseases or conditions that affect the lungs themselves, such as emphysema, chronic bronchitis, asthma or severe pneumonia. Blockage of the airway due to swelling, a foreign object, or vomit can also induce acute respiratory acidosis.

DID YOU KNOW?

People taking statins to reduce blood cholesterol are advised not to drink grapefruit juice. Some components of grapefruit juice bind to the enzymes that break down statins in the liver. This inhibits the enzymes and leads to increased concentrations of statins in the body.

Nitrogenous compounds

The body cannot store excess amino acids. However, amino acids contain almost as much energy as carbohydrates. Therefore, it would be wasteful simply to excrete excess amino acids. Instead they are transported to the liver and the potentially toxic amino group is removed (deamination). The amino group initially forms the very soluble and highly toxic compound, ammonia. This is converted to a less soluble and less toxic compound called urea, which can be transported to the kidneys for excretion. The remaining keto acid can be used directly in respiration to release its energy or it may be converted to a carbohydrate or fat for storage.

Deamination: amino acid + oxygen → keto acid + ammonia

Formation of urea: ammonia + carbon dioxide →   urea   + water

$$2NH_3 \quad + \quad\quad CO_2 \quad \rightarrow (NH_2)_2CO + H_2O$$

Questions

1. List the excretory products found in sweat.

2. What effects could the build-up of excess salts have in the blood?

3. Explain how the waste products of cell metabolism can affect enzyme action.

4. Explain why the majority of waste eliminated as faeces is not considered to be excretion.

5. Suggest why fish can excrete nitrogenous waste in the form of ammonia, while mammals convert it to urea.

By the end of this topic, you should be able to demonstrate and apply your knowledge and understanding of:

* the structure and functions of the mammalian liver
* the examination and drawing of stained sections to show the histology of liver tissue

Blood supply to the liver

The liver cells (hepatocytes) carry out many hundreds of metabolic processes, so the liver has an important role in homeostasis (see topics 5.1.2 and 5.2.3). It is therefore essential that the liver has a good supply of blood. The internal structure of the liver ensures that as much blood as possible flows past as many liver cells as possible. This enables the liver cells to remove excess or unwanted substances from the blood and return substances to the blood to ensure concentrations are maintained.

Blood flow to and from the liver

The liver is supplied with blood from two sources:

* **The hepatic artery**: Oxygenated blood from the heart travels from the aorta via the hepatic artery into the liver. This supplies the oxygen that is essential for aerobic respiration (see Chapter 5.7). The liver cells are very active, because they carry out many metabolic processes. Many of these processes require energy, in the form of ATP, so it is important that the liver has a good supply of oxygen for aerobic respiration.

* **The hepatic portal vein**: Deoxygenated blood from the digestive system enters the liver via the hepatic portal vein. This blood is rich in the products of digestion. The concentrations of various substances will be uncontrolled as they have just entered the body from the products of digestion in the intestines. The blood may also contain toxic compounds that have been absorbed from the intestine. It is important that such substances do not continue to circulate around the body before their concentrations have been adjusted.

Blood leaves the liver via the hepatic vein. The hepatic vein rejoins the vena cava and the blood returns to the body's normal circulation.

A fourth vessel is connected to the liver. However, it is not a blood vessel – it is the bile duct. Bile is a secretion from the liver which has functions in digestion and excretion. The bile duct carries bile from the liver to the gall bladder, where it is stored until required to aid the digestion of fats in the small intestine. Bile also contains some excretory products such as bile pigments like bilirubin, which will leave the body with the faeces.

Histology of the liver

Structure of the liver

The cells, blood vessels and chambers inside the liver are arranged to ensure the greatest possible contact between the blood and the liver cells. The liver is divided into lobes which are further divided into lobules. The lobules are cylindrical.

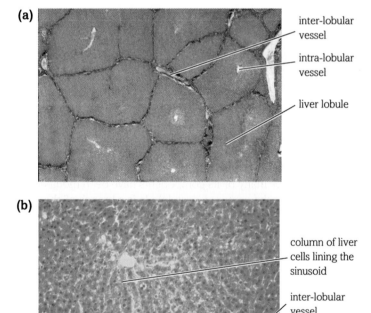

(a)
- inter-lobular vessel
- intra-lobular vessel
- liver lobule

(b)
- column of liver cells lining the sinusoid
- inter-lobular vessel

Figure 2 (a) Liver lobules (×20) and (b) the arrangement of cells in the lobule (×45).

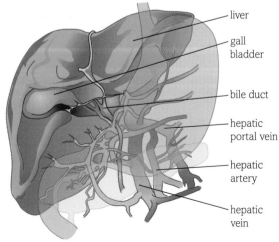

- liver
- gall bladder
- bile duct
- hepatic portal vein
- hepatic artery
- hepatic vein

Figure 1 The supply of blood to the liver.

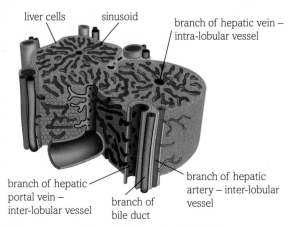

Figure 3 Arrangement of liver cells into cylindrical lobules.

As the hepatic artery and hepatic portal vein enter the liver, they split into smaller and smaller vessels. These vessels run between and parallel to the lobules – they are known as inter-lobular vessels. At intervals, branches from the hepatic artery and the hepatic portal vein enter the lobules. The blood from the two blood vessels is mixed and passes along a special chamber called a sinusoid, which is lined with liver cells. As the blood flows along the sinusoid it is in close contact with the liver cells. These cells are able to remove substances from the blood and return other substances to the blood.

DID YOU KNOW?

Portal vessels are unusual because they have capillaries at each end. There are also portal vessels in parts of the brain to facilitate hormonal communication.

Specialised macrophages called Kupffer cells move about within the sinusoids. Their primary function appears to be to breakdown and recycle old red blood cells. One of the products of haemoglobin breakdown is bilirubin, which is one of the bile pigments excreted as part of the bile.

Bile is made in the liver cells and released into the bile canaliculi. The bile canaliculi join together to form the bile duct, which transports the bile to the gall bladder.

When the blood reaches the end of the sinusoid, the concentrations of many of its components have been modified and regulated. At the centre of each lobule is a branch of the hepatic vein known as the intra-lobular vessel. The sinusoids empty into this vessel. The branches of the hepatic vein, from different lobules, join together to form the hepatic vein, which drains blood from the liver.

LEARNING TIP

Many cell specialisations are not obvious – many cells specialise for different functions by having larger numbers of particular organelles rather than by adopting a specialised shape.

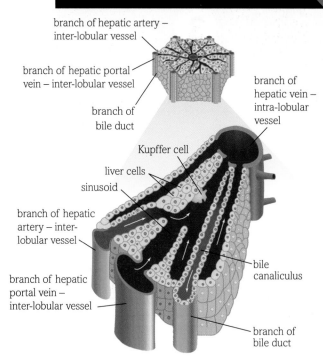

Figure 4 Arrangement of cells in a lobule.

Liver cells

Liver cells, or hepatocytes, appear to be relatively unspecialised. They have a simple cuboidal shape with many microvilli on their surface. However, their many metabolic functions include protein synthesis, transformation and storage of carbohydrates, synthesis of cholesterol and bile salts, detoxification and many other processes. This means that their cytoplasm must be very dense and is specialised in the numbers of certain organelles that it contains.

Questions

1. Explain why the liver has two supplies of blood.

2. Why might blood in the hepatic portal vein contain toxins?

3. Explain why the concentrations of substances in the blood of the hepatic portal vein may change during the day.

4. Use a flow diagram to describe the route taken by blood as it flows from the hepatic artery to the hepatic vein.

5. Describe how the structure of the liver ensures that blood flows past as many liver cells as possible.

6. Suggest which organelles may be particularly common in the cytoplasm of hepatocytes.

(3) Liver function

By the end of this topic, you should be able to demonstrate and apply your knowledge and understanding of:

* the structure and functions of the mammalian liver

Many metabolic functions

The liver is metabolically very active and carries out a wide range of functions, including:

* control of blood glucose levels, amino acid levels, lipid levels
* synthesis of bile, plasma proteins, cholesterol
* synthesis of red blood cells in the fetus
* storage of vitamins A, D and B12, iron, glycogen
* detoxification of alcohol, drugs
* breakdown of hormones
* destruction of red blood cells.

Storage of glycogen

The liver stores sugars in the form of glycogen. It is able to store approximately 100–120 g of glycogen, which makes up about 8% of the fresh weight of the liver. The glycogen forms granules in the cytoplasm of the hepatocytes. This glycogen can be broken down to release glucose into the blood as required. The control of blood glucose concentrations is described in topic 5.4.4.

Detoxification

One important role of the liver is to detoxify substances that may cause harm. Some of these compounds, such as hydrogen peroxide, are produced in the body. Others, such as alcohol, may be consumed as a part of our diet or may be taken for health or recreational reasons, for example, medicines and recreational drugs.

Toxins can be rendered harmless by oxidation, reduction, methylation or by combination with another molecule. Liver cells contain many enzymes that render toxic molecules less toxic.

These include:

* **catalase**, which converts hydrogen peroxide to oxygen and water. Catalase has a particularly high turnover number (the number of molecules of hydrogen peroxide that one molecule of catalase can render harmless in one second) of five million.
* **cytochrome P450**, which is a group of enzymes used to breakdown drugs including cocaine and various medicinal drugs. The cytochromes are also used in other metabolic reactions such as electron transport during respiration. Their role in metabolising drugs can interfere with other metabolic roles and cause the unwanted side effects of some medicinal drugs.

Detoxification of alcohol

Alcohol, or ethanol, is a drug that depresses nerve activity. In addition, alcohol contains chemical potential energy, which can be used for respiration.

Alcohol is broken down in the hepatocytes by the action of the enzyme ethanol dehydrogenase. The resulting compound is ethanal. This is dehydrogenated further by the enzyme ethanal dehydrogenase. The final compound produced is ethanoate (acetate). This acetate is combined with coenzyme A to form acetyl coenzyme A, which enters the process of aerobic respiration. The hydrogen atoms released from alcohol are combined with another coenzyme, called NAD, to form reduced NAD (see topic 5.7.4).

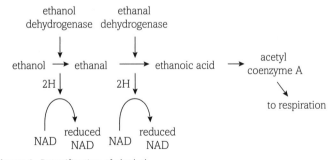

Figure 1 Detoxification of alcohol.

NAD is also required to oxidise and breakdown fatty acids for use in respiration (see topic 5.7.8). If the liver has to detoxify too much alcohol, it uses up its stores of NAD and has insufficient left to deal with the fatty acids. These fatty acids are then converted back to lipids and stored as fat in the hepatocytes, causing the liver to become enlarged. This is a condition known as 'fatty liver', which can lead to alcohol-related hepatitis or to cirrhosis.

Formation of urea

Every day we each need 40–60 g of protein. However, most people in developed countries eat far more than this. Excess amino acids cannot be stored, because the amino groups make them toxic. However, the amino acid molecules contain a lot of energy, so it would be wasteful to excrete the whole molecule. Therefore excess amino acids undergo treatment in the liver to remove and excrete the amino component. This treatment consists of two processes: deamination followed by the **ornithine cycle**.

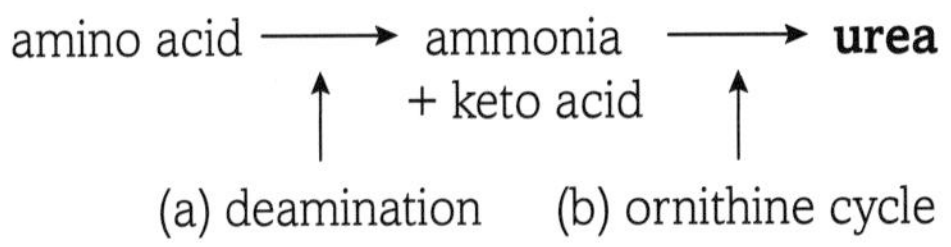

amino acid $\longrightarrow$ ammonia $\longrightarrow$ **urea**
$\quad\quad\quad\quad\quad\quad\uparrow$ + keto acid $\quad\uparrow$
$\quad\quad\quad\quad$ (a) deamination $\quad$ (b) ornithine cycle

Figure 2 Formation of urea.

Deamination

The process of deamination removes the amino group and produces ammonia. Ammonia is very soluble and highly toxic. Therefore, ammonia must not be allowed to accumulate. Deamination also produces an organic compound, a keto acid, which can enter respiration directly (see topic 5.2.1) to release its energy.

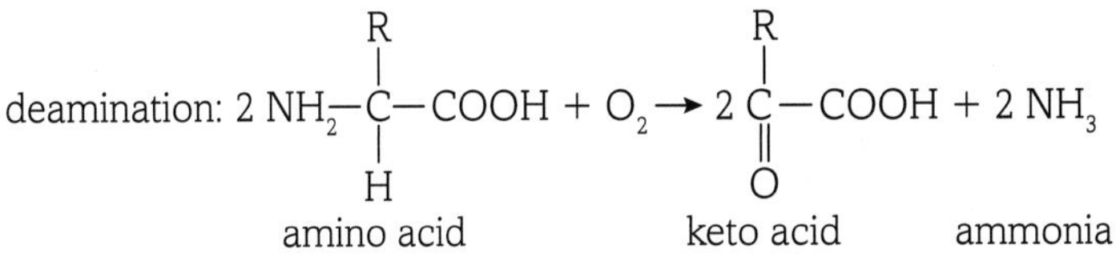

deamination: $2\,NH_2\!-\!\underset{\underset{H}{|}}{\overset{\overset{R}{|}}{C}}\!-\!COOH + O_2 \longrightarrow 2\,\underset{\underset{O}{\|}}{\overset{\overset{R}{|}}{C}}\!-\!COOH + 2\,NH_3$
$\quad\quad\quad\quad\quad\quad\quad$ amino acid $\quad\quad\quad\quad$ keto acid $\quad\quad$ ammonia

Figure 3 Deamination.

The ornithine cycle

Because ammonia is so soluble and toxic, it must be converted to a less toxic form very quickly. The ammonia is combined with carbon dioxide to produce urea. This occurs in the ornithine cycle. Ammonia and carbon dioxide combine with the amino acid ornithine to produce citrulline. This is converted to arginine by addition of further ammonia. The arginine is then re-converted to ornithine by the removal of urea.

Urea is both less soluble and less toxic than ammonia. It can be passed back into the blood and transported around the body to the kidneys. In the kidneys the urea is filtered out of the blood and concentrated in the urine. Urine can be safely stored in the bladder until it is released from the body.

The ornithine cycle can be summarised as:

$$2NH_3 \quad + \quad CO_2 \quad \rightarrow \quad CO(NH_2)_2 + H_2O$$
ammonia + carbon dioxide $\rightarrow$ $\quad$ urea $\quad$ + water

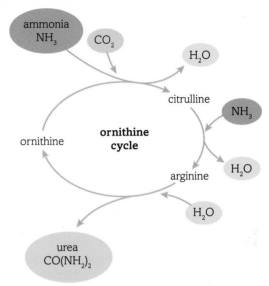

Figure 4 The ornithine cycle.

Questions

1. How can the sugars released from glycogen be used?

2. Explain why substances such as alcohol and excess amino acids should not simply be excreted.

3. Explain why it is essential that ammonia is converted to urea as quickly as possible.

4. Suggest why the hepatocytes contain many mitochondria.

5. Suggest why the hepatocytes contain many ribosomes.

4 Kidney structure

By the end of this topic, you should be able to demonstrate and apply your knowledge and understanding of:

* the structure, mechanisms of action and functions of the mammalian kidney

* the dissection, examination and drawing of the external and internal structure of the kidney

* the examination and drawing of stained sections to show the histology of nephrons (continued in topics 5.2.5 and 5.2.6)

> **KEY DEFINITIONS**
> **nephron:** the functional unit of the kidney.
> **ultrafiltration:** filtration of the blood at a molecular level under pressure.

The structure of the kidney

Most people have two kidneys. These are positioned on each side of the spine, just below the lowest rib. Each kidney is supplied with blood from a renal artery and is drained by a renal vein.

The role of the kidneys is excretion. The kidneys remove waste products from the blood and produce urine. The urine passes out of the kidney down the ureter to the bladder where it can be stored until it is released.

In a longitudinal section you can see that the kidney consists of three regions surrounded by a tough capsule.

* The outer region is called the cortex.
* The inner region is called the medulla.
* The centre is the pelvis, which leads into the ureter.

Fine structure of the kidney

The bulk of each kidney consists of tiny tubules called **nephrons**. Each kidney contains about one million nephrons. Each nephron starts in the cortex at a cup-shaped structure called the Bowman's capsule. The remainder of the nephron is a coiled tubule that passes through the cortex, forms a loop down into the medulla and back to the cortex, before joining a collecting duct that passes back down into the medulla.

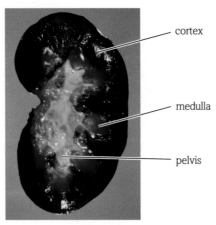

Figure 1 A kidney in longitudinal section.

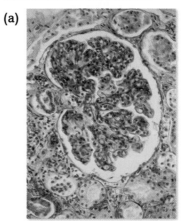

(a)

(b)

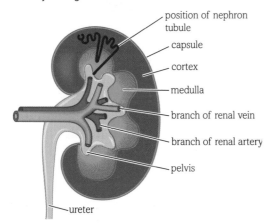

Figure 2 A drawing of a kidney in longitudinal section.

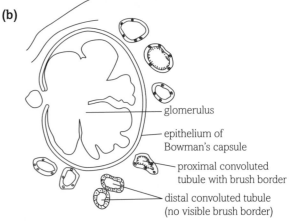

Figure 3 (a) A section through the cortex showing Bowman's capsule and (b) how this might be drawn from a slide.

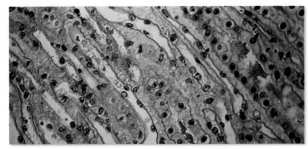

Figure 4 A section through the medulla showing tubules.

Blood supply and filtering

The renal artery splits to form many afferent arterioles, which each lead to a knot of capillaries called the glomerulus. Blood from the glomerulus continues into an efferent arteriole which carries the blood to more capillaries surrounding the rest of the tubule. These capillaries eventually flow together into the renal vein.

Each glomerulus is surrounded by the Bowman's capsule. Fluid from the blood is pushed into the Bowman's capsule by the process of **ultrafiltration**.

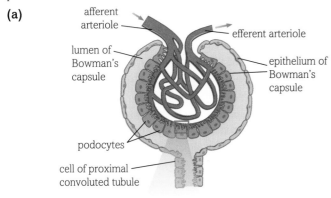

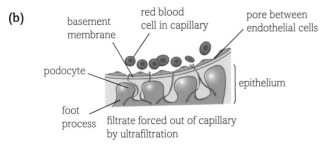

Figure 5 Detail of (a) the glomerulus and (b) Bowman's capsule. Part (b) is an enlargement of the area outlined in blue in part (a).

The filter is the barrier between the blood in the capillary and the lumen of the Bowman's capsule. This barrier consists of three layers, all adapted to enable ultrafiltration:

- **The endothelium of the capillary:** There are narrow gaps between the cells of the endothelium of the capillary wall. The cells of the endothelium also contain pores, called fenestrations. The gaps allow blood plasma and the substances dissolved in it to pass out of the capillary.

- **The basement membrane:** This membrane consists of a fine mesh of collagen fibres and glycoproteins. This mesh acts as a filter to prevent the passage of molecules with a relative molecular mass of greater than 69 000. This means that most proteins (and all blood cells) are held in the capillaries of the glomerulus.

- **The epithelial cells of the Bowman's capsule:** These cells, called podocytes, have a specialised shape – they have many finger-like projections, called major processes. On each major process are minor processes or foot processes that hold the cells away from the endothelium of the capillary. These projections ensure that there are gaps between the cells. Fluid from the blood in the glomerulus can pass between these cells into the lumen of the Bowman's capsule.

The Bowman's capsule leads into the rest of the tubule, which has three parts:

- proximal convoluted tubule
- loop of Henle
- distal convoluted tubule.

The fluid from many nephrons enters the collecting ducts, which pass down through the medulla to the pelvis at the centre of the kidney.

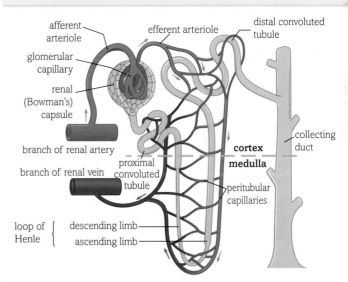

Figure 6 A nephron.

Questions

1. Eight to ten litres of blood pass into the kidneys every hour. Suggest why so much blood must pass through the kidneys.

2. Suggest why the proximal and distal tubules are convoluted.

3. Explain why each nephron is surrounded by capillaries.

4. Explain why reabsorption from the nephron must be selective.

5. Why is it essential that most of the water is reabsorbed from urine?

By the end of this topic, you should be able to demonstrate and apply your knowledge and understanding of:

* the structure, mechanisms of action and functions of the mammalian kidney
* the dissection, examination and drawing of the external and internal structure of the kidney
* the examination and drawing of stained sections to show the histology of nephrons (continued in topic 5.2.6)

Ultrafiltration

Ultrafiltration is the filtering of blood at the molecular level. Blood flows into the glomerulus through the afferent arteriole, which is wider than the efferent arteriole that carries blood away from the glomerulus. The difference in diameters ensures that the blood in the capillaries of the glomerulus maintains a pressure higher than the pressure in the Bowman's capsule. This pressure difference tends to push fluid from the blood into the Bowman's capsule that surrounds the glomerulus.

What is filtered out of the blood?

Blood plasma containing dissolved substances is pushed under pressure from the capillary into the lumen of the Bowman's capsule. The blood plasma contains the following substances:

* water
* amino acids
* glucose
* urea
* inorganic mineral ions (sodium, chloride, potassium).

The concentrations of dissolved solutes will depend upon the water balance in the organism and are therefore variable. The effect of ultrafiltration can be seen in Table 1.

Substance	Concentration in plasma of glomerulus (g dm^{-1})	Concentration in glomerular filtrate (g dm^{-1})	Concentration in urine (g dm^{-1})
Proteins	80.0	0.005	Variable but higher than 0.005
Amino acids	0.5	0.5	0.0
Glucose	1.0	1.0	0.0
Urea	0.3	0.3	Typically 8–10
Mineral ions	7.2	7.2	Typically 3–4

Table 1 A comparison of solute concentrations in blood plasma, glomerular filtrate and urine.

What is left in the capillary?

Blood cells and proteins are left in the capillary. The presence of proteins means that the blood has a very low (very negative) water potential. This ensures that some of the fluid is retained in the blood, and this contains some of the water and dissolved substances listed above. The very low water potential of the blood in the capillaries is important to help reabsorb water at a later stage.

The function of nephrons

As the fluid from the Bowman's capsule passes along the nephron tubule, its composition is altered by selective reabsorption – substances are absorbed back into the tissue fluid and blood capillaries surrounding the nephron.

* In the proximal convoluted tubule, the fluid is altered by the reabsorption of all the sugars, most mineral ions and some water. In total, about 85% of the fluid is reabsorbed here. The cells of these tubules have a highly folded surface producing a brush border which increases the surface area.
* In the descending limb of the loop of Henle, the water potential of the fluid is decreased by the addition of mineral ions and the removal of water.
* In the ascending limb of the loop of Henle, the water potential is increased as mineral ions are removed by active transport.
* In the collecting duct, the water potential is decreased again by the removal of water. The final product in the collecting duct is urine.

This process ensures that the final product (urine) has a low water potential. The urine therefore has a higher concentration of solutes than is found in the blood and tissue fluid. Urine passes into the pelvis and down the ureter to the bladder (see topic 5.2.6).

Selective reabsorption

Reabsorption involves active transport and cotransport. The cells lining the proximal convoluted tubule are specialised to achieve this reabsorption.

* The cell surface membrane in contact with the tubule fluid is highly folded to form microvilli. The microvilli increase the surface area for reabsorption.
* The cell surface membrane also contains special cotransporter proteins that transport glucose or amino acids, in association with sodium ions, from the tubule into the cell.

- The opposite membrane of the cell, close to the tissue fluid and blood capillaries, is also folded to increase its surface area. This membrane contains sodium/potassium pumps that pump sodium ions out of the cell and potassium ions into the cell.
- The cell cytoplasm has many mitochondria. This indicates that an active, or energy-requiring, process is involved, because many mitochondria will produce a lot of ATP.

The mechanism of reabsorption

Figure 1 explains how glucose and amino acids are selectively reabsorbed. The movement of sodium ions and glucose into the cell is driven by the concentration gradient created by pumping sodium ions out of the cell. The sodium ions move into the cell by facilitated diffusion but they cotransport glucose or amino acids against their concentration gradient. This is sometimes called secondary active transport.

The movement of these substances reduces the water potential of the cells so that water is drawn in from the tubule by osmosis. As the substances move through to the blood, the water follows.

DID YOU KNOW?

Most proteins are not filtered out of the blood, so proteins in the urine can indicate kidney disease. Some small proteins may enter the urine and can be used to diagnose conditions, for example, hCG is a polypeptide hormone that is detected in a pregnancy test.

LEARNING TIP

The details of membrane structure and cell specialisation are AS level topics – but they could be tested at A level standard in a question about the kidney.

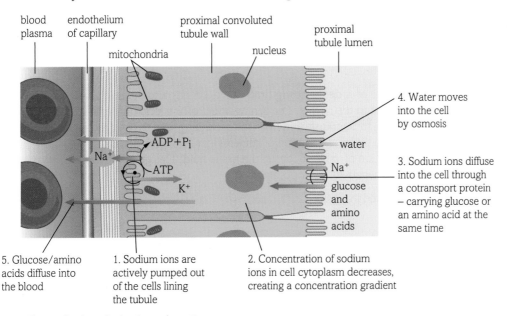

Figure 1 The mechanism of selective reabsorption.

Larger molecules, such as small proteins that may have entered the tubule, can be reabsorbed by endocytosis (Book 1, topic 2.5.4).

Questions

1. Explain what is meant by ultrafiltration.

2. Explain why it is not correct to say that the pressure in the glomerulus is raised.

3. Explain why the glomerular filtrate contains no large proteins.

4. Explain why the concentrations of glucose and amino acids are the same in the glomerular filtrate as in the blood plasma.

5. Explain why the concentration of proteins in urine is higher than that in the glomerular filtrate.

6. What may happen if water is not reabsorbed from the nephron?

By the end of this topic, you should be able to demonstrate and apply your knowledge and understanding of:

* the structure, mechanisms of action and functions of the mammalian kidney

* the dissection, examination and drawing of the external and internal structure of the kidney

* the examination and drawing of stained sections to show the histology of nephrons

Reabsorption of water

Each minute about 125 cm³ of fluid is filtered from the blood and enters the nephrons. After selective reabsorption in the proximal convoluted tubule, about 45 cm³ of fluid is left. By the time this fluid reaches the bladder, the volume has dropped to about 1.5 cm³.

The loop of Henle

The loop of Henle consists of a **descending limb** that descends into the medulla and an **ascending limb** that ascends back out to the cortex. The arrangement of the loop of Henle allows mineral ions (sodium and chloride ions) to be transferred from the ascending limb to the descending limb. The overall effect is to increase the concentration of mineral ions in the tubule fluid, which has a similar effect upon the concentration of mineral ions in the tissue fluid. This gives the tissue fluid in the medulla a very low (very negative) water potential.

The mechanism is explained in Figure 1.

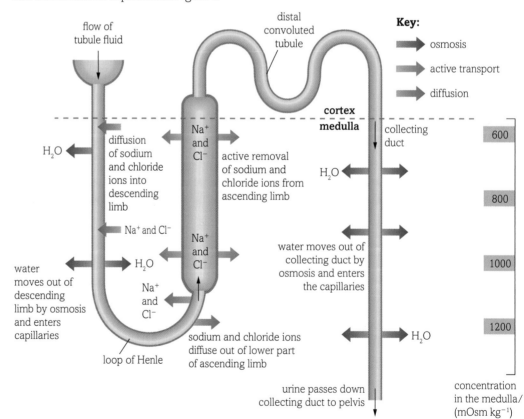

Figure 1 Reabsorption of water in the loop of Henle and the collecting duct.

As mineral ions enter the descending limb, the concentration of the fluid in the descending limb rises. This means that its water potential decreases (becomes more negative). It becomes increasingly more negative the deeper the tubule descends into the medulla.

As the fluid rises up the ascending limb, mineral ions leave the fluid. At the base this movement is by diffusion. However, higher up the ascending limb active transport is used to move mineral ions out. The upper portion of the ascending limb is also impermeable to water.

The effect of these ionic movements is to create a higher water potential in the fluid of the ascending limb. It also decreases the water potential in the tissue fluid of the medulla. The water potential of the tissue fluid becomes lower (much more negative) towards the bottom of the loop of Henle.

As fluid passes down the collecting duct, it passes through tissues with an ever-decreasing water potential. Therefore, there is always a water potential gradient between the fluid in the collecting duct and that in the tissues. This allows water to be moved out of the collecting duct and into the tissue fluid by osmosis.

The arrangement of the loop of Henle is known as a hairpin countercurrent multiplier system. The overall effect of this arrangement is to increase the efficiency of transfer of mineral ions from the ascending limb to the descending limb, in order to create the water potential gradient seen in the medulla.

The collecting duct

From the top of the ascending limb the tubule fluid passes along a short distal convoluted tubule, where active transport is used to adjust the concentrations of various mineral ions. From here the fluid flows into the collecting duct. At this stage the tubule fluid still contains a lot of water – it has a high water potential. The collecting duct carries the fluid back down through the medulla to the pelvis. Remember that the tissue fluid in the medulla has a low water potential that becomes even lower deeper into the medulla. As the tubule fluid passes down the collecting duct, water moves by osmosis from the tubule fluid into the surrounding tissue. It then enters the blood capillaries by osmosis, and is carried away.

The amount of water that is reabsorbed depends on the permeability of the collecting duct walls. Only about 1.5–2.0 dm^3 of fluid (urine) reaches the pelvis each day. By the time the urine reaches the pelvis, it has a low (very negative) water potential and the concentration of minerals and urea is higher than in blood.

Concentration changes in the tubule fluid

Figure 2 shows the changes in concentration of various substances as the fluid passes along the nephron and collecting duct.

1. Glucose decreases in concentration as it is selectively reabsorbed from the proximal tubule.

2. Sodium ions diffuse into the descending limb of the loop of Henle, causing the concentration to rise. They are then pumped out of the ascending limb, so the concentration falls.

3. The urea concentration rises as water is withdrawn from the tubule. Urea is also actively moved into the tubule.

4. Sodium ions are removed from the tubule, but their concentration rises as water is removed from the tubule, and potassium ions increase in concentration as water is removed. Potassium ions are also actively transported into the tubule to be removed in urine.

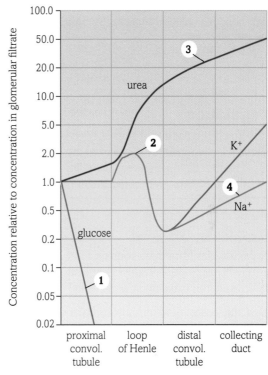

Figure 2 The changes in concentration as fluid passes along the nephron.

Questions

1. Why must the collecting duct pass back through a region of low water potential?

2. Why is it important for terrestrial mammals to reabsorb as much water as possible?

3. Explain why the kidney of a camel has very long loops of Henle.

4. Suggest why the kidney of a beaver has short loops of Henle.

5. Look at Figure 2.
   (a) Explain why the concentration of urea rises as the fluid passes along the distal tubule.
   (b) Explain why the concentration of sodium stays constant as the fluid passes along the proximal tubule.

By the end of this topic, you should be able to demonstrate and apply your knowledge and understanding of:

* the control of the water potential of the blood

KEY DEFINITIONS

antidiuretic hormone (ADH): a hormone that controls the permeability of the collecting duct walls.
osmoreceptor: a sensory receptor that detects changes in water potential.

Osmoregulation

Osmoregulation is the control of the water potential in the body. Water potential is the tendency of water to move from one place to another. Osmoregulation involves controlling levels of both water and salt in the body. The correct water balance between cells and the surrounding fluids must be maintained to prevent water entering cells and causing lysis or leaving cells and causing crenation (see Book 1, topic 2.5.3).

The body gains water from three sources: food, drink and metabolism (e.g. respiration). Water is lost from the body in urine, sweat, water vapour in exhaled air, and faeces.

These gains and losses of water must be balanced. The kidneys act as an effector to control the water content of the body and the salt concentration in the body fluids. On a cool day or when you have drunk a lot of fluid, the kidneys will produce a large volume of dilute urine. Alternatively, on a hot day or when you have drunk very little, the kidneys will produce smaller volumes of more concentrated urine.

The mechanism of osmoregulation

The kidneys alter the volume of urine produced by altering the permeability of the collecting ducts. The walls of the collecting ducts can be made more or less permeable according to the needs of the body:

* If you need to conserve less water (on a cool day or when you have drunk a lot of fluid), the walls of the collecting ducts become less permeable. This means that less water is reabsorbed and a greater volume of urine will be produced.
* If you need to conserve more water (on a hot day or when you have drunk very little), the collecting duct walls are made more permeable so that more water can be reabsorbed into the blood. You will produce a smaller volume of urine.

DID YOU KNOW?

The water potential of the blood is not kept completely level – it oscillates around the normal level according to the uptake and loss of water from the body. The feeling of thirst is a sign that the water potential has become too low.

Altering the permeability of the collecting duct

The cells in the walls of the collecting duct respond to the level of **antidiuretic hormone (ADH)** in the blood. These cells have membrane-bound receptors for ADH. The ADH binds to these receptors and causes a chain of enzyme-controlled reactions inside the cell (an example of cell signalling). The end result of these reactions is to cause vesicles containing water-permeable channels (aquaporins) to fuse with the cell surface membrane. This makes the walls more permeable to water.

When the level of ADH in the blood rises, more water-permeable channels are inserted. This allows more water to be reabsorbed, by osmosis, into the blood. Less urine is produced and the urine has a lower water potential.

If the level of ADH in the blood falls, then the cell surface membrane folds inwards (invaginates) to create new vesicles that remove water-permeable channels from the membrane. This makes the walls less permeable and less water is reabsorbed, by osmosis, into the blood. More water passes on down the collecting duct to form a greater volume of urine which is more dilute (higher water potential).

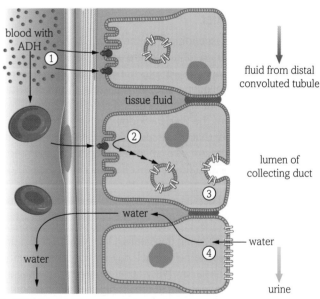

1. ADH detected by cell surface receptors
2. Enzyme-controlled reactions
3. Vesicles containing water-permeable channels (aquaporins) fuse to membrane
4. More water can be reabsorbed

Figure 1 The effect of ADH on the wall of the collecting duct.

Adjusting the concentration of ADH in the blood

The hypothalamus in the brain contains specialised cells called **osmoreceptors**. These are the sensory receptors that detect the stimulus – they monitor the water potential of the blood. These cells respond to the effects of osmosis. When the water potential of the blood is low (very negative), the osmoreceptor cells lose water by osmosis and shrink. As a result they stimulate neurosecretory cells in the hypothalamus.

The neurosecretory cells are specialised neurones (nerve cells) that produce and release ADH. The ADH is manufactured in the cell body, which lies in the hypothalamus. ADH moves down the axon to the terminal bulb in the posterior pituitary gland, where it is stored in vesicles. When the neurosecretory cells are stimulated by the osmoreceptors, they carry action potentials down their axons and cause the release of ADH by exocytosis (see Book 1, topic 2.5.4), as shown in Figure 2.

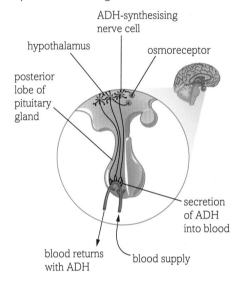

Figure 2 Osmoreceptors in the hypothalamus.

ADH enters the blood capillaries running through the posterior pituitary gland. It is transported around the body and acts on the cells of the collecting ducts (its target cells). Once the water potential of the blood rises again, less ADH is released. ADH is slowly broken down – it has a half-life of about 20 minutes. Therefore the ADH present in the blood is broken down and the collecting ducts will receive less stimulation.

Figure 3 shows how the level of ADH is controlled by negative feedback.

> **DID YOU KNOW?**
> Alcohol inhibits the release of ADH – so drinking alcohol reduces reabsorption from the collecting ducts and makes you need to go to the toilet.

Questions

1. A simple flow diagram of a response is: stimulus → receptor → coordination → effector → response. Draw a flow diagram to show the mechanism of response to a change in water potential in the blood.

2. Why must ADH be broken down?

3. Suggest where ADH is broken down.

4. Explain why negative feedback is an important part of osmoregulation.

5. Describe how aquaporins are inserted into the cell surface membrane of cells in the collecting duct.

6. How are neurosecretory cells different from ordinary nerve cells?

7. Describe how the osmoreceptors respond to an increase in the water potential of the blood.

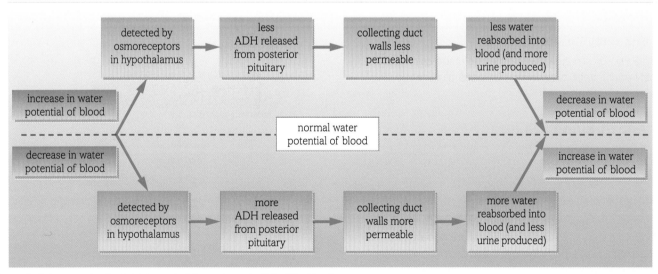

Figure 3 Using negative feedback to control the water potential of the blood.

By the end of this topic, you should be able to demonstrate and apply your knowledge and understanding of:

* * the effects of kidney failure and its potential treatments
* * how excretory products can be used in medical diagnosis

> **KEY DEFINITIONS**
>
> **glomerular filtration rate (GFR):** the rate at which fluid enters the nephrons.
> **monoclonal antibodies:** antibodies made from one type of cell – they are specific to one complementary molecule.
> **renal dialysis:** a mechanism used to artificially regulate the concentrations of solutes in the blood.

Kidney failure

If the kidneys fail completely they are unable to regulate the levels of water and electrolytes (substances that form charged particles in water) in the body or to remove waste products such as urea from the blood. This will rapidly lead to death.

Assessing kidney function

Kidney function can be assessed by estimating the **glomerular filtration rate (GFR)** and by analysing the urine for substances such as proteins. Proteins in the urine indicate the filtration mechanism has been damaged.

The GFR is a measure of how much fluid passes into the nephrons each minute. A normal reading is in the range 90–120 $cm^3 min^{-1}$. A figure below 60 $cm^3 min^{-1}$ indicates that there may be some form of chronic kidney disease. A figure below 15 $cm^3 min^{-1}$ indicates kidney failure and a need for immediate medical attention.

Causes of kidney failure

The possible causes of kidney failure include diabetes mellitus (both Type 1 and Type 2 sugar diabetes), heart disease, hypertension and infection.

Treatment of kidney failure

The main treatments for kidney failure are renal dialysis and kidney transplant.

Renal dialysis

Renal dialysis is the most common treatment for kidney failure. Waste products, excess fluid and mineral ions are removed from the blood by passing it over a partially permeable dialysis membrane that allows the exchange of substances between the blood and dialysis fluid. The dialysis fluid contains the correct concentrations of mineral ions, urea, water and other substances found in blood plasma. Any substances in excess in the blood diffuse across the membrane into the dialysis fluid. Any substances that are too low in concentration diffuse into the blood from the dialysis fluid.

The two types of renal dialysis are:

* **Haemodialysis**: blood from an artery or vein is passed into a machine that contains an artificial dialysis membrane shaped to form many artificial capillaries, which increases the surface area for exchange. Heparin is added to avoid clotting. The artificial capillaries are surrounded by dialysis fluid, which flows in the opposite direction to the blood (a countercurrent). This improves the efficiency of exchange. Any bubbles are removed before the blood is returned to the body via a vein.

Haemodialysis is usually performed at a clinic two or three times a week for several hours at each session. Some patients learn to carry it out at home.

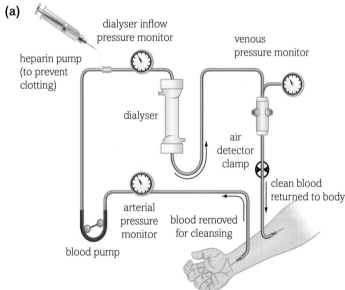

(a)
dialyser inflow pressure monitor
venous pressure monitor
heparin pump (to prevent clotting)
dialyser
air detector clamp
clean blood returned to body
arterial pressure monitor
blood removed for cleansing
blood pump

(b)
blood pumped in at a higher pressure than the dialysis fluid
header
tube sheet
solution outlet
dialyser – contains fine capillary tubes of dialysis membrane
fibres
jacket
solution inlet

blood flows through artificial capillaries, while the dialysis fluid flows along the outside in the opposite direction

Figure 1 (a) Haemodialysis and (b) detail of a dialyser.

- **Peritoneal dialysis (PD)**: the dialysis membrane is the body's own abdominal membrane (peritoneum). First, a surgeon implants a permanent tube in the abdomen. Dialysis solution is poured through the tube and fills the space between the abdominal wall and organs. After several hours, the used solution is drained from the abdomen.

PD can be carried out at home or work. Because the patient can walk around while having dialysis, the method is sometimes called ambulatory PD.

Dialysis must be combined with a carefully monitored diet.

Kidney transplant

A kidney transplant is the best life-extending treatment for kidney failure. This involves major surgery. While the patient is under anaesthesia, the surgeon implants the new organ into the lower abdomen and attaches it to the blood supply and the bladder. Patients are given immunosuppressant drugs to help prevent their immune system recognising the new organ as a foreign object and rejecting it.

Many patients feel much better immediately after the transplant.

The advantages and disadvantages of kidney transplants are shown in Table 1.

Advantages	Disadvantages
Freedom from time-consuming renal dialysis	Need to take immunosuppressant drugs
Feeling physically fitter	Need for major surgery under general anaesthetic
Improved quality of life – able to travel	Need for regular checks for signs of rejection
Improved self-image – no longer have a feeling of being chronically ill	Side effects of immunosuppressant drugs – fluid retention, high blood pressure, susceptibility to infections

Table 1 Advantages and disadvantages of kidney transplants.

Urine analysis

Molecules with a relative molecular mass of less than 69 000 can enter the nephron. Any metabolic product or other substance in the blood can therefore be passed into the urine if it is small enough. If these substances are not reabsorbed further down the nephron they can be detected in urine.

For example, urine can be tested for:

- glucose in the diagnosis of diabetes
- alcohol to determine blood alcohol levels in drivers
- many recreational drugs (tests may be carried out as random tests at work – especially where there are safety issues related to the type of work)
- human chorionic gonadotrophin (hCG) in pregnancy testing
- anabolic steroids, to detect improper use in sporting competitions.

DID YOU KNOW?

Marijuana can be detected in the urine for up to seven days after a single use. In a habitual user, it can be detected for over a month after the last use.

Pregnancy testing

Once a human embryo is implanted in the uterine lining, it produces a hormone called human chorionic gonadotrophin (hCG). hCG is a relatively small glycoprotein, with a molecular mass of 36 700, that can be found in urine as early as six days after conception. Pregnancy-testing kits use **monoclonal antibodies** which bind to hCG in urine (see Figure 2).

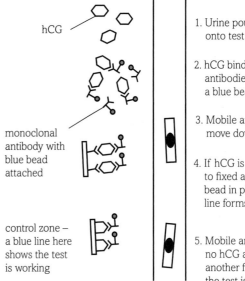

hCG

monoclonal antibody with blue bead attached

control zone – a blue line here shows the test is working

1. Urine poured onto test stick
2. hCG binds to mobile antibodies attached to a blue bead
3. Mobile antibodies move down test stick
4. If hCG is present, it binds to fixed antibodies holding bead in place – a blue line forms
5. Mobile antibodies with no hCG attached bind to another fixed site to show the test is working

Figure 2 How monoclonal antibodies work in a pregnancy test. A negative test shows one blue line in the control zone; a positive test shows two blue lines.

Testing for anabolic steroids

Anabolic steroids increase protein synthesis within cells, which results in the build-up of cell tissue, especially in the muscles. Non-medical uses for anabolic steroids are controversial, because they can give advantage in competitive sports and they have dangerous side effects.

All major sporting bodies ban the use of anabolic steroids. Anabolic steroids have a half-life of about 16 hours and remain in the blood for many days. They are relatively small molecules and can enter the nephron easily. Testing for anabolic steroids involves analysing a urine sample in a laboratory using **gas chromatography**.

Questions

1. Explain why renal dialysis must be combined with a carefully monitored diet.

2. What components of the diet must be monitored in a patient undergoing dialysis?

3. Explain why dialysis fluid must be (a) changed regularly, (b) kept at 37 °C and (c) sterile.

4. Create a table of advantages and disadvantages of renal dialysis as a treatment for kidney failure.

THINKING BIGGER

POLLUTION IN RIVERS

For some decades, scientists have been concerned about the pollution being poured into rivers and streams. While modern approaches to industry and agriculture have reduced many types of pollution, there are still many issues to solve. The following extract is from an article written on CARE2, an online community that calls itself the world's largest network for good.

MALE FISH TURNING FEMALE DUE TO POLLUTION

Canadian researchers who studied fish in Alberta's Red Deer and Oldman rivers found a peculiar effect of pollution on a native minnow called the long-nosed dace. Males they sampled contained elevated levels of a protein marker normally only found in females. Also, '…up to 44 per cent of the male fish had eggs in their testes.' In normal circumstances only females produce eggs. One of the study's co-authors Lee Jackson said, '… it tells us the fish are exposed to oestrogen or something that looks like oestrogen to the fish.' Mr. Jackson is the executive director of a research facility which creates new approaches for treating wastewater.

The researchers also analysed water samples from the river locations where they examined the minnows. They found synthetic oestrogens from birth control pills and hormone therapy drugs in the river water. In addition, synthetic and natural hormones were found in the water, which could be from agricultural run-off and cattle ranching. The researchers took samples at 15 locations along 600 kilometres of river. Their study indicated the disruptive chemicals could be present in a span covering the entire 600 kilometres of river water.

The ratio of females to male minnows in the rivers could also be shifting much toward greater numbers of females. If this ratio remains unbalanced, fewer numbers of new fish will be born, as there will not be enough males to fertilise fish eggs. With so many chemicals in the minnow's habitat, it could be difficult to pinpoint exactly which chemical is causing the gender bending, or if it is a combination of all of them.

The effect, called fish feminisation, is not unique to the Alberta minnows. In fact, it has been documented in Europe, and the United States as well.

The Canadian researchers are going to continue their studies to deepen their knowledge of what is taking place in the Red Deer and Oldman rivers, with the ultimate goal of perhaps influencing policy decisions that can stop the problem.

Long-nosed dace eat aquatic insects and are prey for larger fish, so they are an important part of the food chain.

Recently in Colorado a similar situation was partly remedied by upgrading a wastewater treatment plant. Fish exposed to the water previously had been feminised in about seven days, but after the treatment plant's upgrade, it took more than 28 days.

One of the researchers involved in the Colorado study said that it is not only about changing wastewater technology, it is also about modifying consumer behaviour, 'We excrete natural and synthetic oestrogens and use shampoos, detergents and cosmetics containing a variety of hormone disrupters that wind up in waterways. All of these different chemicals we are putting into the environment have the potential to alter the biology of animals and to affect ecosystems.'

He said consumers can refuse to buy milk made with growth hormones and antibacterial soaps. Also they can reduce their use of shampoos and detergents.

Source
- www.care2.com/greenliving/male-fish-turning-female-due-to-pollution.html. By Jake Richardson.

Figure 1 A water treatment plant.

Where else will I encounter these themes?

Let's start by considering the nature of the writing in the article:

1. Why does the writer use language such as 'gender bending' and 'fish feminisation'?
2. The writer suggests that some of the hormones may be coming from 'synthetic and natural hormones... which could be from agricultural run-off and cattle ranching.' Why does the writer phrase this as only a suggestion?
3. The writer says the fish are being exposed to 'oestrogen or something that looks like oestrogen to fish'. Explain what he means by 'something that looks like oestrogen'.

Now we will look at the biology in or connected to this article. Do not worry if you are not ready to answer these questions yet. You may like to return to the questions once you have covered other topics later in the book. Use the timeline at the bottom of the page to help you to put this work in context with what you have already learned and what is ahead in your course.

4. Oestrogen is a steroid hormone:
    a. Describe how oestrogen enters cells.
    b. Describe how oestrogen alters the metabolism of the cell.
5. Is it likely that long-nosed dace is the only species affected? Explain your answer.
6. Define the term 'excretion'.
7. The writer suggests that the oestrogen could be coming from pills used for birth control or hormone replacement therapy. How are the hormones released from people's bodies?
8. The water could be tested for the presence of oestrogen using monoclonal antibodies. What other tests use monoclonal antibodies?
9. In other articles, it is suggested that oestrogens in wastewater could eventually affect the human population. Explain how this could occur.

Activity

The article suggests that treatment of wastewater must be upgraded to deal with hormones. Treatment of wastewater to remove hormones could be carried out by introducing microorganisms that have been genetically modified.

Produce a presentation to show how genetically modified microorganisms could be used to treat wastewater.

Ensure that your presentation explains:

- what is meant by genetic modification
- how bacteria can be genetically modified
- how the modified bacteria can be upscaled (increased in numbers)
- how the bacteria would remove oestrogen from the water.

The water must be tested after treatment to ensure that it is safe. This can be done using immobilised monoclonal antibodies. Extend your presentation to explain how immobilised monoclonal antibodies can be used to test water.

Ensure that your presentation explains:

- what a monoclonal antibody is
- how monoclonal antibodies can be produced
- how the monoclonal antibodies can be immobilised
- how a test stick might work when placed in water.

Practice questions

1. Which row correctly identifies the functions of parts of the nephron? [1]

	Proximal convoluted tubule	Distal convoluted tubule	Loop of Henle	Collecting duct
A	selective reabsorption	adjusts concentrations	concentrates the salts in the medulla	reabsorbs water
B	filters blood	selective reabsorption	adjusts concentrations	concentrates the salts in the medulla
C	reabsorbs water	filters blood	selective reabsorption	adjusts concentrations
D	concentrates the salts in the medulla	reabsorbs water	filters blood	selective reabsorption

2. How does antidiuretic hormone affect the kidney? [1]

A. Decreases permeability of collecting ducts to water.

B. Increases volume of blood passing into the nephron.

C. Increases permeability of collecting ducts to water.

D. Increases concentration of salts in the medulla.

3. Statements i–vi refer to excretory organs and their products:

(i) The liver excretes carbon dioxide

(ii) The kidneys excrete urea

(iii) The lungs excrete oxygen

(iv) The lungs excrete carbon dioxide

(v) The liver excretes bile pigments

(vi) The skin excretes carbon dioxide.

Which row is correct? [1]

A. All statements are correct.

B. Statements (ii), (iii), (v) and (vi) are correct.

C. Statements (ii), (iv) and (v) are correct.

D. Statements (ii), (iv), (v) and (vi) are correct.

4. What substances leave the blood in the renal capsule? [1]

A. Blood plasma only

B. Blood plasma and any substances that have a molecular weight below 69 000

C. Blood plasma proteins and white blood cells

D. Whole blood

5. Which row correctly describes the function of each vessel leading into and out of the liver? [1]

	Hepatic artery	Hepatic vein	Hepatic portal vein	Bile duct
A	Carries oxygenated blood into the liver	Carries deoxygenated blood away from the liver	Transports products from liver to digestive system	Carries waste products to the gall bladder
B	Carries deoxygenated blood away from the liver	Carries oxygenated blood into the liver	Transports products of digestion to the liver	Carries waste products to the gall bladder
C	Carries oxygenated blood into the liver	Carries deoxygenated blood away from the liver	Transports products of digestion to the liver	Carries waste products to the bladder
D	Carries oxygenated blood into the liver	Carries deoxygenated blood away from the liver	Transports products of digestion to the liver	Carries waste products to the gall bladder

[Total: 5]

6. Figure 1 shows a series of reactions that occur in the cells of the liver.

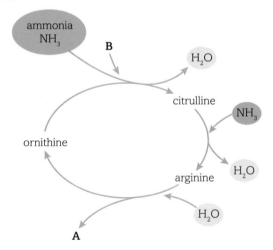

Figure 1

(a) Name this series of reactions. [1]

(b) Describe the source of the ammonia (NH_3). [2]

(c) Name substances A and B. [2]

(d) Substance A is an excretory product. Describe the route taken from the liver to the excretory organ where substance A is excreted from the body. [8]

[Total: 13]

7. The liver can detoxify substances such as alcohol. This process involves the reduction of NAD.

   (a) List three other homeostatic functions of the liver. [3]

   (b) Describe the metabolic pathway used to detoxify alcohol in the hepatocytes. You may use an annotated flow diagram. [5]

   (c) Explain why drinking too much alcohol can lead to fat accumulating in the liver. [3]

   [Total: 11]

8. (a) What is meant by the term 'ultrafiltration'? [2]

   (b) The glomerulus and Bowman's capsule are adapted to perform ultrafiltration of the blood:

   (i) How does fluid from the blood leave the capillaries of the glomerulus? [2]

   (ii) Describe how the wall of the Bowman's capsule is adapted to allow filtration to occur. [3]

   (iii) What structure filters out larger molecules? [1]

   (c) Ultrafiltration is performed under pressure. Explain how the pressure in the glomerulus is maintained. [2]

   [Total: 10]

9. The table below shows the composition of fluids in parts of the kidney.

Substance	Concentration in plasma of glomerulus (g dm^{-1})	Concentration in glomerular filtrate (g dm^{-1})	Concentration in urine (g dm^{-1})
protein	80.0	0.0	0.0
amino acids	0.5	0.5	0.0
glucose	1.0	1.0	0.0
urea	0.3	0.3	9.0
mineral ions	7.2	7.2	3.5

   (a) Explain why there are no proteins in the glomerular filtrate. [2]

   (b) Explain why amino acids are found in the glomerular filtrate but not in the urine. [2]

   (c) (i) Calculate the percentage increase in urea concentration between the glomerular filtrate and the urine. Show your working. [2]

   (ii) Describe how this increase is achieved. [1]

   (d) Describe the mechanism that maintains the water potential of the blood, despite a person sweating on a hot day when little water is available to drink. [7]

   [Total: 14]

10. (a) Glucose is selectively reabsorbed from the fluid in the proximal convoluted tubule.

    Describe how the cells of the proximal convoluted tubule reabsorb all the glucose. [6]

    (b) What is the role of the loop of Henle? [1]

    (c) Figure 2 shows the relative length of the loop of Henle in three mammals. Explain the differences. [6]

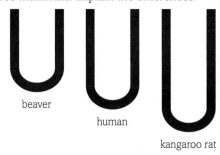

beaver

human

kangaroo rat

Figure 2

[Total: 13]

11. A Formula 1 driver who weighs 70 kg may lose 6 kg of water during a race.

    (a) Calculate the percentage of body mass lost during a race. Show your working and give your answer to two decimal places. [2]

    (b) Suggest how this water is lost. [1]

    (c) What is the effect on the water potential of the driver's blood as water is lost during a race? [1]

    (d) Urine production almost stops towards the end of a race. Explain why less urine is produced. [9]

    [Total: 13]

Communication, homeostasis and energy

NEURONAL COMMUNICATION

Introduction

The nervous system forms a complete communication network around the body. Sensory receptors detect changes in the environment and convert the stimuli to electrical impulses. These impulses, or action potentials, travel rapidly along sensory neurones to the central nervous system (CNS). They act as input to the control centres in the CNS. The CNS collects all the input from a variety of receptors and coordinates a response. It sends action potentials down motor neurones to the effectors. This is the output from the control centres.

Neurones have a resting state in which the plasma membrane is polarised. Action potentials consist of a series of ionic movements across the plasma membrane that cause depolarisation followed by repolarisation. The action potential is transmitted along the neurone by localised ion movements called local currents. In this chapter you will learn about how the resting potential is maintained and how action potentials are produced and transmitted.

Neurones communicate with one another at synapses, where the signal is transmitted as a neurotransmitter chemical. Synapses have a number of important functions in nervous communication.

All the maths you need

To unlock the puzzles of this chapter you need the following maths:

- Recognise and make use of appropriate units in calculations
- Estimate results
- Use an appropriate number of significant figures
- Construct and interpret frequency tables and diagrams, bar charts and histograms
- Understand and use the symbols: $=, <, \leqslant, \geqslant, >, \sim$
- Translate information between graphical, numerical and algebraic forms

What have I studied before?

- The eye as a sensory receptor
- The main parts of the nervous system
- The gross structure of neurones
- Nerves communicate via a neurotransmitter at synapses
- Nervous responses are rapid
- Voluntary responses are controlled by the brain
- The need for communication
- The structure of the plasma membrane and its role as a selectively permeable barrier
- How substances move across membranes
- Active transport and diffusion
- The role of the plasma membrane in cell signalling

What will I study later?

- How the nervous system is organised
- Somatic and autonomic nervous systems
- The structure of the brain
- Reflex actions
- The neuromuscular junction
- The action of muscles
- The coordination of responses by the nervous and endocrine systems

What will I study in this chapter?

- The role of sensory receptors as transducers of energy
- The structure and function of neurones
- The generation and transmission of action potentials
- The structure and roles of synapses

1 Roles of sensory receptors

By the end of this topic, you should be able to demonstrate and apply your knowledge and understanding of:

* the roles of mammalian sensory receptors in converting different types of stimuli into nerve impulses

---

KEY DEFINITIONS

Pacinian corpuscle: a pressure sensor found in the skin.
sensory receptors: cells/sensory nerve endings that respond to a stimulus in the internal or external environment of an organism and can create action potentials.
transducer: a cell that converts one form of energy into another – in this case to an electrical impulse.

DID YOU KNOW?

Sensory receptors detect a change in the environment. So if a stimulus is constant it will not continue to cause a response. This is the basis of some forms of behaviour where organisms become habituated to a stimulus.

Sensory receptors

Sensory receptors are specialised cells that can detect changes in our surroundings. Most are energy **transducers** that convert one form of energy to another.

Each type of transducer is adapted to detect changes in a particular form of energy. This may be a change in light levels, a change in pressure on the skin or one of many other energy changes. Other receptors detect the presence of chemicals.

Each change in the environment, whether it is a change in the energy level or the presence of a new chemical, is called a stimulus. Whatever the stimulus, the sensory receptors respond by creating a signal in the form of electrical energy. This is a called a nerve impulse.

Table 1 shows some different receptors and the energy changes that they detect.

Stimulus (change in environment)	Sensory receptor	Energy change involved
Change in light intensity	Light sensitive cells (rods and cones) in the retina	Light to electrical
Change in temperature	Temperature receptors in the skin and hypothalamus	Heat to electrical
Change in pressure on the skin	Pacinian corpuscles in the skin	Movement to electrical
Change in sound	Vibration receptors in the cochlea of the ear	Movement to electrical
Movement	Hair cells in inner ear	Movement to electrical
Change in length of muscle	Muscle spindles in skeletal muscles	Movement to electrical
Chemicals in the air	Olfactory cells in epithelium lining the nose	These receptors detect the presence of a chemical and create an electrical nerve impulse
Chemicals in food	Chemical receptors in taste buds on tongue	

Table 1 Sensory receptors and the energy changes that they detect.

Pacinian corpuscles

A **Pacinian corpuscle** is a pressure sensor that detects changes in pressure on the skin.

The corpuscle is an oval-shaped structure that consists of a series of concentric rings of connective tissue wrapped around the end of a nerve cell. When pressure on the skin changes this deforms the rings of connective tissue, which push against the nerve ending.

The corpuscle is sensitive only to changes in pressure that deform the rings of connective tissue. Therefore, when pressure is constant they stop responding.

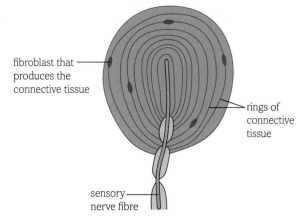

fibroblast that produces the connective tissue

rings of connective tissue

sensory nerve fibre

Figure 1 A Pacinian corpuscle.

Generating nerve impulses
Changing membrane permeability

As you may recall from your AS Biology work, all cell surface membranes contain proteins. Some proteins are channels that allow the movement of ions (charged particles) across the membrane by facilitated diffusion. Others are transport proteins that can actively move ions across the membrane against their concentration gradient; this requires the use of energy in the form of ATP.

LEARNING TIP

Remember that lipid bilayers are not permeable to charged particles. Such particles (ions) must pass through protein channels to cross the membrane.

If the channel proteins are permanently open then ions can diffuse across the membrane and will do so until their concentrations on either side of the membrane are in equilibrium. If the channels can be closed then the action of the active pumps can create a concentration gradient across the membrane.

Cells associated with the nervous system have specialised channel proteins. Some of these, called **sodium channels**, are specific to sodium ions (Na^+). Others, called **potassium channels**, are specific to potassium ions (K^+). These channels also possess a gate that can open or close the channel.

The sodium channels are sensitive to small movements of the membrane, so when the membrane is deformed by the changing pressure the sodium channels open. This allows sodium ions to diffuse into the cell, producing a **generator potential** (also called a receptor potential).

The membranes also contain **sodium/potassium pumps** that actively pump sodium ions out of the cell and potassium ions into the cell. Three sodium ions are pumped out for every two potassium ions pumped into the cell. When the channel proteins are all closed, the sodium/potassium pumps work to create a concentration gradient. The concentration of sodium ions outside the cell increases, while the concentration of potassium ions inside the cell increases. The membrane is more permeable to potassium ions, so some of these leak out of the cell. The membrane is less permeable to sodium ions, so few of these are able to leak into the cell.

The result of these ionic movements is a potential gradient across the cell membrane. The cell is negatively charged inside compared with outside. This negative potential is enhanced by the presence of negatively charged anions inside the cell.

Creating a nerve impulse

When the cell is inactive the cell membrane is said to be **polarised**, that is negatively charged inside compared with the outside.

A nerve impulse is created by altering the permeability of the nerve cell membrane to sodium ions. This is achieved by opening the sodium ion channels. As the sodium ion channels open, the membrane permeability is increased and sodium ions can move across the membrane down their concentration gradient into the cell. The movement of ions across the membrane creates a change in the potential difference (charge) across the membrane. The inside of the cell becomes less negative (compared with the outside) than usual. This is called **depolarisation**. The change in potential across a receptor membrane is often called a generator potential.

If a small stimulus is detected only a few sodium channels will open. The larger the stimulus (the change in energy levels in the environment) the more gated channels will open. If enough gates are opened and enough sodium ions enter the cell, the potential difference across the cell membrane changes significantly and will initiate an impulse or **action potential**.

Questions

1. Why do membranes need special channel proteins to enable the movement of ions?

2. Explain why a constant sound will often become unnoticeable after a short time.

3. Why is energy required to produce a concentration gradient?

4. Explain why a concentration gradient is needed to ensure sodium ions move rapidly into the cell.

5. What is meant by facilitated diffusion?

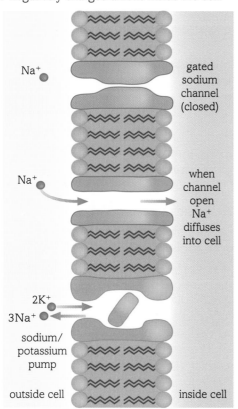

Na^+

gated sodium channel (closed)

Na^+

when channel open Na^+ diffuses into cell

$2K^+$

$3Na^+$

sodium/ potassium pump

outside cell   inside cell

Figure 2 The action of the sodium/potassium pump and sodium ion channels.

② Structure and function of neurones

By the end of this topic, you should be able to demonstrate and apply your knowledge and understanding of:

* the structure and functions of sensory, relay and motor neurones

Function of neurones

Once a stimulus has been detected and its energy has been converted to a depolarisation of the receptor cell membrane, the impulse must be transmitted to other parts of the body. The impulse is transmitted along neurones as an action potential. The action potential is carried as a rapid depolarisation of the membrane caused by the influx of sodium ions.

There are a number of different types of neurone. These include:

* **motor neurones** that carry an action potential from the central nervous system (CNS) to an effector such as a muscle or gland

* **sensory neurones** that carry the action potential from a sensory receptor to the CNS

* **relay neurones** that connect sensory and motor neurones.

Structure of neurones

All neurones have a similar basic structure that enables them to transmit the action potential. Neurones are specialised cells with the following features.

* Many are very long so that they can transmit the action potential over a long distance.

* The cell surface (plasma) membrane has many gated ion channels that control the entry or exit of sodium, potassium or calcium ions.

* Sodium/potassium pumps use ATP to actively transport sodium ions out of the cell and potassium ions into the cell.

* Neurones maintain a potential difference across their cell surface (plasma) membrane.

* A cell body contains the nucleus, many mitochondria and ribosomes.

* Numerous dendrites connect to other neurones. The dendrites carry impulses towards the cell body.

* An axon carries impulses away from the cell body.

* Neurones are surrounded by a fatty layer that insulates the cell from electrical activity in other nerve cells nearby. This fatty layer is composed of Schwann cells closely associated with the neurone.

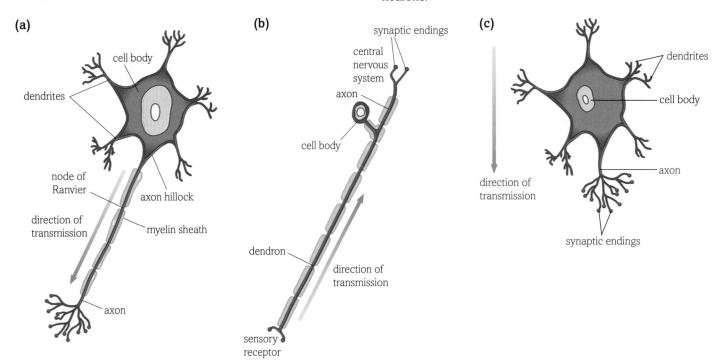

Figure 1 (a) A motor neurone, (b) a sensory neurone and (c) a relay neurone.

Differences between types of neurone

- Motor neurones have their cell body in the CNS and have a long axon that carries the action potential out to the effector.

- Sensory neurones have a long dendron carrying the action potential from a sensory receptor to the cell body, which is positioned just outside the CNS. They then have a short axon carrying the action potential into the CNS.

- Relay neurones connect the sensory and motor neurones together. They have many short dendrites and a short axon. The number of dendrites and the number of divisions of the axon is variable. Relay neurones are an essential part of the nervous system, which conduct impulses in coordinated pathways.

Myelinated and non-myelinated neurones

Around one-third of the peripheral neurones in vertebrates are **myelinated neurones** – that is, they are insulated by an individual **myelin sheath**. The remainder of the peripheral neurones and the neurones found in the CNS are not myelinated.

Myelinated neurones

Most sensory and motor neurones are associated with many Schwann cells, which make up a fatty sheath called the myelin sheath. These Schwann cells are wrapped tightly around the neurone so the sheath actually consists of several layers of membrane and thin cytoplasm from the Schwann cell.

At intervals of 1–3 mm along the neurone there are gaps in the myelin sheath. These are called the **nodes of Ranvier**. Each node is very short (about 2–3 μm long).

Because the myelin sheath is tightly wrapped around the neurone it prevents the movement of ions across the neurone membranes. Therefore, movement of ions across the membrane can only occur at the nodes of Ranvier. This means that the impulse, or action potential, jumps from one node to the next. This makes conduction much more rapid.

Non-myelinated neurones

Non-myelinated neurones are also associated with Schwann cells, but several neurones may be enshrouded in one loosely wrapped Schwann cell. This means that the action potential moves along the neurone in a wave rather than jumping from node to node as seen in myelinated neurones.

> **DID YOU KNOW?**
> You can notice the difference in speed of conduction when you stub your toe or drop something on your foot. You feel the contact very quickly, but a greater sensation of pain arrives almost a second later – this is conducted by non-myelinated neurones.

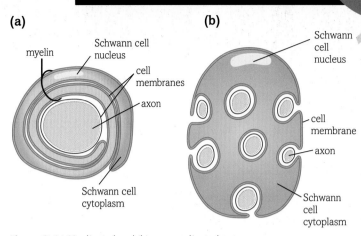

Figure 2 (a) Myelinated and (b) non-myelinated neurones.

Advantages of myelination

Myelinated neurones can transmit an action potential much more quickly than non-myelinated neurones can. The typical speed of transmission in myelinated neurones is 100–120 m s^{-1}. A non-myelinated neurone may only reach transmission speeds of 2–20 m s^{-1}.

Myelinated neurones carry action potentials from sensory receptors to the CNS and from the CNS to effectors. They carry action potentials over long distances – the longest neurone in a human can be about 1 m in length. The increased speed of transmission means that the action potential reaches the end of the neurone much more quickly. This enables a more rapid response to a stimulus.

Non-myelinated neurones tend to be shorter and carry action potentials only over a short distance. They are often used in coordinating body functions such as breathing, and the action of the digestive system. Therefore the increased speed of transmission is not so important.

> ## Questions
>
> **1** Draw a table to compare and contrast sensory and motor neurones.
>
> **2** Suggest why neurones need to contain a large number of mitochondria.
>
> **3** Draw a table to summarise the differences between myelinated and non-myelinated neurones.

By the end of this topic, you should be able to demonstrate and apply your knowledge and understanding of:

* the generation and transmission of nerve impulses in mammals (continued in topic 5.3.4)

Neurones at rest

When a neurone is not transmitting an **action potential** it is said to be at rest. In fact, it is actively pumping ions across its cell surface (plasma) membrane. Just like the sensory receptor described in topic 5.3.1, sodium/potassium ion pumps use ATP to pump three sodium ions out of the cell for every two potassium ions that are pumped in. The gated sodium ion channels are kept closed. However, some of the potassium ion channels are open, and therefore the plasma membrane is more permeable to potassium ions than to sodium ions. Potassium ions tend to diffuse out of the cell. The cell cytoplasm also contains large organic anions (negatively charged ions). Hence, the interior of the cell is maintained at a negative potential compared with the outside. The cell membrane is said to be polarised. The potential difference across the cell membrane is about −60 mV. This is called the **resting potential**. Note that in myelinated neurones, the ion exchanges described occur only at the nodes of Ranvier.

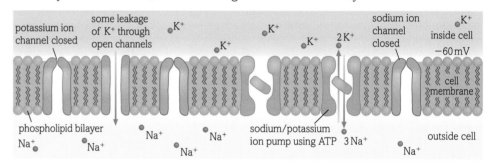

Figure 1 A neurone at rest.

Generating an action potential

While the neurone is at rest it maintains a concentration gradient of sodium ions across its plasma membrane – the concentration is higher outside then inside. Equally the concentration of potassium ions is higher inside than outside.

If some of the sodium ion channels are opened, then sodium ions will quickly diffuse down their concentration gradient into the cell from the surrounding tissue fluid. This causes a depolarisation of the membrane.

In the generator region of a neurone the gated channels are opened by the action of the **synapse** (a nerve junction) (see topic 5.3.5). When a few gated channels open they allow a few sodium ions into the cell and produce a small depolarisation. This is known as a generator potential. It may go no further. However, when more gated channels are opened the generator potentials are combined to produce a larger depolarisation. If the depolarisation reaches a particular magnitude it passes a threshold and will cause an action potential.

Most of the sodium ion channels in a neurone are opened by changes in the potential difference across the membrane – they are called **voltage-gated channels**. When there are sufficient generator potentials to reach the threshold potential they cause the voltage-gated channels to open. This is an example of **positive feedback** – a small depolarisation of the membrane causing a change that increases the depolarisation further.

The opening of voltage-gated sodium ion channels allows a large influx of sodium ions and the depolarisation reaches +40 mV on the inside of the cell. Once this value is reached the neurone will transmit the action potential. The action potential is self-perpetuating – once it starts at one point in the neurone, it will continue along to the end of the neurone. All action potentials are the same magnitude (+40 mV). Therefore they are referred to as an 'all-or-nothing' response.

Stages of an action potential

The following stages of an action potential are shown in Figure 2.

1. The membrane starts in its resting state – polarised with the inside of the cell being −60 mV compared to the outside. There is a higher concentration of sodium ions outside than inside and a higher concentration of potassium ions inside than outside.

2. Sodium ion channels open and some sodium ions diffuse into the cell.

3. The membrane depolarises – it becomes less negative with respect to the outside and reaches the threshold value of −50 mV.

4. Positive feedback causes nearby voltage-gated sodium ion channels to open and many sodium ions flood in. As more sodium ions enter, the cell becomes positively charged inside compared with outside.

5. The potential difference across the plasma membrane reaches +40 mV. The inside of the cell is positive compared with the outside.

6. The sodium ion channels close and potassium channels open.

7. Potassium ions diffuse out of the cell bringing the potential difference back to negative inside compared with the outside – this is called repolarisation.

8. The potential difference overshoots slightly, making the cell hyperpolarised.

9. The original potential difference is restored so that the cell returns to its resting state.

DID YOU KNOW?
Autoimmune diseases such as multiple sclerosis are caused by demyelination of the motor neurones – the neurones lose their myelin sheath and are unable to conduct impulses properly.

Refractory period

After an action potential the sodium and potassium ions are in the wrong places. The concentrations of these ions inside and outside the cell must be restored by the action of the sodium/potassium ion pumps. For a short time after each action potential it is impossible to stimulate the cell membrane to reach another action potential. This is known as the refractory period and allows the cell to recover after an action potential. It also ensures that action potentials are transmitted in only one direction.

Questions

1. Explain why a neurone is active while it is said to be resting.

2. Why is it essential to maintain a concentration gradient across the cell membrane?

3. What is the role of the organic anions inside the neurone?

4. What is the difference between the sodium channels in the generator region and those elsewhere along the neurone?

5. Explain why it is not possible to stimulate a neurone immediately after an action potential.

6. Explain the role of positive feedback in the generation of an action potential.

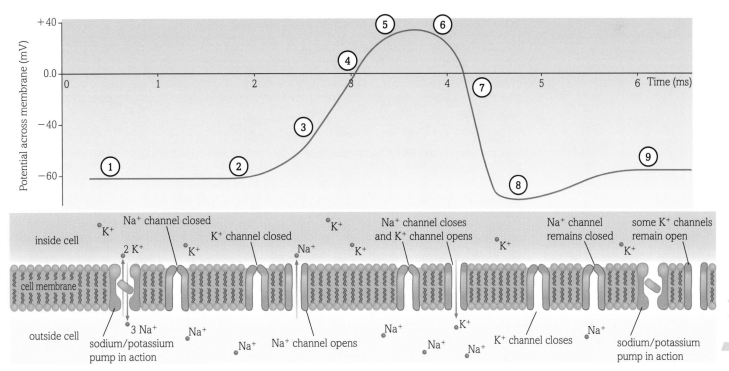

Figure 2 The ionic movements during an action potential.

Nerve impulses: transmission

By the end of this topic, you should be able to demonstrate and apply your knowledge and understanding of:

* the generation and transmission of nerve impulses in mammals

In topic 5.3.3 we saw how an action potential can be generated. We also saw how the action potential consists of a series of movements of ions across the cell surface membrane of the neurone. The role of the neurone is to transmit information in the form of action potentials to other parts of the body. So how does that action potential travel along the neurone?

Local currents

The opening of sodium ion channels at one particular point of the neurone upsets the balance of sodium and potassium ions set up by the action of the sodium/potassium pumps. When sodium ions are allowed to flood into the neurone causing depolarisation, this creates **local currents** in the cytoplasm of the neurone. Sodium ions begin to move along the neurone towards regions where their concentration is still lower. These local currents cause a slight depolarisation of the membrane and cause sodium ion channels further along the membrane to open (positive feedback).

The steps in the formation of local currents and the transmission of a nerve impulse (see Figure 1) are as follows.

1. When an action potential occurs the sodium ion channels open at that point in the neurone.
2. The open sodium ion channels allow sodium ions to diffuse across the membrane from the region of higher concentration outside the neurone into the neurone. The concentration of sodium ions inside the neurone rises at the point where the sodium ion channels are open.
3. Sodium ions continue to diffuse sideways along the neurone, away from the region of increased concentration. This movement of charged particles is a current called a local current.
4. The local current causes a slight depolarisation further along the neurone which affects the voltage-gated sodium ion channels, causing them to open. The open channels allow rapid influx of sodium ions causing a full depolarisation (action potential) further along the neurone. The action potential has therefore moved along the neurone.

The action potential will continue to move in the same direction until it reaches the end of the neurone – it will not reverse direction, because the concentration of sodium ions behind the action potential is still high.

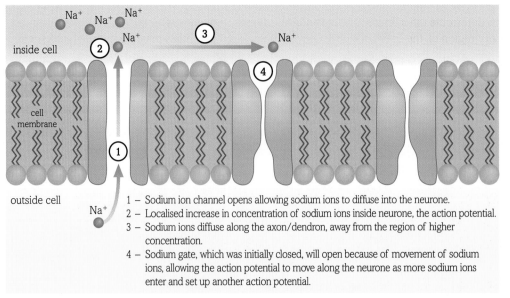

1 – Sodium ion channel opens allowing sodium ions to diffuse into the neurone.
2 – Localised increase in concentration of sodium ions inside neurone, the action potential.
3 – Sodium ions diffuse along the axon/dendron, away from the region of higher concentration.
4 – Sodium gate, which was initially closed, will open because of movement of sodium ions, allowing the action potential to move along the neurone as more sodium ions enter and set up another action potential.

Figure 1 How local currents are formed.

Saltatory conduction

As described in topic 5.3.2, the myelin sheath is an insulating layer of fatty material, composed of Schwann cells wrapped tightly around the neurone. Sodium and potassium ions cannot diffuse through this fatty layer. In between the Schwann cells are small gaps – the nodes of Ranvier. Therefore, the ionic movements that create an action potential cannot occur over much of the length of the neurone: they occur only at the nodes of Ranvier. In myelinated neurones the local currents are therefore elongated and sodium ions diffuse along the neurone from one node of Ranvier to the next. This means that the action potential appears to jump from one node to the next. This is called **saltatory conduction** (Latin, meaning 'to jump').

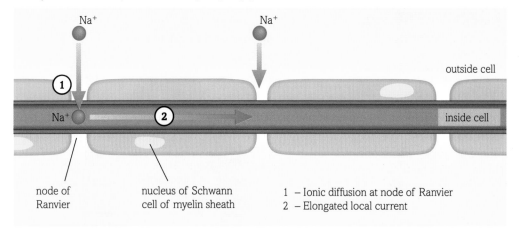

node of Ranvier

nucleus of Schwann cell of myelin sheath

1 – Ionic diffusion at node of Ranvier
2 – Elongated local current

Figure 2 Saltatory conduction.

Advantages of saltatory conduction

The myelin sheath means that action potentials can only occur at the gaps between the Schwann cells that make up the myelin sheath. Effectively the action potential jumps from one node of Ranvier to the next. This speeds up the transmission of the action potential along the neurone. Myelinated neurones conduct action potentials more quickly than non-myelinated neurones. A myelinated neurone can conduct an action potential at up to 120 m s^{-1}.

Frequency of transmission

The impulse carried by a neurone is an action potential. All action potentials are the same intensity: each one produces a depolarisation of +40 mV. This is the 'all-or-nothing' rule (topic 5.3.3).

Although the size of the action potential is unrelated to the intensity of the stimulus that caused the action potential, we can still detect stimuli of different intensities, such as loud or quiet sounds. Our brains determine the intensity of the stimulus from the frequency of action potentials arriving in the sensory region of the brain. A higher frequency of action potentials means a more intense stimulus.

When a stimulus is at higher intensity more sodium channels are opened in the sensory receptor. This produces more generator potentials. As a result there are more frequent action potentials in the sensory neurone. Therefore there are more frequent action potentials entering the central nervous system.

LEARNING TIP

The flow of sodium ions along the axon in a local current is much more rapid than the movement of an action potential involving exchange of ions across the membrane. Therefore a myelinated neurone will transmit the impulse much more quickly along the neurone.

DID YOU KNOW?

Frequency of transmission is not the only way that the intensity of a stimulus can be transmitted. Similar stimuli, but of different intensity, may stimulate different receptors. For example, the skin contains a range of receptors that detect touch and pressure. Free nerve endings in the epidermis detect the lightest of touches. Meissner's corpuscles are very close to the surface of the skin and detect touch and light pressure. Pacinian corpuscles are deeper and do not detect touch – they need substantial pressure to be stimulated.

Questions

1. What causes the sodium ion channels to open?

2. Explain how the myelin sheath causes saltatory conduction.

3. Suggest why saltatory conduction makes conduction more rapid.

4. Explain why the maximum frequency of action potentials is limited.

(5) Synapses 1

By the end of this topic, you should be able to demonstrate and apply your knowledge and understanding of:

* the structure and roles of synapses in neurotransmission (continued in topic 5.3.6).

The structure of a cholinergic synapse

A synapse is a junction between two or more neurones, where one neurone can communicate with, or signal to, another neurone. Between the two neurones is a small gap called the **synaptic cleft**, which is approximately 20 nm wide.

As described in topic 5.3.3, an action potential travels along a neurone as a series of ionic movements across the neurone membrane. This action potential cannot bridge the gap between two neurones. Instead, the action potential in the pre-synaptic neurone causes the release of a chemical (the **neurotransmitter**) that diffuses across the synaptic cleft and generates a new action potential in the post-synaptic neurone. Synapses that use acetylcholine as the neurotransmitter are called **cholinergic synapses**.

The pre-synaptic bulb

The pre-synaptic neurone ends in a swelling called the **pre-synaptic bulb** (or pre-synaptic knob). This bulb contains a number of specialised features (see Figure 1):

* many mitochondria – indicating that an active process needing ATP is involved

* a large amount of smooth endoplasmic reticulum, which packages the neurotransmitter into vesicles

* large numbers of vesicles containing molecules of a chemical called **acetylcholine**, the transmitter that will diffuse across the synaptic cleft

* a number of voltage-gated calcium ion channels on the cell surface membrane.

The post-synaptic membrane

The post-synaptic membrane contains specialised sodium ion channels that can respond to the neurotransmitter (see Figure 2). These channels consist of five polypeptide molecules. Two of these polypeptides have a special receptor site that is specific to acetylcholine. The receptor sites have a shape that is complementary to the shape of the acetylcholine molecule. When acetylcholine is present in the synaptic cleft it binds to the two receptor sites and causes the sodium ion channel to open.

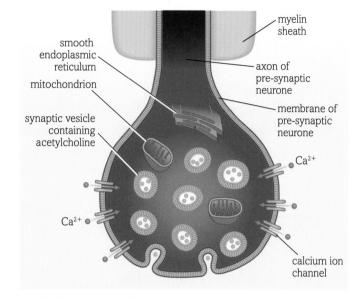

Figure 1 The pre-synaptic bulb.

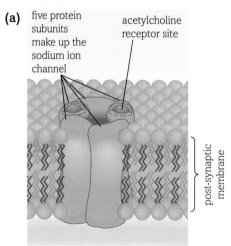

Figure 2 (a) The post-synaptic membrane contains sodium ion channels. (b) The channels are opened by acetylcholine.

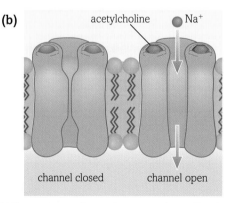

Transmission across the synapse

The sequence of events in the transmission of a signal across the synaptic cleft is as follows (see Figure 3).

1. An action potential arrives at the synaptic bulb.

2. The voltage-gated calcium ion channels open.

3. Calcium ions diffuse into the synaptic bulb.

4. The calcium ions cause the synaptic vesicles to move to, and fuse with, the pre-synaptic membrane.

5. Acetylcholine is released by exocytosis.

6. Acetylcholine molecules diffuse across the cleft.

7. Acetylcholine molecules bind to the receptor sites on the sodium ion channels in the post-synaptic membrane.

8. The sodium ion channels open.

9. Sodium ions diffuse across the post-synaptic membrane into the post-synaptic neurone.

10. A generator potential or excitatory post-synaptic potential (EPSP) is created (see topic 5.3.6).

11. If sufficient generator potentials combine then the potential across the post-synaptic membrane reaches the threshold potential.

12. A new action potential is created in the post-synaptic neurone.

Once an action potential is achieved it will pass down the post-synaptic neurone.

LEARNING TIP
Remember that acetylcholine molecules are released from the vesicles by exocytosis and diffuse across the synaptic cleft. Your description should make it clear that vesicles do not pass across the synaptic cleft.

The role of acetylcholinesterase

If acetylcholine is left in the synaptic cleft it will continue to open the sodium ion channels in the post-synaptic membrane and will continue to cause action potentials. **Acetylcholinesterase** is an enzyme found in the synaptic cleft. It hydrolyses the acetylcholine to ethanoic acid (acetic acid) and choline. This stops the transmission of signals, so that the synapse does not continue to produce action potentials in the post-synaptic neurone.

The ethanoic acid and choline are recycled. They re-enter the synaptic bulb by diffusion and are recombined to acetylcholine using ATP from respiration in the mitochondria. The recycled acetylcholine is stored in synaptic vesicles for future use.

DID YOU KNOW?
Organophosphate insecticides work by permanently inhibiting acetylcholinesterase, causing convulsions and paralysis. Farmers who are over-exposed to these pesticides can suffer conditions associated with poor synaptic function.
Nicotine acts as a stimulant by activating the acetylcholine receptor sites.

Questions

1. Suggest why the pre-synaptic neurone ends in a bulb.

2. Explain why the pre-synaptic bulb contains:

   (a) many mitochondria

   (b) a lot of smooth endoplasmic reticulum.

3. Explain why the calcium ion channels are voltage-gated.

4. Why is it important that the synaptic cleft contains the enzyme acetylcholinesterase?

5. What deductions can you draw about the shape of the nicotine molecule?

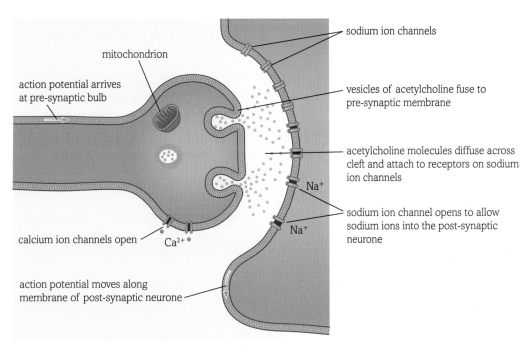

mitochondrion

action potential arrives at pre-synaptic bulb

calcium ion channels open — Ca^{2+}

action potential moves along membrane of post-synaptic neurone

sodium ion channels

vesicles of acetylcholine fuse to pre-synaptic membrane

acetylcholine molecules diffuse across cleft and attach to receptors on sodium ion channels

Na^+

sodium ion channel opens to allow sodium ions into the post-synaptic neurone

Na^+

Figure 3 Transmission across a synapse.

By the end of this topic, you should be able to demonstrate and apply your knowledge and understanding of:

* the structure and roles of synapses in neurotransmission

> **KEY DEFINITION**
>
> **summation:** occurs when the effects of several excitatory post-synaptic potentials (EPSPs) are added together.

Action potentials and cell signalling

In topic 5.3.4 we saw that an action potential is an all-or-nothing response. Once the action potential starts, it will be conducted along the entire length of the neurone. The action potential does not vary in size or intensity. At the end of the neurone the pre-synaptic membrane releases neurotransmitter molecules into the synaptic cleft. The post-synaptic neurone responds to these molecules, which is an example of cell signalling. In cholinergic synapses the signal sent to the next neurone consists of molecules of acetylcholine.

These processes are the same in all neurones with cholinergic synapses. In topic 5.3.4 we also saw that a more intense stimulus is transmitted as more frequent action potentials.

Synapses and nervous communication

The main role of synapses is to connect two neurones together so that a signal can be passed from one to the other. However, nerve junctions can be much more complex than a simple connection between two neurones. Nerve junctions often involve several neurones – this could be several neurones from different places converging on one neurone, or it could be one neurone sending signals out to several neurones that diverge to different effectors.

When one action potential passes down an axon to the synapse it will cause a few vesicles to move to, and fuse with, the pre-synaptic membrane. The relatively small number of acetylcholine molecules diffusing across the cleft produces a small depolarisation. This is an **excitatory post-synaptic potential (EPSP)**. This, on its own, will not be sufficient to cause an action potential in the post-synaptic neurone.

It may take several EPSPs to reach the threshold and cause an action potential. The effects of several EPSPs combine together to increase the membrane depolarisation until it reaches the threshold. This combined effect is known as **summation**.

Summation can result from several action potentials in the same pre-synaptic neurone (**temporal summation**), or from action potentials arriving from several different pre-synaptic neurones (**spatial summation**).

(a)

One action potential in the pre-synaptic neurone does not produce an action potential in the post-synaptic neurone – it requires a series of action potentials in the pre-synaptic neurone.

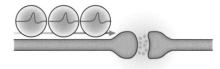

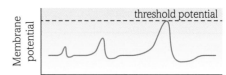

Small EPSPs (excitatory post-synaptic potentials) in post-synaptic neurone do not create an action potential until they act together.

(b)

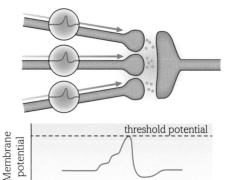

Several pre-synaptic neurones may each contribute to producing an action potential in the post-synaptic neurone.

Figure 1 (a) Temporal summation and (b) spatial summation.

In addition, some pre-synaptic neurones can produce **inhibitory post-synaptic potentials (IPSPs)**. These can reduce the effect of summation and prevent an action potential in the post-synaptic neurone.

> **DID YOU KNOW?**
>
> In many synapses in the brain, EPSPs and IPSPs compete with each other and determine whether or not the post-synaptic membrane will generate an action potential. GABA (γ-aminobutyric acid) and glycine are common neurotransmitters involved in IPSPs. An IPSP can be achieved by opening chloride ion channels that allow chloride ions into the post-synaptic neurone or by opening potassium ion channels that allow potassium ions out of the cell. In both cases a temporary hyperpolarisation is produced.

Control of communication

Because nerve junctions may involve several neurones, this enables synapses to control the communication passed along the nervous system:

- Several pre-synaptic neurones might converge on one post-synaptic neurone. This can allow action potentials from different parts of the nervous system to contribute to generating an action potential in one post-synaptic neurone – so creating a particular response. This is spatial summation. This could be useful where several different stimuli are warning us of danger.

- The combination of several EPSPs could be prevented from producing an action potential by one IPSP.

- One pre-synaptic neurone might diverge to several post-synaptic neurones. This can allow one action potential to be transmitted to several parts of the nervous system. This is useful in a reflex arc. One post-synaptic neurone elicits the response, while another informs the brain.

- Synapses ensure that action potentials are transmitted in the correct direction – only the pre-synaptic bulb contains vesicles of acetylcholine. Therefore, if an action potential happens to start half way along a neurone and ends at the post-synaptic membrane, it will not cause a response in the next cell.

- Synapses can filter out unwanted low-level signals. If a low-level stimulus creates an action potential in the pre-synaptic neurone it is unlikely to pass across a synapse to the next neurone, because several vesicles of acetylcholine must be released to create an action potential in the post-synaptic neurone.

- Low-level action potentials can be amplified by summation. If a low-level stimulus is persistent it will generate several successive action potentials in the pre-synaptic neurone. The release of many vesicles of acetylcholine over a short period of time will enable the post-synaptic EPSPs to combine together to produce an action potential.

- After repeated stimulation a synapse may run out of vesicles containing the neurotransmitter. The synapse is said to be fatigued. This means the nervous system no longer responds to the stimulus – we have become **habituated** to it. It explains why we soon get used to a smell or a background noise. It may also help to avoid overstimulation of an effector, which could cause damage.

- The creation and strengthening of specific pathways within the nervous system is thought to be the basis of conscious thought and memory. Synaptic membranes are adaptable. In particular, the post-synaptic membrane can be made more sensitive to acetylcholine by the addition of more receptors. This means that a particular post-synaptic neurone is more likely to fire an action potential, creating a specific pathway in response to a stimulus.

Questions

1. Explain the difference between an action potential and a cell signal.

2. What is meant by the term cholinergic synapse?

3. Explain what is meant by summation.

4. Explain how modifying the post-synaptic membrane could mean that one post-synaptic neurone is more likely to fire an action potential than another.

5. Suggest an example where the organism could benefit from preventing an action potential by creating an IPSP.

THINKING BIGGER

AUTOIMMUNE DISEASE

Multiple sclerosis, or MS, is a disease of the nervous system. Most healthy neurones are insulated by myelin, a fatty substance which aids the flow of action potentials. In MS, the myelin is damaged or breaks down. This distorts or even blocks the flow of action potentials, resulting in the many symptoms of MS. The following newspaper article reports on a potential cure for MS.

COULD A CURE FOR MS AND DIABETES BE ON THE WAY?

SCIENTISTS have discovered how to "switch off" autoimmune diseases such as multiple sclerosis and Type 1 diabetes in a breakthrough which could improve the lives of millions of people.

British researchers have revealed how to stop cells from attacking healthy body tissue.

A team at the University of Bristol have discovered how cells convert from being aggressive, allowing the body's immune system to destroy its own tissue by mistake, to actually protecting against disease.

It is hoped the discovery will lead to the widespread use of a very targeted immunotherapy treatment for many autoimmune disorders, including multiple sclerosis (MS), Type 1 diabetes, Graves' disease and systemic lupus erythematosus (SLE).

MS alone affects around 100 000 people in the UK and around 400 000 have Type 1 diabetes.

Professor David Wraith, of the university's School of Cellular and Molecular Medicine, led the "exciting" research – which was funded by the Wellcome Trust.

He said: "Insight into the molecular basis of antigen-specific immunotherapy opens up exciting new opportunities to enhance the selectivity of the approach while providing valuable markers with which to measure effective treatment."

"These findings have important implications for the many patients suffering from autoimmune conditions that are currently difficult to treat."

In the study, published in the journal *Nature*, scientists were able to selectively target the cells that cause autoimmune disease by dampening down their aggression against the body's own tissue, while converting them into cells capable of protecting against disease.

This type of conversion has previously been applied to allergies, in a treatment known as "allergic desensitisation", but its application to autoimmune disorders has only recently been appreciated.

The researchers have now revealed how the administration of fragments of the proteins that are normally the target for the attack leads to correction of the autoimmune response.

Their work also shows that effective treatment can be achieved by gradually increasing the dose of antigenic fragment injected.

In order to analyse how this type of immunotherapy works, the scientists looked inside the immune cells themselves to see which genes and proteins were turned off by the treatment.

They found changes in gene expression that help explain how effective treatment leads to conversion of aggressor into protector cells.

The outcome is to reinstate self-tolerance, where an individual's immune system ignores its own tissues while remaining fully armed to protect against infection.

Researchers say that by specifically targeting the cells at fault, the immunotherapeutic approach avoids the need for immune-suppressive drugs.

These drugs are often associated with side effects such as infections, development of tumours and disruption of natural regulatory mechanisms.

The treatment approach is currently undergoing clinical development through biotechnology company Apitope, a spin-out from the University of Bristol.

Figure 1 A representation of damaged myelin (top) and healthy myelin (bottom).

Source
● http://www.express.co.uk/news/uk/506643/MS-Breakthrough-Cure-Diabetes

Where else will I encounter these themes?

Book 1 | 5.1 | 5.2 | 5.3 | YOU ARE HERE | 5.4 | 5.5 | 5.6

Let's start by considering the nature of the writing in the article.

1. After reading the article, do you think it is written for non-scientists? Use some examples of text to justify your answer.
2. What is the correct name for the 'fragments of the proteins that are normally the target for the attack'?

Now we will look at the biology in, or connected to, this article. Don't worry if you are not ready to give answers to these questions yet. You may like to return to the questions once you have covered other topics later in the book. Use the timeline at the bottom of the page to help you to put this work in context with what you have already learned and what is ahead in your course.

3. What is meant by the resting potential of a neurone?
4. Explain what is meant by an action potential?
5. Describe the effect of myelination on nerve action and explain how this effect is achieved.
6. Suggest how damage to the myelin sheath will affect nervous conduction and why this disruption to the nervous conduction affects the contraction of skeletal muscle.
7. What is meant by the terms:
   a. 'allergy'
   b. 'autoimmune disease'?

Activity

Design and write a leaflet for someone who has just been diagnosed with MS. The leaflet should aim to explain what MS does to the nerves and how this can cause the symptoms described above. The leaflet could then go on to explain how a process of desensitisation may prevent the immune cells from attacking the myelin sheath.

Your leaflet should contain:

- information about healthy neurones
- the role of neurones in the body
- the role of the immune system
- the effect of MS on the neurones
- an explanation of why this prevents normal conduction
- an explanation of how desensitisation could work (extension).

1. Figure 1 shows an action potential. Which row in the table below correctly identifies the stages of the action potential? [1]

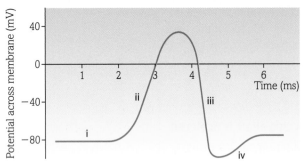

Figure 1

	i	ii	iii	iv
A	resting potential	hyperpolarisation	depolarisation	repolarisation
B	resting potential	depolarisation	repolarisation	hyperpolarisation
C	resting potential	repolarisation	hyperpolarisation	depolarisation
D	hyperpolarisation	depolarisation	repolarisation	resting potential

2. Figure 1 shows an action potential. Which row in the table below correctly identifies the ionic movements associated with the action potential? [1]

	i	ii	iii	iv
A	Na⁺ out, K⁺ in	K⁺ in	Na⁺ out	Na⁺ out, K⁺ in
B	Na⁺ out, K⁺ in	Na⁺ in	K⁺ out	K⁺ out
C	Na⁺ out, K⁺ in	Na⁺ in	K⁺ out	Na⁺ out, K⁺ in
D	Na⁺ out, K⁺ in	Na⁺ in	K⁺ out	Na⁺ out

3. Which row below correctly describes the structure of neurones? [1]

A. Both motor neurones and sensory neurones have a long dendron.

B. Both motor neurones and sensory neurones have a long axon.

C. Motor neurones have a long axon, sensory neurones have a long dendron.

D. Sensory neurones have a long axon, motor neurones have a long dendron.

4. What does the frequency of action potentials passing along a sensory neurone inform the brain about? [1]

A. The intensity of the stimulus.

B. How excited the animal is.

C. The type of stimulus.

D. The type of sensory receptor involved.

5. The roles of synapses include: [1]

A. reversing the action potential

B. cutting out low level responses

C. connecting several neurones together

D. initiating a response.

[Total: 5]

6. Table 1 shows the rate of conduction of an action potential in a number of different neurones in a human.

Type of neurone	Diameter of axon (μm)	Speed of conduction (ms⁻¹)
myelinated	15	120
myelinated	6	12
myelinated	1.5	5
unmyelinated	1.5	2

Table 1

(a) Describe the effect of reducing the diameter of a neurone on the rate of conduction. [2]

(b) Calculate how many times faster a myelinated neurone can conduct an action potential compared with an unmyelinated neurone of the same size. Show your working. [2]

(c) Explain why a myelinated neurone conducts more quickly than an unmyelinated neurone. [3]

[Total: 7]

7. (a) What is the role of a sensory receptor? [2]

(b) Figure 2 shows a sensory receptor.

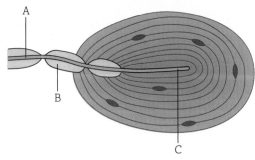

Figure 2

(i) Name the sensory receptor shown in Figure 2. [1]

(ii) Identify structures A and B. [2]

(iii) Describe the changes to the membrane at point C when the sensory receptor in Figure 2 is activated to produce an action potential. [7]

[Total: 12]

8. Figure 3 shows the mechanism of synaptic transmission.

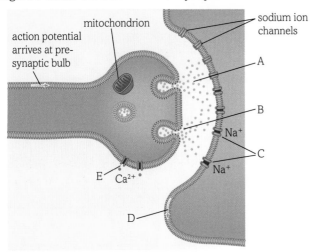

Figure 3

(a) Write the letters A–E in the correct sequence to describe the sequence of events that occurs after arrival of the action potential at the synapse. [5]

(b) Describe the process occurring at the following points.
   (i) A
   (ii) B
   (iii) C
   (iv) E [8]

[Total: 13]

9. Synapses have many important roles in communication.

(a) What is meant by an excitatory post-synaptic potential (EPSP)? [2]

(b) Explain what is meant by summation. [2]

(c) Explain the difference between spatial summation and temporal summation. [2]

(d) What is meant by an inhibitory post-synaptic potential (IPSP)? [2]

(e) Explain what is meant by the term 'all or nothing' when applied to creating a nervous impulse. [2]

[Total: 10]

10. (a) Complete the following paragraph. [6]

A resting neurone maintains a potential across the plasma membrane of 60 mV with the inside being .... compared with the outside. This potential is maintained by sodium/potassium pumps which use .... to pump three sodium ions out of the cell for every two potassium ions pumped in. There are also large organic .... inside the cell that help to maintain the potential difference across the membrane. An action potential occurs when ion channels open allowing .... ions to flood into the cell causing .... The membrane is repolarised by movement of potassium ions out of the cell when potassium ion channels open. After an action potential the membrane becomes .... due to too many potassium ions leaving the cell. This ensures that the neurone can rest briefly between action potentials.

(b) Describe how an action potential is transmitted along an axon. [6]

[Total: 12]

Communication, homeostasis and energy

HORMONAL COMMUNICATION

Introduction

Hormones are cell signalling molecules released from specific glands called endocrine glands. Hormones are released directly into the blood and are transported around the body. Each hormone has a specific target tissue and, in the case of non-steroid hormones, the cells in the target tissue must have specific membrane-bound receptors that detect the presence of the hormone. The presence of the hormone induces specific responses inside the cell.

In this chapter you will learn about certain hormones and the ways in which they bring about their effects:

- adrenaline from the adrenal glands
- adrenocorticoid hormones from the adrenal glands
- insulin and glucagon from the islets of Langerhans in the pancreas.

When hormonal control is defective, the balance inside the body can be disrupted and homeostasis fails. The effects of defective hormonal control are considered through an examination of diabetes. You will learn how blood glucose concentrations are normally controlled by the action of insulin and glucagon acting by negative feedback. You will also learn how medical technology can be used in overcoming defects in hormonal control systems.

All the maths you need

To unlock the puzzles of this chapter you need the following maths:

- Recognise and make use of appropriate units in calculations
- Estimate results
- Use an appropriate number of significant figures
- Construct and interpret frequency tables and diagrams, bar charts and histograms
- Understand and use the symbols: $=$, $<$, $\leq$, $\geq$, $>$, $\sim$
- Translate information between graphical, numerical and algebraic forms

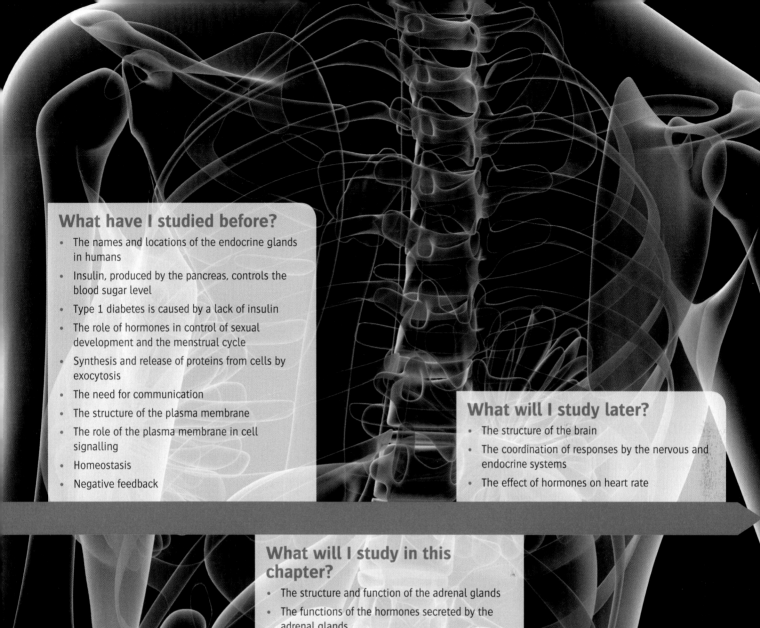

What have I studied before?

- The names and locations of the endocrine glands in humans
- Insulin, produced by the pancreas, controls the blood sugar level
- Type 1 diabetes is caused by a lack of insulin
- The role of hormones in control of sexual development and the menstrual cycle
- Synthesis and release of proteins from cells by exocytosis
- The need for communication
- The structure of the plasma membrane
- The role of the plasma membrane in cell signalling
- Homeostasis
- Negative feedback

What will I study later?

- The structure of the brain
- The coordination of responses by the nervous and endocrine systems
- The effect of hormones on heart rate

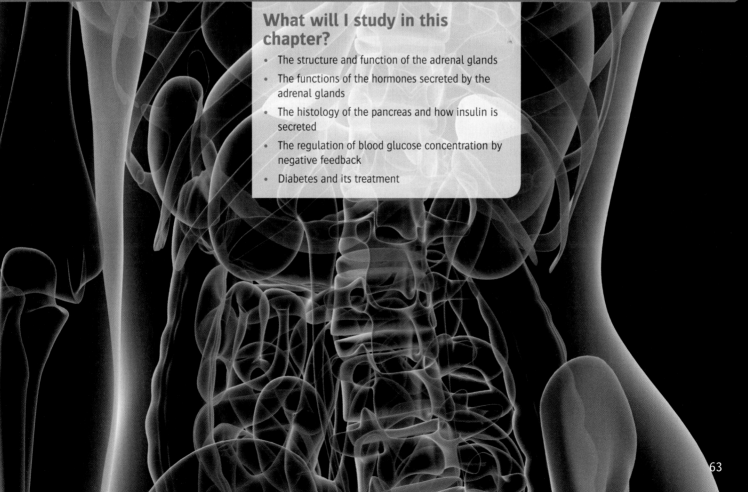

What will I study in this chapter?

- The structure and function of the adrenal glands
- The functions of the hormones secreted by the adrenal glands
- The histology of the pancreas and how insulin is secreted
- The regulation of blood glucose concentration by negative feedback
- Diabetes and its treatment

By the end of this topic, you should be able to demonstrate and apply your knowledge and understanding of:

* endocrine communication by hormones

KEY DEFINITIONS

endocrine system: a communication system using hormones as signalling molecules.

hormones: molecules (proteins or steroids) that are released by endocrine glands directly into the blood. They act as messengers, carrying a signal from the endocrine gland to a specific target organ or tissue.

target cells: for non-steroid hormones, cells that possess a specific receptor on their plasma (cell surface) membrane. The shape of the receptor is complementary to the shape of the hormone molecule. Many similar cells together form a target tissue.

Signalling using hormones

The **endocrine system** is another system, in addition to the nervous system, used for communication around the body. The endocrine system uses the blood circulatory system to transport its signals. The signals released by the endocrine system are molecules called **hormones**. The blood system transports materials all over the body, therefore any hormone released into the blood will be transported throughout the body.

Types of hormone

There are two types of hormone:

* protein and peptide hormones, and derivatives of amino acids (e.g. adrenaline, insulin and glucagon)
* steroid hormones (e.g. oestrogen and testosterone).

These two types of hormone work in different ways.

Proteins are not soluble in the phospholipid membrane and do not enter the cell. Protein hormones need to bind to the cell surface membrane and release a second messenger inside the cell.

Steroid hormones, however, can pass through the membrane and enter the cell and the nucleus, to have a direct effect on the DNA in the nucleus.

Endocrine glands

Hormones are released directly into the blood from **endocrine glands**. The endocrine glands are ductless glands – they consist of groups of cells that manufacture and release the hormone directly into the blood in capillaries running through the gland.

Figure 1 shows the major endocrine organs in the human body.

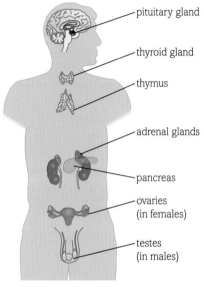

Figure 1 The endocrine organs in the human body.

pituitary gland
thyroid gland
thymus
adrenal glands
pancreas
ovaries (in females)
testes (in males)

Endocrine glands have groups of cells with associated capillaries, but no visible ducts (see topics 5.4.2 and 5.4.3 for examples of endocrine glands).

LEARNING TIP

There are two types of gland in the mammalian body: endocrine glands, which release hormones, and exocrine glands, which do not. Exocrine glands consist of groups of cells surrounding a small duct. These cells secrete their products into this duct, which then leads to the site where the secretion is required. For example, the salivary glands secrete saliva into a duct, which carries the saliva into the mouth.

Detecting the signal

Hormones always have a specific function – they are transported all over the body, but they have an effect in one type of tissue only. The cells receiving an endocrine signal are called **target cells**. These cells may be grouped together in a target tissue such as the epithelium of the collecting ducts, as described in topic 5.2.7. Alternatively they may be more widely dispersed in a number of tissues, such as the receptors for adrenalin found in the central nervous system and the tissues innervated by the peripheral nervous system including the heart, smooth muscle and skeletal muscle. For non-steroid hormones, the target cells must possess a specific receptor on their plasma membrane that is complementary in shape to the shape of the signalling molecule (hormone). The hormone binds to this receptor and initiates changes in the cell.

If all the cells in the body possess such a receptor then all the cells can respond to the signal. However, each hormone is different from all the others. This means that a hormone can be carried in the blood without affecting cells that do not possess the correct specific receptor. Only those specific target cells that possess the correct receptor will respond to the hormone.

DID YOU KNOW?

There are two types of adrenergic receptor. **Alpha receptors** are excitatory in smooth muscles and gland cells, but cause relaxation of intestinal smooth muscles. **Beta receptors** produce an inhibitory response, although in heart muscle the effect is excitatory. Beta blockers are drugs used to inhibit the response of these receptors to adrenalin in the control of certain heart conditions.

First and second messengers

Non-steroid hormones are known as **first messengers**. They are signalling molecules outside the cell that bind to the cell surface membrane and initiate an effect inside the cell. They usually cause the release of another signalling molecule in the cell, which is called the **second messenger**. The second messenger stimulates a change in the activity of the cell.

Many non-steroid hormones act via a G protein in the membrane. The G protein is activated when the hormone binds to the receptor. The G protein in turn activates an effector molecule – usually an enzyme that converts an inactive molecule into the active second messenger. In many cells the effector molecule is the enzyme **adenyl cyclase**, which converts ATP to cyclic AMP (cAMP), as shown in Figure 2. cAMP is the second messenger. This second messenger may act directly on another protein (such as an ion channel), or it may initiate a cascade of enzyme-controlled reactions that alter the activity of the cell.

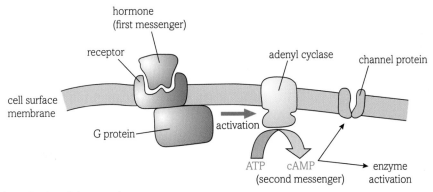

Figure 2 The activation of the second messenger.

Questions

1. What is the difference between exocrine and endocrine glands?

2. Compile a list of endocrine organs and the hormones they release (you may need to use your GCSE knowledge to recall the hormones).

3. What is the relationship between a first messenger and the receptor it combines with?

4. Explain why target cells and tissues must have specific receptors on their cell surface membranes.

5. Explain the difference between first messengers and second messengers.

6. Why are steroid hormones able to enter the cell?

Adrenal glands

By the end of this topic, you should be able to demonstrate and apply your knowledge and understanding of:

∗ the structure and functions of the adrenal glands

KEY DEFINITIONS

adrenal cortex: the outer layer of the adrenal gland.
adrenal gland: one of a pair of glands lying above the kidneys, which release adrenaline and a number of other hormones known as corticoids (or corticosteroids) such as aldosterone.
adrenaline: a hormone released from the adrenal glands, which stimulates the body to prepare for fight or flight.
adrenal medulla: the inner layer of the adrenal gland.

The structure of the adrenal glands

The adrenal glands are a good example of an endocrine gland. They are found lying anterior to (just above) the kidneys – one on each side of the body.

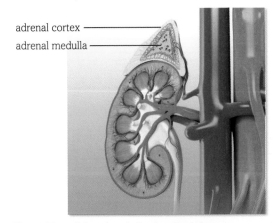

adrenal cortex ————
adrenal medulla ————

Figure 1 The position of the adrenal gland on the kidney.

Each gland is divided into the outer **adrenal cortex** and the inner **adrenal medulla**. Both regions are well supplied with blood vessels and produce hormones which are secreted directly into the blood vessels.

The adrenal cortex

The adrenal gland has an outer capsule surrounding three distinct layers of cells, which are the:

- **zona glomerulosa** – the outermost layer, which secretes **mineralocorticoids** such as aldosterone

- **zona fasciculata** – the middle layer, which secretes glucocorticoids such as cortisol

- **zona reticularis** – the innermost layer, which is thought to secrete precursor molecules that are used to make sex hormones.

The adrenal medulla

The adrenal medulla is found at the centre of the adrenal gland and secretes **adrenaline** and noradrenaline.

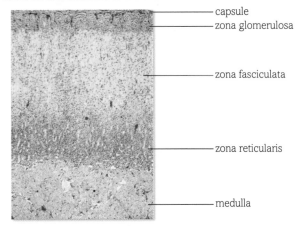

capsule
zona glomerulosa

zona fasciculata

zona reticularis

medulla

Figure 2 A section through the adrenal gland.

The functions of hormones from the adrenal glands

Hormones from the adrenal cortex

The adrenal cortex uses cholesterol to produce a range of hormones. These hormones are steroid based and are able to enter cells directly by dissolving into the cell surface membrane. The steroid hormones enter the nucleus and have a direct effect on the DNA to cause protein synthesis.

The action of steroid hormones can be summarised as follows.

1. The steroid hormone passes through the cell membrane of the target cell.

2. The steroid hormone binds with a specific receptor (with a complementary shape) in the cytoplasm.

3. The receptor–steroid hormone complex enters the nucleus of the target cell and binds to another specific receptor on the chromosomal material.

4. Binding stimulates the production of messenger RNA (mRNA) molecules, which code for the production of proteins.

Hormones from the adrenal cortex have a variety of roles in the body.

- Mineralocorticoids (e.g. aldosterone) from the zona glomerulosa help to control the concentrations of sodium and potassium in the blood. As a result they also contribute to maintaining blood pressure. Aldosterone acts on the cells of the distal tubules and collecting ducts in the kidney. It increases absorption of sodium ions, decreases absorption of potassium ions, and increases water retention so increasing blood pressure.

- Glucocorticoids (e.g. cortisol) from the zona fasciculata help to control the metabolism of carbohydrates, fats and proteins in the liver. Cortisol is released in response to stress or as a result of a low blood glucose concentration. It stimulates the production of glucose from stored compounds (especially glycogen, fats and proteins) in the liver.

- Cortisol may also be released by the zona reticularis. However, if the correct enzymes are not present for the release of cortisol, then the zona reticularis releases precursor androgens into the blood. These are taken up by the ovaries or testes and converted to sex hormones (e.g. testosterone in males and oestrogen in females). The sex hormones help development of the secondary sexual characteristics and regulate the production of gametes.

Adrenaline from the adrenal medulla

Adrenaline is released from the adrenal medulla into the blood and is transported throughout the body.

Adrenaline is a polar molecule derived from the amino acid tyrosine. This means that it cannot enter cells through the plasma membrane like a steroid hormone can. Therefore, it must be detected by specialised receptors on the plasma membrane of the target cells (see topic 5.4.1 for more detail). Many cells and tissues have adrenaline receptors. Therefore the effects of adrenaline are widespread. The role of adrenaline is to prepare the body for activity, which includes the following effects:

- relaxing smooth muscle in the bronchioles
- increasing stroke volume of the heart
- increasing heart rate
- causing general vasoconstriction to raise blood pressure
- stimulating conversion of glycogen to glucose
- dilating the pupils
- increasing mental awareness
- inhibiting the action of the gut
- causing body hair to stand erect.

DID YOU KNOW?

Adrenaline is known as the 'fight or flight' hormone. It prepares the body either to defend itself or run away to escape danger. However, its roles are wider than this and can be described as arousal, flight, fight and flirt.

An adrenaline junkie is someone who appears to be addicted to the effects of adrenaline. They appear to enjoy stressful activities or risky situations that cause the release of adrenaline giving them a 'high'.

Questions

1. Why are the adrenal cortex and the medulla well supplied with blood vessels?

2. What is the difference between the hormones released by the adrenal cortex and the hormones released by the medulla?

3. Describe how steroid hormones cause the production of new enzymes in their target cells.

4. Explain how increasing heart rate and stroke volume prepares the body for activity.

5. Explain why adrenaline causes an increase in blood glucose levels.

The pancreas and release of insulin

By the end of this topic, you should be able to demonstrate and apply your knowledge and understanding of:

* the histology of the pancreas
* the examination and drawing of stained sections of the pancreas to show the histology of the endocrine tissues
* how blood glucose concentration is regulated (continued in topic 5.4.4)

KEY DEFINITIONS

beta cells: cells found in the islets of Langerhans that secrete the hormone insulin.

glucagon: a hormone that causes an increase in blood glucose concentration.

insulin: the hormone, released from the pancreas, that causes blood glucose levels to go down.

The pancreas

The pancreas is a small organ lying below the stomach. It is unusual in that it has both exocrine and endocrine functions. The two main secretions of the pancreas are:

* pancreatic juices containing enzymes which are secreted into the small intestine
* hormones which are secreted from the islets of Langerhans into the blood.

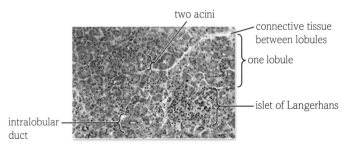

Figure 1 A section through a part of the pancreas as seen using a light microscope (×210).

Exocrine function

Exocrine glands secrete substances into a duct. Most cells in the pancreas synthesise and release digestive enzymes. This is the exocrine function of the pancreas. The exocrine cells are in small groups surrounding tiny tubules. Each group of cells is called an **acinus** (plural acini). The acini are grouped together into small lobules separated by connective tissue. The cells of the acini secrete the enzymes they synthesise into the tubule at the centre of the group. The tubules from the acini join to form intralobular ducts that eventually combine to make up the pancreatic duct. The pancreatic duct carries the fluid containing the enzymes into the first part of the small intestine (duodenum).

The fluid from the pancreatic duct contains the following enzymes:

* pancreatic amylase – a carbohydrase which digests amylose to maltose
* trypsinogen – an inactive protease which will be converted to the active form trypsin when it enters the duodenum
* lipase – which digests lipid molecules.

The fluid also contains sodium hydrogencarbonate, which makes it alkaline. This helps to neutralise the contents of the digestive system that have just left the acid environment of the stomach.

Endocrine function

Dispersed in small patches among the lobules of acini are the islets of Langerhans.

The islets of Langerhans contain the **alpha (α) cells** and **beta (β) cells** that make up the endocrine tissue in the pancreas. The alpha cells secrete **glucagon** (see topic 5.4.4) and the beta cells secrete **insulin**.

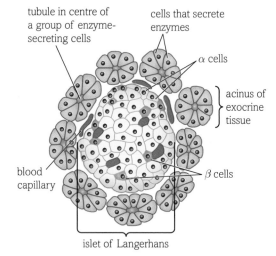

Figure 2 A drawing of a section through the pancreas.

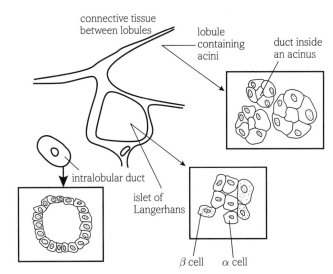

Figure 3 A tissue plan showing alpha and beta cells in the islets of Langerhans.

Releasing insulin

When insulin is secreted from the beta cells in the islets of Langerhans, it brings about effects that reduce the blood glucose concentration. If the blood glucose concentration is too high then it is important that insulin is released from the beta cells. However, if the blood glucose concentration drops too low it is important that insulin secretion stops.

The control of blood glucose concentration is described in the next topic. Here we consider how the insulin is released from the beta cells.

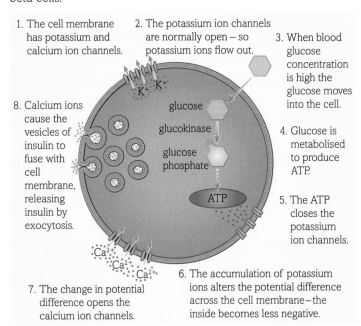

Figure 4 The mechanism of insulin secretion.

1. The cell membranes of the beta cells contain both calcium ion channels and potassium ion channels.

2. The potassium ion channels are normally open and the calcium ion channels are normally closed. Potassium ions diffuse out of the cell making the inside of the cell more negative; at rest the potential difference across the cell membrane is about −70 mV.

3. When glucose concentrations outside the cell are high, glucose molecules move into the cell.

4. The glucose is quickly used in metabolism to produce ATP. This involves the enzyme glucokinase.

5. The extra ATP causes the potassium channels to close.

6. The potassium can no longer diffuse out and this alters the potential difference across the cell membrane – it becomes less negative inside.

7. This change in potential difference opens the calcium ion channels.

8. Calcium ions enter the cell and cause the secretion of insulin by making the vesicles containing insulin move to the cell surface membrane and fuse with it, releasing insulin by exocytosis.

DID YOU KNOW?

Glucokinase acts as a glucose sensor and triggers a change in the metabolism of glucose in the cells. Some people have a mutated version of the gene for glucokinase. This gene produces an enzyme that is not as sensitive to glucose and leads to a raised glucose concentration in the blood.

Questions

1 Distinguish clearly between the exocrine and endocrine functions of the pancreas.

2 Suggest how the cells of the acini are specialised to perform their function.

3 Suggest why the protease trypsin is released in the inactive form trypsinogen.

4 Why is it essential that insulin is not secreted in large amounts continuously?

5 Compare the secretion of insulin with the secretion of a neurotransmitter at a synapse.

By the end of this topic, you should be able to demonstrate and apply your knowledge and understanding of:

* how blood glucose concentration is regulated

Blood glucose concentration

The concentration of blood glucose is carefully regulated. The normal blood concentration of glucose is between 4 and 6 mmol dm^{-3}.

If a person's blood glucose concentration is allowed to drop below 4 mmol dm^{-3} and remain too low for long periods the person is said to be hypoglycaemic. The main problem caused by **hypoglycaemia** is an inadequate delivery of glucose to the body tissues and, in particular, to the brain. Mild hypoglycaemia may simply cause tiredness and irritability. However, in severe cases there may be impairment of brain function and confusion, which may lead on to seizures, unconsciousness and even death.

If blood glucose concentration is allowed to rise too high for long periods this is known as **hyperglycaemia**. Permanently high blood glucose concentrations can lead to significant organ damage. A blood glucose concentration that is consistently higher than 7 mmol dm^{-3} is used as the diagnosis for **diabetes mellitus** (see topic 5.4.5).

The cells in the islets of Langerhans constantly monitor the concentration of glucose in the blood. If the concentration rises or falls away from the acceptable concentration then the alpha and beta cells in the islets of Langerhans detect the change and respond by releasing the relevant hormone: insulin if blood glucose is high, and glucagon if it is too low.

These hormones act on the cells in the liver (hepatocytes) which can store glucose in the form of glycogen. When there is excess glucose in the blood it is converted to glycogen. If glucose is needed to raise the blood concentration then glycogen is converted back to glucose.

If blood glucose rises too high

A high blood glucose concentration is detected by the beta cells in the islets of Langerhans. The beta cells respond by secreting insulin into the blood (as described in topic 5.4.3). Insulin travels throughout the body in the circulatory system. The target cells are the liver cells, muscle cells and some other body cells including those in the brain.

Human insulin is a small protein of 51 amino acids, therefore it is unable to pass through the cell surface membrane. The target cells possess the specific membrane-bound receptors for insulin. When insulin binds to the insulin receptor, this activates the enzyme tyrosine kinase which is associated with the receptor on the inside of the membrane. Tyrosine kinase causes **phosphorylation** of inactive enzymes in the cell. This activates the enzymes leading to a cascade of enzyme-controlled reactions inside the cell.

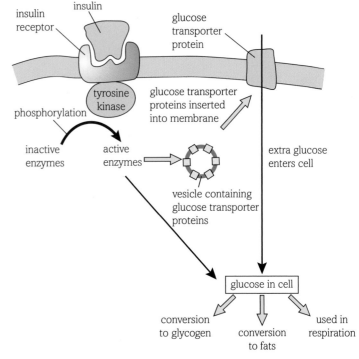

Figure 1 The action of insulin on the liver cells.

Insulin has several effects on the cell:

* More transporter proteins specific to glucose are placed into the cell surface membrane. This is achieved by causing vesicles containing these transporter proteins to fuse with the membrane.

* More glucose enters the cell.

* Glucose in the cell is converted to glycogen for storage (**glycogenesis**).

* More glucose is converted to fats.

* More glucose is used in respiration.

The increased uptake of glucose, through the specific transporter proteins, reduces the blood glucose concentration.

If blood glucose drops too low

A low blood glucose concentration is detected by the alpha cells in the islets of Langerhans. The alpha cells then secrete the hormone glucagon into the blood.

Glucagon is a small protein containing 29 amino acids. Its target cells are the hepatocytes (liver cells), which possess the specific receptor for glucagon. When the blood passes these cells the glucagon binds to the receptors. This stimulates a G protein (see topic 5.4.1) inside the membrane, which activates the adenyl cyclase inside each cell. The adenyl cyclase converts ATP to cAMP, which activates a series of enzyme-controlled reactions in the cell (this mechanism is described in topic 5.4.1). The effects of glucagon include the following.

- Glycogen is converted to glucose (**glycogenolysis**) by phosphorylase A, which is one of the enzymes activated in the cascade.

- More fatty acids are used in respiration.

- Amino acids and fats are converted into additional glucose, by **gluconeogenesis**.

The overall effect of these changes is to increase the blood glucose concentration.

Negative feedback

The concentration of blood glucose is controlled by a negative feedback mechanism involving both the hormones insulin and glucagon. The hormones are antagonistic – they have opposite effects on blood glucose concentration. One of their effects is to inhibit the effects of the opposing hormone.

The action of these hormones in negative feedback is summarised in Figure 2. It should be noted that the concentration of glucose in the blood will not remain constant – it will fluctuate around the required concentration. When the glucose concentration rises too high the release of insulin will act to bring the concentration down. But if the concentration falls too low the release of glucagon will act to raise the concentration again.

LEARNING TIP

It is easy to mix up the spellings of the terms 'glucose', 'glucagon' and 'glycogen'. Add in the terms 'glycogenesis', 'gluconeogenesis' and 'glycogenolysis' and things get very confusing. Write these terms out on a separate page of your notebook along with their meanings. Make sure you can spell them accurately.

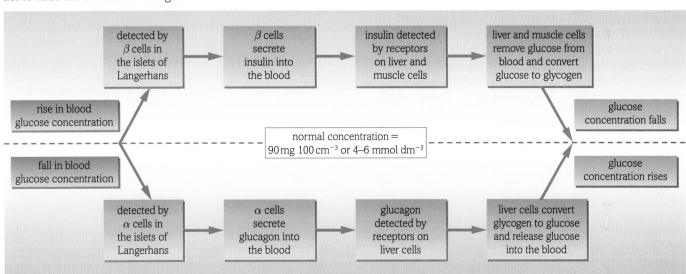

Figure 2 The control of blood glucose concentration by negative feedback.

Questions

1. The normal concentration of glucose in the blood is 4–6 mmol dm^{-3}. Explain what is meant by mmol dm^{-3}.

2. Why do hepatocytes have specialised receptors for both insulin and glucagon?

3. State which molecules named in this topic are first messengers and which are second messengers.

4. Explain why negative feedback is used to regulate the glucose concentration.

5. Why does the concentration of glucose in the blood fluctuate?

5 Diabetes

By the end of this topic, you should be able to demonstrate and apply your knowledge and understanding of:

* the differences between Type 1 and Type 2 diabetes mellitus

* the potential treatments for diabetes mellitus

> **KEY DEFINITIONS**
>
> **diabetes mellitus:** a condition in which blood glucose concentrations cannot be controlled effectively.
> **stem cells:** unspecialised cells that have the potential to develop into any type of cell.

Diabetes mellitus

Diabetes mellitus is a condition in which the body is no longer able to produce sufficient insulin to control its blood glucose concentration. This can lead to prolonged very high concentrations of glucose (hyperglycaemia) after a meal rich in sugars and other carbohydrates, as shown in Figure 1, which compares blood insulin and glucose concentrations in a normal and a diabetic person. It can also lead to the concentration dropping too low (hypoglycaemia) after exercise or after fasting.

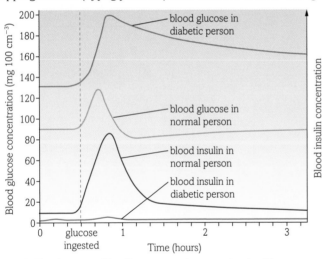

Figure 1 The glucose and insulin concentrations seen in a healthy person and a diabetic.

Type 1 diabetes

Type 1 diabetes is also known as insulin-dependent diabetes, or as juvenile-onset diabetes because it usually starts in childhood. It is thought to be the result of an **autoimmune response** in which the body's immune system attacks and destroys the beta cells. Type 1 diabetes may also result from a viral attack.

In a healthy person, glucose is absorbed into the blood and any excess is converted to glycogen in the liver and muscles. This glycogen can then be used to release glucose when blood glucose concentration falls.

A person with Type 1 diabetes is no longer able to synthesise sufficient insulin and cannot store excess glucose as glycogen. Excess glucose in the blood is not removed quickly, leaving a prolonged period of high concentration. However, when the blood glucose falls, there is no store of glycogen that can be used to release glucose. Therefore, the blood glucose concentration falls too low. This is when a diabetic can suffer a 'hypo' – a period of hypoglycaemia.

Type 2 diabetes

Type 2 diabetes is also known as non-insulin-dependent diabetes. A person with Type 2 diabetes can produce insulin, but not enough. Also, as people age, their responsiveness to insulin declines. This is probably because the specific receptors on the surface of the liver and muscle cells become less responsive and the cells lose their ability to respond to the insulin in the blood.

In Type 2 diabetes the blood glucose concentration is almost permanently raised, which can damage the major organs and circulation.

Certain factors seem to bring an earlier onset of Type 2 diabetes. These include:

* obesity
* lack of regular exercise
* a diet high in sugars, particularly refined sugars
* being of Asian or Afro-Caribbean origin
* family history.

> **DID YOU KNOW?**
>
> Anyone who lives long enough will eventually become diabetic, as their pancreas secretes less insulin and their liver cells become less responsive – but this may not be until they are about 120 years old!

Treating diabetes

The effects of diabetes are severe and become worse with time. It is therefore important that diabetes is diagnosed as early as possible so that treatment can be started.

Treating Type 1 diabetes

Type 1 diabetes is usually treated using insulin injections. The blood glucose concentration must be monitored and the correct dose of insulin administered to keep the glucose concentration fairly stable.

Alternatives to insulin injections include:

- insulin pump therapy – a small device constantly pumps insulin (at a controlled rate) into the bloodstream through a needle that is permanently inserted under the skin
- islet cell transplantation – healthy beta cells from the pancreas of a deceased donor are implanted into the pancreas of someone with Type 1 diabetes
- a complete pancreas transplant.

In addition, recent research has shown that it may be possible to treat Type 1 diabetes using **stem cells** to grow new islets of Langerhans in the pancreas. Stem cells are not yet differentiated and can be induced to develop into a variety of cell types.

The most common sources of stem cells are bone marrow and the placenta. However, scientists have found precursor cells in the pancreas of adult mice. These cells are capable of developing into a variety of cell types and may be true stem cells. If similar cells can be found in the human pancreas then they could be used to produce new beta cells in patients with Type 1 diabetes. This would give the patient freedom from daily insulin injections.

Treating Type 2 diabetes

Type 2 diabetes is usually treated by changes in lifestyle. A type 2 diabetic will be advised to lose weight, exercise regularly and carefully monitor their diet, taking care to match carbohydrate intake and use. This may be supplemented by medication that reduces the amount of glucose the liver releases to the bloodstream or that boosts the amount of insulin released from the pancreas.

In severe cases, further treatment may include insulin injections or the use of other drugs that slow down the absorption of glucose from the digestive system.

The source of insulin for treating diabetes

Insulin used to be extracted from the pancreas of animals – usually from pigs as this matches human insulin most closely. However, more recently, insulin has been produced by *Escherichia coli* bacteria that have undergone **genetic modification** to manufacture human insulin.

The advantages of using insulin from genetically modified bacteria include the following.

- It is an exact copy of human insulin, therefore it is faster acting and more effective.
- There is less chance of developing tolerance to the insulin.
- There is less chance of rejection due to an immune response.
- There is a lower risk of infection.
- It is cheaper to manufacture the insulin than to extract it from animals.
- The manufacturing process is more adaptable to demand.
- Some people are less likely to have moral objections to using the insulin produced from bacteria than to using that extracted from animals.

Questions

1. Insulin is a protein. Why must insulin be injected rather than being taken orally?

2. Explain the meaning of the terms hyperglycaemia and hypoglycaemia.

3. How does regular exercise help to control Type 2 diabetes?

4. What are the advantages of an insulin pump over daily injections?

5. Explain why the insulin produced by genetically engineered bacteria can be considered to be human insulin.

THINKING BIGGER

DIAGNOSING DIABETES

The enzyme glucokinase found in β cells in the pancreas acts as a trigger to release insulin from the cell. Scientists have found a mutated version of the gene, which codes for a slightly different version of the glucokinase gene. People who possess this gene have slightly raised blood glucose concentrations (hyperglycaemia), which can often mean they are incorrectly diagnosed with diabetes. The following piece is a selection from an article about testing people with this modified glucokinase gene for diabetes.

IMPACT OF USING FPG OR HBA1C FOR DIABETES DIAGNOSIS IN *GCK* MODY

Introduction

There is a contemporary drive to focus on haemoglobinA1c (HbA1c) rather than fasting plasma glucose (FPG) or oral glucose tolerance test (OGTT) in the diagnosis of diabetes. Maturity onset diabetes of the young (MODY) is a rare form of diabetes caused by a mutation in a single gene. Patients with MODY due to a heterozygous inactivating mutation in the *GCK* gene are frequently misdiagnosed with Type 1 (T1D) or Type 2 (T2D) diabetes, yet they have lifelong mild fasting hyperglycaemia, pharmacological treatment is rarely required and the development of complications is atypical. These patients have traditionally been identified as suitable for genetic testing if they have a FPG in the range of 5.5–8.4 mmol/l or a small (<3 mmol/l) OGTT 2-hour increment.

The range of HbA1c has not been systematically studied in *GCK* families. The current recommendation that an HbA1c of ≥48 mmol/mol is diagnostic of diabetes could mean many more patients with *GCK* mutation-related hyperglycaemia will be misdiagnosed with either T1D or T2D compared with using FPG or 2-hour OGTT values and be inappropriately treated.

We aimed to study HbA1c in families with a *GCK* mutation. The objectives were to: 1) determine how effective HbA1c is as a discriminator between those with a mutation, those without a mutation (controls) and those with other young-onset diabetes; 2) to establish age-related HbA1c reference ranges in order to identify levels to help select those patients most likely to have hyperglycaemia due to a *GCK* mutation; 3) to investigate how effective upper and lower reference ranges are in distinguishing between GCK-MODY, controls and young-onset T1D and T2D and 4) to investigate whether changing from a glucose-based diagnostic criteria to an HbA1c-based diagnostic criteria will alter the proportion of patients with *GCK* mutations who are diagnosed with diabetes.

Discussion

In known *GCK* families, either FPG or HbA1c may be used to distinguish between individuals likely or unlikely to have a mutation.

However, the care and screening provided for patients with a mutation should be based on their clinical characteristics and should not vary if the HbA1c or fasting glucose is just above or just below the criteria for diabetes either by FPG or HbA1c.

Previous work has shown that 38% of patients with a *GCK* mutation would be classified as having diabetes using FPG alone (>7.0 mmol/l) and that only 24% of these would also have a 2-hour OGTT value in the range where diabetes was diagnosed. Population screening for diabetes now frequently uses HbA1c rather than FPG. Our results suggest that more individuals with a *GCK* mutation will be diagnosed with diabetes using an HbA1c ≥48 mmol/mol (6.5%) rather than using FPG of ≥7.1 mmol/l (68% vs. 48%). Using both HbA1c *and* FPG would mean only 19% of subjects with the *GCK* mutations would be diagnosed with diabetes.

If the correct diagnosis of a *GCK* mutation is not made, patients may receive unnecessary treatment for hyperglycaemia, more aggressive treatment of risk factors such as high blood pressure and cholesterol and an increase in screening for complications. Since studies to date have not established that patients with a *GCK* mutation have an increased risk of macrovascular disease, such treatment is unlikely to be beneficial. Serious microvascular complications have not been reported in children and young adults with a mutation and they are very rare, even after a lifetime of untreated *GCK*-related hyperglycaemia.

It is important that those with stable, mild hyperglycaemia (either by FPG or HBA1c) be considered for *GCK* mutation testing to prevent inappropriate management as T1D or T2D. This should also avoid the adverse impact of making a diagnosis of diabetes on insurance and employment opportunities.

Source

● http://www.plosone.org/article/info%3Adoi%2F10.1371%2Fjournal.pone.0065326#pone-0065326-g005

Where else will I encounter these themes?

Let's start by considering the nature of the writing in the article.

1. This article was presented in an open access forum for peer review. Select some examples from the text to demonstrate that this article is written for scientists to read.
2. What is meant by the terms 'microvascular' and 'macrovascular'?

Now we will look at the biology in, or connected to, this article. Don't worry if you are not ready to give answers to these questions yet. You may like to return to the questions once you have covered other topics later in the book. Use the timeline at the bottom of the page to help you to put this work in context with what you have already learned and what is ahead in your course.

3. Explain the differences between Type 1 and Type 2 diabetes.
4. Glucokinase activates glucose by phosphorylation. Explain why a slightly different version of the enzyme may not be as active as a normal type.
5. In the article the writer suggests that people with stable mild hyperglycaemia should be tested for the GCK mutation.
   a. What is meant by stable mild hyperglycaemia?
   b. Why should family members also be tested?
6. The difference between the normal GCK gene and the modified version is likely to be a point mutation. What is a point mutation?
7. Explain how a point mutation can result in a slightly different protein.
8. Explain why point mutations sometimes make no difference to the protein coded for.
9. If one parent has the mutated GCK gene, what is the chance that their second child will have the gene?

Activity

Plan a story board for an animation that explains the role of glucokinase in the β cells and how a modified version of the gene could lead to higher glucose concentrations in the blood. Your poster should include the following information:

- how the structure of protein is coded in the DNA
- how a mutation could lead to a slightly different enzyme molecule
- how the enzyme (glucokinase) acts in the β cell
- how the action of glucokinase causes the release of insulin
- how insulin affects the liver cells
- how blood glucose concentration is regulated through negative feedback.

5.7　6.1　6.2　6.3　6.4　6.5　6.6

Practice questions

1. Which of the following is not a response to the release of adrenaline? [1]

   A. Body hair stands erect.

   B. Increase in heart rate.

   C. Decrease in action of the muscles in the digestive system.

   D. Constriction of the bronchioles.

2. Which row correctly describes a hormone produced by the endocrine gland and its function? [1]

	Endocrine gland	Hormone released	Function of hormone
A	Adrenal gland	Glucocorticoid	Regulation of carbohydrate metabolism
B	Pituitary gland	Antidiuretic hormone	Decrease absorption of water from the collecting ducts
C	Pancreas	Glucagon	Decrease blood glucose concentration
D	Pancreas	Lipase	Digestion of lipid molecules

3. Read the following statements about the endocrine system.
   (i) All hormones are proteins.
   (ii) The target cell for a hormone possesses a receptor protein that is complementary in shape to the hormone.
   (iii) Hormones are called first messengers.
   (iv) Second messengers act inside cells.
   (v) Hormones are transported in the blood.
   (vi) Peptide hormones are released from the endocrine cells by exocytosis.

   Which statements are correct? [1]

   A. All the statements are correct.

   B. (ii), (iii), (v) and (vi) are correct.

   C. All the statements except (i) are correct.

   D. All the statements except (i) and (vi) are correct.

4. Which statement about the pancreas is correct? [1]

   A. α cells release insulin.

   B. β cells release glucagon.

   C. β cells contain glucokinase.

   D. α cells store glucose as glycogen.

5. Which statement about adrenaline is correct? [1]

   A. Adrenaline is a steroid hormone.

   B. Adrenaline is released from the adrenal medulla.

   C. Adrenaline consists of 41 amino acids.

   D. There are many different forms of adrenaline that each target a different tissue.

   [Total: 5]

6. Figure 1 shows a section through the pancreas.

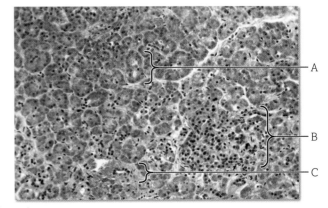

Figure 1

   (a) Identify the components labelled A and C. [2]

   (b) Name two types of cell found in component B and state the function of each. [4]

   (c) The cells in component B have certain adaptations in common. Explain why the cells in component B contain:
   (i) many ribosomes
   (ii) obvious Golgi apparatus. [2]

   [Total: 8]

7. (a) State three differences between Type 1 and Type 2 diabetes. [3]

   (b) Suggest what advice a medical professional might give to a person with Type 2 diabetes. [3]

   (c) Explain the potential advantages of treating Type 1 diabetes with stem cells. [3]

   [Total: 9]

8. Table 1 shows the effect of various foods on the blood glucose concentration and on the insulin concentration in the blood.

Food	Main component of food	Effect on blood glucose concentration	Insulin concentration in blood 1 hour after consumption
Pasta	Starch	Large increase	High
Bacon	Protein and fat	No effect	Low
Milk	Fat, lactose and water	Small increase	Moderate
Fibre-rich cereal	Cellulose	No effect	Low
Sugary sweets	Sucrose	Moderate increase	Moderate

Table 1

(a) Starch and cellulose are both polymers of glucose. Explain why they have such different effects on the blood glucose concentration. [4]

(b) Describe the response of the cells in the pancreas to eating a lot of pasta. [7]

(c) Which food would be best for a marathon runner to eat the night before a marathon?

Justify your answer. [2]

(d) Select two foods that could be recommended to a person with Type 2 diabetes who is controlling their diabetes by diet alone. Justify your choices. [2]

[Total: 15]

9. (a) Insulin and glucagon are described as antagonistic. Explain why. [2]

(b) Figure 2 shows how insulin affects a liver cell.

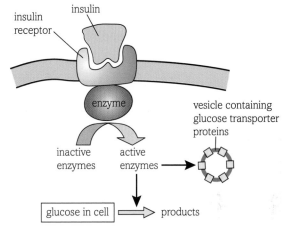

Figure 2

Describe the effects that insulin has on the liver cells. [5]

(c) Identify three risk factors that are linked to Type 2 diabetes. [3]

[Total: 10]

MODULE 5
Communication, homeostasis and energy

CHAPTER 5.5
PLANT AND ANIMAL RESPONSES

Introduction

Plant responses to environmental changes are just as important as those of animals. Plants respond by changes in their growth. Some of these changes can be very rapid, for example the closing of a Venus flytrap to trap food or the closing of a leaf on the sensitive *Mimosa* plant. Other growth responses are gradual and result in more permanent changes to the plant. These include the growth of shoots towards light and the growth of roots towards gravity.

Plant responses are coordinated by plant growth substances or hormones. Many of these plant hormones have important commercial uses such as producing seedless fruit, acting as a selective weedkiller and stimulating the growth of roots in cuttings.

In animals, responding to changes in the environment is a much more complex process and often involves coordination between the neuronal and endocrine systems. Most animals are continuously responding to changes in their environment through coordinated muscular action. Many processes may be unconscious and these are coordinated by the hind brain through a set of motor pathways known as the autonomic nervous system. Conscious activity is coordinated by the higher parts of the brain, such as the cerebrum.

All the maths you need

To unlock the puzzles of this chapter you need the following maths:

- Recognise and make use of appropriate units in calculations
- Recognise and use expressions in decimal form and standard form
- Use ratios, fractions and percentages
- Use an appropriate number of significant figures
- Construct and interpret frequency tables and diagrams, bar charts and histograms
- Understand simple probability
- Find arithmetic means
- Understand the terms mean, median and mode
- Select and use a statistical test
- Understand measure of dispersion, including standard deviation and range
- Translate information between graphical, numerical and algebraic forms
- Plot two variables from experimental data

What have I studied before?

- Plants respond to the environment
- Plant hormones can be used in agriculture
- Auxin released in the tip of a plant can become unevenly distributed
- The nervous system is used to detect changes in our environment
- The nervous system consists of the central nervous system and the peripheral nervous system
- A simple reflex arc
- The need for communication
- The action of neurones and the roles of synapses
- The endocrine system
- The role of membranes as selectively permeable barriers

What will I study later?

- How ATP for muscular contraction is produced
- Production of cuttings from plants
- Artificial cloning through micropropagation and tissue culture

What will I study in this chapter?

- The range of plant responses to changes in the environment
- The role of hormones in plant responses
- How plant hormones can be used commercially
- The organisation of the nervous system into somatic and autonomic systems
- The structure of the brain and the function of parts of the brain
- Reflex actions
- Coordinated responses using both the neuronal and endocrine systems, including control of heart rate
- The different types of muscle
- The structure and action of skeletal muscle

Plant responses to the environment

By the end of this topic, you should be able to demonstrate and apply your knowledge and understanding of:

* the types of plant responses
* the roles of plant hormones

Plant responses to external stimuli

It seems obvious that animals respond to the **biotic** (living) and **abiotic** (non-living) components of their environment. Plants also respond to external stimuli.

Types of stimuli

Responding to the environment may help plants to survive long enough to reproduce. For example, in higher temperatures, plants may deposit thicker layers of wax on their leaves; in very windy conditions, they may have vascular tissue which is more heavily lignified. Plants show specific responses to the threat of herbivores, employing the following chemical defences:

* **Tannins** are toxic to microorganisms and larger herbivores. In leaves, they are found in the upper epidermis, and make the leaf taste bad. In roots, they prevent infiltration by pathogenic microorganisms.

* **Alkaloids** are derived from amino acids. In plants, scientists think they are a feeding deterrent to animals, tasting bitter. They are located in growing tips and flowers, and peripheral cell layers of stems and roots.

* **Pheromones** are chemicals which are released by one individual and which can affect the behaviour or physiology of another.

Types of response

Tropisms are directional growth responses of plants. They include:

* phototropism – shoots grow towards light (they are positively phototropic), which enables them to photosynthesise

* geotropism – roots grow towards the pull of gravity. This anchors them in the soil and helps them to take up water, which is needed for support (to keep cells turgid), as a raw material for photosynthesis and to help cool the plant. There will also be minerals, such as nitrate in the water, needed for the synthesis of amino acids

* **chemotropism** – on a flower, pollen tubes grow down the style, attracted by chemicals, towards the ovary where fertilisation can take place (see Figure 1)

* **thigmotropism** – shoots of climbing plants, such as ivy, wind around other plants or solid structures to gain support.

LEARNING TIP

Don't confuse '-tropic' with '-trophic'. The first describes a growth response towards or away from a stimulus (e.g. photo*tropic* response). Tro*phic* is connected with how living things feed (e.g. plants are auto*trophic*).

KEY DEFINITIONS

abiotic components: components of an ecosystem that are non-living.

alkaloids: organic nitrogen-containing bases that have important physiological effects on animals; includes nicotine, quinine, strychnine and morphine.

biotic components: components of an ecosystem that are living.

pheromone: any chemical substance released by one living thing, which influences the behaviour or physiology of another living thing.

tannins: phenolic compounds, located in cell vacuoles or in surface wax on plants.

tropism: a directional growth response in which the direction of the response is determined by the direction of the external stimulus.

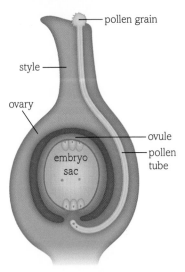

Figure 1 Longitudinal section showing a pollen tube growing from a pollen grain towards the ovary of a flower.

If a plant responds towards a stimulus, it is a *positive tropic* response. If a plant responds away from a stimulus, it is still a *tropic* (directional) response, but *negative* rather than positive.

Non-directional responses to external stimuli are called **nastic responses**. The sensitive plant, *Mimosa pudica*, responds to touch with a sudden folding of the leaves (see Figure 2). This response is an example of **thigmonasty**.

Figure 2 Thigmonastic response of the sensitive plant.

Control of responses – the role of plant hormones

Hormones coordinate plant responses to environmental stimuli. Like animal hormones, plant hormones are chemical messengers that can be transported away from their site of manufacture to act in other parts (target cells or tissues) of the plant. They are not produced in endocrine glands, but by cells in a variety of tissues in the plant.

When hormones reach their target cells, they bind to receptors on the plasma membrane. Specific hormones have specific shapes, which can only bind to specific receptors with complementary shapes on the membranes of particular cells. This specific binding makes sure that the hormones only act upon the correct tissues. Some hormones can have different effects on different tissues; some can amplify each other's effects, and some can even cancel out each other's effects. Hormones can influence cell division, cell elongation or cell differentiation. You can see some hormones and their effects in Table 1.

Hormone	Effects
Cytokinins	Promote cell division Delay leaf senescence Overcome apical dominance Promote cell expansion
Abscisic acid	Inhibits seed germination and growth Causes stomatal closure when the plant is stressed by low water availability
Auxins e.g. IAA (indole-3-acetic acid)	Promote cell elongation Inhibit growth of side-shoots Inhibit leaf abscission (leaf fall)
Gibberellins	Promote seed germination and growth of stems
Ethene	Promotes fruit ripening

Table 1 Plant hormones and their effects.

Hormones move around the plant in any of the following ways:

- active transport
- diffusion
- mass flow in the phloem sap or in xylem vessels.

Questions

1. How might thigmonastic responses in plants help them survive?

2. What is the difference between a tropic and a nastic response?

3. Suggest why plant growth regulators are called hormones, although they are not produced in endocrine glands.

4. Explain why only certain tissues in a plant respond to a particular plant hormone.

5. Ethene is a gas. It is produced by flowers, stems and leaves and causes ripening of fruit and ageing of flowers. Ripe fruit also produces it. Bananas produce lots of it. Why is it not advisable to place a vase of freshly cut flowers next to a fruit bowl containing bananas?

By the end of this topic, you should be able to demonstrate and apply your knowledge and understanding of:

* the experimental evidence for the role of auxins in the control of apical dominance
* the experimental evidence for the role of gibberellin in the control of stem elongation and seed germination

Auxins

Auxins are plant hormones which are responsible for regulating plant growth.

If you break the shoot tip (apex) off a plant, the plant starts to grow side branches from lateral buds that were previously dormant (Figure 1).

Researchers suggested that auxins from the apical bud prevent lateral buds from growing. When the tip is removed, auxin levels in the shoot drop and the buds grow.

To test their hypothesis, scientists applied a paste containing auxins to the cut end of the shoot, and the lateral buds did not grow.

However, scientists' manipulation of the plants could have had an unexpected effect; upon exposure to oxygen, cells on the cut end of the stem could have produced a hormone that promoted lateral bud growth. Because of this, scientists applied a ring of auxin transport inhibitor below the apex of the shoot. The lateral buds grew.

Based on this result, scientists suggested that a normal auxin level in lateral buds inhibits growth, whereas low auxin levels promote growth. However, the two variables – auxin levels and growth inhibition – may have no effect on each other, but could both be affected by a third variable. Years later, a different scientist remarked that auxin levels in lateral buds of the kidney bean actually increased when the shoot tip was cut off. Scientists now think that two other hormones are involved:

* Abscisic acid inhibits bud growth. High auxin in the shoot may keep abscisic acid levels high in the bud. When the tip (the source of auxin) is removed, abscisic acid levels drop and the bud starts to grow.
* Cytokinins promote bud growth – directly applying cytokinin to buds can override the **apical dominance** effect. High levels of auxin make the shoot apex a sink for cytokinins produced in the roots – most of the cytokinin goes to the shoot apex. When the apex is removed, cytokinin spreads evenly around the plant.

Gibberellins

Gibberellins are plant hormones which are responsible for control of stem elongation and seed germination.

Stem elongation

In Japan, a fungus causes a disease which makes rice grow very tall. The fungal compounds involved are gibberellins and include gibberellic acid (GA_3).

> **DID YOU KNOW?**
> The Japanese call this disease Bakanae: the 'foolish seedling disease'!

Scientists tested gibberellic acid on many different plants. When they applied it to dwarf varieties of plants such as maize and peas, or to rosette plants (like cabbages), they grew taller (Figure 2).

This suggests that gibberellic acid is responsible for plant stem growth. However, think about whether the experiment actually investigated a natural phenomenon – just because GA_3 *can* cause stem

Figure 1 Apical dominance: the apical bud has been removed from the plant on the right.

Figure 2 Effect of gibberellins on cabbage growth: plants treated with gibberellins have grown taller.

elongation, it does not mean that it *does* so in nature. An experiment like this needs to work within concentrations of gibberellins naturally found in plants, and in parts of the plant that gibberellin molecules normally reach.

Researchers found a way to meet these criteria. They compared GA_1 levels (another member of the gibberellin family) of tall pea plants (homozygous for the dominant *Le* allele) and dwarf pea plants (homozygous for the recessive *le* allele), which were otherwise genetically identical (Figure 3). They found that plants with higher GA_1 levels were taller.

However, to show that GA_1 *directly* causes stem growth, researchers need to know how GA_1 is formed (see Figure 4). They worked out that the *Le* gene was responsible for producing the enzyme that converted GA_{20} to GA_1.

Then the researchers chose a pea plant with a mutation that blocks gibberellin production between *ent*-kaurene and GA_{12}-aldehyde in the pathway shown in Figure 4. Those plants produce no gibberellin and grow to only about 1 cm tall. The researchers grafted (a bit like transplanting) a shoot onto a homozygous *le* plant (which cannot convert GA_{20} to GA_1), and it grew tall.

Such a shoot, with no GA_{20} of its own, does have the enzyme to convert GA_{20} to GA_1, and it can use the unused GA_{20} from the normal plant. Because the shoot grew tall, this confirmed that GA_1 causes stem elongation.

Further studies have shown that gibberellins cause growth in the internodes by stimulating cell elongation (by loosening cell walls) and cell division (by stimulating production of a protein that controls the cell cycle).

LEARNING TIP
Try representing the results of these experiments on gibberellin as a branching tree, to help you understand and remember how results were obtained, and the decisions the researchers made as their work continued.

Seed germination

Gibberellins also promote seed germination. When the seed absorbs water, the embryo releases gibberellin, which travels to the aleurone layer in the endosperm region of the seed. The gibberellin enables the production of amylase, which can break down starch into glucose. This provides a substrate for respiration for the embryo, and so it grows. The glucose is also used for protein synthesis.

Questions

1. Draw a flow chart to represent the contribution of auxins, cytokinins and abscisic acid to apical dominance.

2. In deep-water rice plants, lower oxygen levels stimulate production of ethene, which reduces abscisic acid levels. Abscisic acid is an antagonist of gibberellin. Using this information, explain how deep-water rice plants keep their upper foliage above water.

3. Explain how the research above confirms that GA_1 causes stem elongation.

4. Compare and contrast the action of gibberellins and auxins.

5. Explain why a result gained by artificial application of hormones in an experiment may not necessarily reflect what happens inside a plant.

6. Explain why an association or correlation between two variables does not necessarily mean that one directly *causes* a change in the other.

le le   Le Le or Le le

Figure 3 Dwarf and tall pea plants.

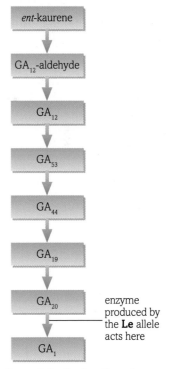

ent-kaurene

GA_{12}-aldehyde

GA_{12}

GA_{53}

GA_{44}

GA_{19}

GA_{20} enzyme produced by the **Le** allele acts here

GA_1

Figure 4 Synthesis pathway for gibberellins.

By the end of this topic, you should be able to demonstrate and apply your knowledge and understanding of:

* practical investigations into phototropism and geotropism
* practical investigations into the effect of plant hormones on growth

Where in a plant does growth occur?

In contrast to animal cells, in plant cells the cell wall limits the cell's ability to divide and expand. Growth only happens in particular places in the plant where there are groups of immature cells that are still capable of dividing, called **meristems** (see Figure 1).

* Apical meristems are at the tips or apices (singular: apex) of roots and shoots, and are responsible for the roots and shoots getting longer.
* Lateral bud meristems are found in the buds. These can give rise to side shoots.
* Lateral meristems forming a cylinder near the outside of roots and shoots are responsible for the roots and shoots getting wider.
* In some plants, intercalary meristems are located between the nodes, where the leaves and buds branch off the stem. Growth between the nodes is responsible for the shoot getting longer.

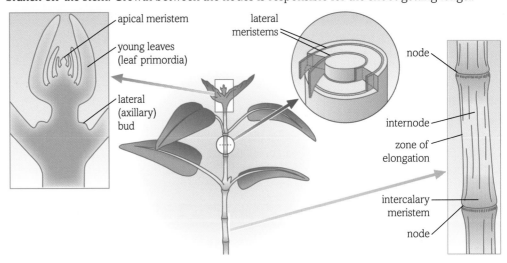

Figure 1 Meristems in a plant.

Investigating phototropic and geotropic responses

Phototropic responses can be investigated using the experiments in Figure 2. The experiment has an experimental plant and a control plant (with 10 replicates). The control plant is illuminated from all sides, while the experimental plant has illumination from just one side. In each plant, the shoots and roots are marked every 2 mm at the start.

Look at the results after several days. The shoot has bent towards the light, because the shady side of the shoot has elongated more than the illuminated side. The mean and standard deviation of the lengths between the marks has increased on the shady side.

We can also investigate **geotropic** responses. A control plant is constantly spun (very slowly) by a machine called a klinostat to ensure the effect of gravity is applied equally to all sides of the plant. For the experimental plant, the klinostat is not switched on, so gravity is only applied to one side.

In the experimental plant, the root bends downwards, because the upper side of the root has elongated more than the lower side. The shoot bends upwards, because the lower side of the shoot has elongated more than the upper side. In the control plant, both root and shoot grow horizontally.

If you are given the chance to conduct these experiments, take care not to handle mains plugs, switches, lamps or the klinostat with wet hands.

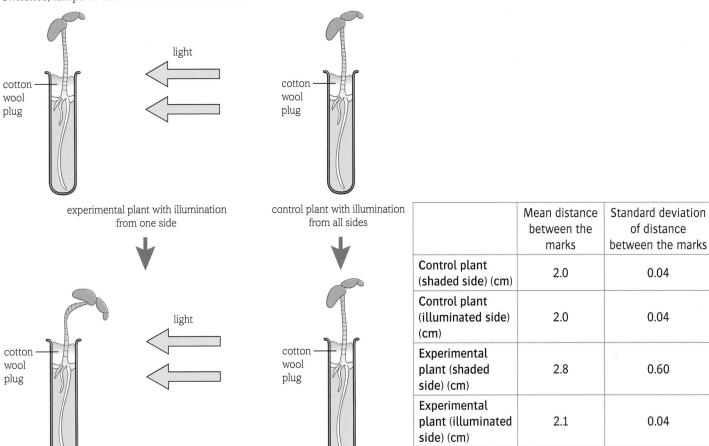

	Mean distance between the marks	Standard deviation of distance between the marks
Control plant (shaded side) (cm)	2.0	0.04
Control plant (illuminated side) (cm)	2.0	0.04
Experimental plant (shaded side) (cm)	2.8	0.60
Experimental plant (illuminated side) (cm)	2.1	0.04

Figure 2 Experiment investigating phototropic responses.

Investigating the effect of plant hormones on phototropisms

Are plant hormones involved?

A series of classic experiments confirmed that a chemical messenger from the shoot tip is responsible for the phototropic responses shown in Figure 2.

Darwin's experiments (see Figure 3) confirmed that the shoot tip was responsible for phototropic responses. Boysen–Jensen's work (see Figure 3) confirmed that water and/or solutes need to be able to move backwards from the shoot tip for phototropism to happen. When a permeable gelatine block was inserted behind the shoot tip, the shoot still showed positive phototropism. When an impermeable mica block was inserted, there was no phototropic response. You can replicate these experiments using cress or cereal seedlings (see Figure 3).

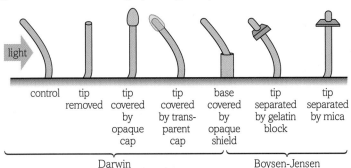

Figure 3 Classic experiments to confirm the role of the shoot tip in controlling phototropic responses.

To demonstrate that a chemical messenger existed and could stimulate a phototropic effect artificially, Went carried out the work shown in Figure 4.

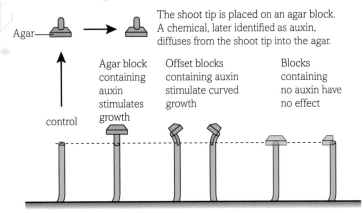

Figure 4 Classic experiments to confirm the role of the shoot tip in producing a chemical messenger that controls phototropic responses.

The chemical messenger: auxin

To confirm the role of auxin as the chemical messenger, agar blocks impregnated with different concentrations of auxin give the same result. In fact, using a series of blocks of different concentrations of auxin (i.e. indole-3-acetic acid; IAA) created by serial dilution (see feature below) gives shoot curvature in proportion to the amount of auxin.

Creating a serial dilution

- Set up five 10 ml screw-top bottles, or boiling tubes with bungs.
- Using a syringe, add 10 ml of auxin solution at 100 parts per million to the first boiling tube (tube 1).
  Using a separate syringe, add 9 ml of distilled water to the other four tubes (tubes 2, 3, 4 and 5).
- Remove 1 ml of auxin solution from tube 1 and add it to tube 2.
- Using a separate syringe each time, repeat the process between tubes 2 and 3, 3 and 4, and 4 and 5.
  Shake vigorously after each transfer (close the lid or insert the bung!).
- All five jars now contain auxin solution, each one 10 times more dilute than the previous one. This is serial dilution.
Note: Indole-3-acetic-acid (IAA) (auxin) is of low hazard but it is good practice to avoid all skin contact with plant hormones and wash hands after use.

Auxins are produced at the apex of the shoot (Figure 5). The auxin travels to the cells in the zone of elongation, causing them to elongate, and making the shoot grow. When light is equal on all sides, the auxin simply promotes shoot growth evenly (you can confirm this using the protocol in the Investigation).

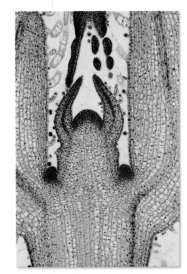

Figure 5 Longitudinal section of a shoot

INVESTIGATION

Effect of IAA on shoot growth
- Take 15 wheat seedlings, measure them, and cut the final 2 mm from the tip. Take 5 further seedlings and measure them, but do not remove the tip.
- Treat 5 seedlings with lanolin to cover the end of the tip (Group A).
- Treat 5 seedlings with lanolin infused with IAA (an auxin) (Group B).
- Leave 5 seedlings untreated (Group C).
- Return all the seedlings to a beaker with their roots surrounded by wet cotton wool.
- After 3 days, measure them all again. The seedlings in Groups A and C will not have grown. Those in Group B and the intact seedlings will have grown.
Note: IAA and lanolin are low hazard but avoid skin contact. If you have a wheat (gluten) allergy, make sure you wash your hands after handling the seeds.

However, combining the conclusions from the experiments above, light shining on one side of the shoot appears to cause auxins to be transported to the shaded side, causing the cells there to elongate more quickly (see Figure 6), making the shoot bend towards the light. The extent to which cells elongate is proportional to the concentration of auxins.

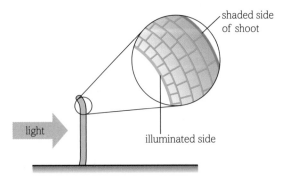

Figure 6 Mechanism of the phototropic response in shoots.

The mechanism of auxin's effect

Auxin increases the stretchiness of the cell wall by promoting the active transport of H^+ by an ATPase enzyme on the plasma membrane, into the cell wall. The resulting low pH provides optimum conditions for wall-loosening enzymes (expansins) to work. These enzymes break bonds within the cellulose (at the same time, the increased hydrogen ions also disrupt hydrogen bonds within cellulose), so the walls become less rigid and can expand as the cell takes in water.

How the light causes redistribution of auxin is still uncertain. Two enzymes have been identified – phototropin 1 and phototropin 2 – whose activity is promoted by blue light. Blue light is the main component of white light that causes the phototropic response. Hence, there is lots of phototropin 1 activity on the light side, but progressively less activity towards the dark side. This gradient is thought to cause the redistribution of auxins through their effect on PIN proteins. These transmembrane proteins can be found dorsally, ventrally or laterally on the plasma membrane of cells, and they control the efflux of auxin from each cell, essentially sending auxin in different directions in the shoot, depending upon their location on the plasma membrane.

The activity of PIN proteins is controlled by a different molecule, called PINOID. One theory suggests that phototropins affect the activity of PINOID, which then affects PIN activity. However, recent evidence from *Arabidopsis* suggests this may only work for pulse-induced phototropism (short bursts of light), with another independent mechanism able to operate in continuous light.

Auxin in geotropic responses of roots

Auxin is also involved in the geotropic responses of roots. In a root lying flat, Went discovered that auxin accumulates on the lower side, where it inhibits cell elongation. The upper side continues to grow and the root bends downwards.

This effect of auxin in roots is in contrast to that in the shoot, where auxin promotes cell elongation on the lower side, making the shoot lying flat bend upwards.

This happens because root and shoot cells in the elongation zone exhibit different responses to the same concentrations of auxin. You can see in Figure 7 that concentrations that stimulate shoot growth inhibit root growth.

LEARNING TIP

Remember that auxin causes cell elongation in shoots, rather than cell division.

DID YOU KNOW?

Arabidopsis is a small flowering plant often used as a model in plant biology.

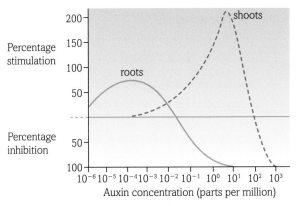

Figure 7 Graph showing effect of auxin concentration on elongation of shoots and roots compared to controls.

Questions

1. What effects do auxins have on cells?

2. Explain how auxin is involved in shoot growth.

3. Why do roots and shoots behave differently in response to particular levels of auxin?

4. Tropic responses are described as growth responses. Discuss whether they really do involve growth.

5. For the Investigation, suggest why there is more than one control.

6. Write a protocol for someone to follow to carry out (a) Darwin's work and (b) Went's work.

By the end of this topic, you should be able to demonstrate and apply your knowledge and understanding of:

* the commercial uses of plant hormones

Auxins

Artificial auxins can be used to prevent leaf and fruit 'drop' and to promote flowering for commercial flower production (surprisingly, in high concentrations auxins can also promote fruit drop). This is useful if there are too many small fruit that will be difficult to sell – the plant then produces fewer, larger fruit. Other uses of auxins are shown in Table 1.

Taking cuttings	Dipping the end of a cutting in rooting powder before planting it encourages root growth (Figure 1). Rooting powder contains auxins and talcum powder!
Seedless fruit	Treating unpollinated flowers with auxin can promote growth of seedless fruit (parthenocarpy). Applying auxin promotes ovule growth, which triggers automatic production of auxin by tissues in the developing fruit, helping to complete the developmental process.
Herbicides	Auxins are used as herbicides to kill weeds. Because they are man-made, plants find them more difficult to break down, and they can act within the plant for longer. They promote shoot growth so much that the stem cannot support itself, buckles and dies.

Table 1 Commercial uses of auxins.

Figure 1 Dipping a cutting into rooting powder.

Cytokinins

Because cytokinins can delay leaf senescence, they are sometimes used to prevent yellowing of lettuce leaves after they have been picked.

Cytokinins are used in tissue culture to help mass-produce plants. Cytokinins promote bud and shoot growth from small pieces of tissue taken from a parent plant. This produces a short shoot with a lot of side branches, which can be split into lots of small plants. Each of these is then grown separately.

Gibberellins

Fruit production

* Gibberellins delay senescence in citrus fruit, extending the time fruits can be left unpicked, and making them available for longer in the shops.

* Gibberellins acting with cytokinins can make apples elongate to improve their shape.

* Without gibberellins, bunches of grapes are very compact; this restricts the growth of individual grapes. With gibberellins, the grape stalks elongate, they are less compacted, and the grapes get bigger.

Brewing

To make beer you need malt, which is usually produced in a malt-house at a brewery (Figure 2). When barley seeds germinate, the aleurone layer of the seed produces amylase enzymes that break down stored starch into maltose. Usually, the genes for amylase production are switched on by naturally occurring gibberellins. Adding gibberellin can speed up the process. Malt is then produced by drying and grinding up the seeds.

Figure 2 Barley grains in a malt-house at a brewery.

Sugar production

Spraying sugar cane with gibberellins stimulates growth between the nodes, making the stems elongate. This is useful because sugar cane stores sugar in the cells of the internodes (the sub-sections of the stems in Figure 3), making more sugar available from each plant.

Figure 3 Sugar cane.

Plant breeding

A plant breeder's job is to produce plants with desired characteristics, by breeding together other plants, usually over many generations. However, in conifer plants this can take a particularly long time, because conifers spend a long time as juveniles before becoming reproductively active. Gibberellins can speed up the process by inducing seed formation on young trees.

Seed companies that want to harvest seeds from biennial plants (which flower only in their second year of life) can add gibberellins to induce seed production.

Stopping plants making gibberellins is also useful. Spraying with gibberellin synthesis inhibitors can keep flowers short and stocky (desirable in plants like poinsettias), and ensures that the internodes of crop plants stay short, helping to prevent lodging. Lodging happens in wet summers – stems bend over because of the weight of water collected on the ripened seed heads, making the crop difficult to harvest.

Ethene

Because ethene is a gas, and cannot be sprayed directly, scientists have developed 2-chloroethylphosphonic acid, which can be sprayed in solution, is easily absorbed, and slowly releases ethene inside the plant. Commercial uses of ethene include:

- speeding up fruit ripening in apples, tomatoes and citrus fruits
- promoting fruit drop in cotton, cherry and walnut
- promoting female sex expression in cucumbers, reducing the chance of self-pollination (pollination makes cucumbers taste bitter) and increasing yield
- promoting lateral growth in some plants, yielding compact flowering stems.

Restricting ethene's effects can also be useful. Storing fruit at a low temperature, with little oxygen and high carbon dioxide levels, prevents ethene synthesis and thus prevents fruit ripening. This means fruits can be stored for longer – essential when shipping unripe bananas from the Caribbean (see Figure 4). Other inhibitors of ethene synthesis, such as silver salts, can increase the shelf life of cut flowers.

Figure 4 Preparing bananas for shipping. The fruit are stored in conditions that prevent ethene synthesis.

Questions

1. Which hormones are involved in:
   (a) slowing down senescence?
   (b) fruit ripening?
   (c) encouraging root growth?

2. Which hormones can help clone daughter plants from a parent plant?

3. Using examples, describe the commercial benefits of restricting levels of (a) ethene and (b) gibberellins.

4. Instead of using hormones to produce seedless fruit, plant breeders sometimes take a different approach and breed genetically parthenocarpic varieties. A good example is bananas, whose cells are triploid, stopping them forming male and female gametes. Suggest one advantage and one disadvantage of each approach.

5. Putting a silver coin into the water of a vase of flowers is said to keep the flowers fresh for longer. Suggest how this might work.

6. Explain, using examples, how hormones can increase the yield of a crop.

5 The mammalian nervous system

By the end of this topic, you should be able to demonstrate and apply your knowledge and understanding of:

* the organisation of the mammalian nervous system

Responding to the environment

A successful organism must be able to respond to changes in the environment. These changes could be in the external environment or in the internal environment. In topic 5.1.1 we saw that, in order to respond, the organism needs a communication system. This communication system must enable:

* detection of changes in the environment
* cell signalling to occur between all parts of the body
* coordination of a range of effectors to carry out responses to the sensory input
* suitable responses.

Many environmental changes require rapid and well-coordinated responses to ensure survival. This may involve a wide array of responses such as coordinated muscle action, control of balance and posture, temperature regulation and coordination with the endocrine system. This is the role of the nervous system.

Divisions of the nervous system

The most obvious division of the nervous system is into the **central nervous system (CNS)** and the **peripheral nervous system (PNS)**.

The PNS is further divided into the **sensory system** and the **motor system**.

The motor system is divided into the **somatic nervous system** and the **autonomic nervous system**.

These divisions help us describe nervous actions and to understand coordination processes.

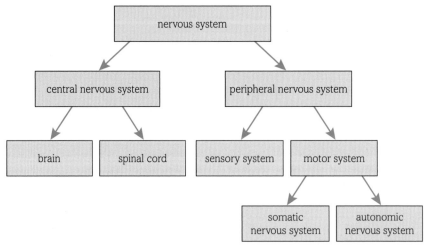

Figure 1 The organisation of the nervous system.

Central nervous system

The central nervous system (CNS) consists of the brain and spinal cord. The human brain contains about 86 billion neurones. Much of the brain is composed of relay neurones, which have multiple connections enabling complex neural pathways. Most of these cells are non-myelinated cells and the tissue looks grey in colour. It is known as grey matter.

The spinal cord also has many non-myelinated relay neurones making up the central grey matter. However, the spinal cord also contains large numbers of myelinated neurones making up an outer region of white matter. These myelinated neurones carry action potentials up and down the spinal cord for rapid communication over longer distances.

The spinal cord is protected by the vertebral column. Between each of the vertebrae, peripheral nerves enter and leave the spinal cord carrying action potentials to and from the rest of the body.

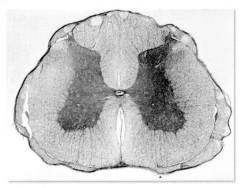

Figure 2 A section through the spinal cord.

Peripheral nervous system

The role of the peripheral nervous system (PNS) is to ensure rapid communication between the sensory receptors, the CNS and the effectors.

The PNS is composed of sensory and motor neurones. These are usually bundled together in a connective tissue sheath to form nerves.

Sensory nervous system

The sensory fibres entering the CNS are dendrons of the sensory neurones. These neurones conduct action potentials from the sensory receptors into the CNS. These neurones have their cell body in the dorsal root leading into the spinal cord and a short axon connecting to other neurones in the CNS.

Motor nervous system

The motor nervous system conducts action potentials from the CNS to the effectors. It is further subdivided according to the functions of the motor nerves:

- The somatic nervous system consists of motor neurones that conduct action potentials from the CNS to the effectors that are under voluntary (conscious) control, such as the skeletal muscles. These neurones are mostly myelinated, so that responses can be rapid. There is always one single motor neurone connecting the CNS to the effector.

- The autonomic nervous system consists of motor neurones that conduct action potentials from the CNS to effectors that are not under voluntary control. This includes the glands, the cardiac muscle and smooth muscle in the walls of the blood vessels, the airways and the wall of the digestive system. The control of many of these effectors does not require rapid responses, and the neurones are mostly non-myelinated. There are at least two neurones involved in the connection between the CNS and the effector. These neurones are connected at small swellings called ganglia.

Autonomic means 'self-governing', and the autonomic nervous system operates to a large extent independently of conscious control. It is responsible for controlling the majority of the homeostatic mechanisms and so plays a vital role in regulating the internal environment of the body.

The autonomic nervous system can be further divided into the **sympathetic** system, which prepares the body for activity, and the **parasympathetic** system, which conserves energy (see Figure 3). The sympathetic and parasympathetic systems differ in both structure and action. They are antagonistic systems, as the action of one system opposes the action of the other. In general, at rest, action potentials pass along the neurones of both systems at a relatively low frequency. This is controlled by subconscious parts of the brain. Changes to internal conditions, or stress, lead to changes in the balance of stimulation between the two systems, leading to an appropriate response.

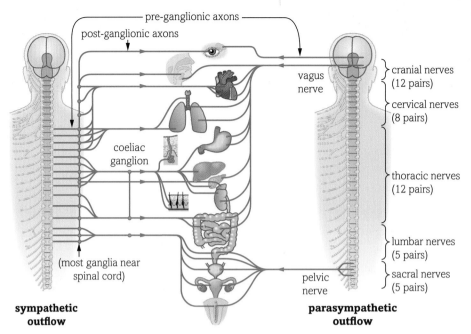

Figure 3 The autonomic nervous system.

Table 1 shows the differences between the sympathetic and parasympathetic systems.

Sympathetic system	Parasympathetic system
Consists of many nerves leading out of the CNS, each leading to a separate effector	Consists of a few nerves leading out of the CNS, which divide up and lead to different effectors
Ganglia just outside the CNS	Ganglia in the effector tissue
Short pre-ganglionic neurones	Long pre-ganglionic neurones (variable in length, dependent upon the position of the effector)
Long post-ganglionic neurones (variable in length, dependent upon the position of the effector)	Short post-ganglionic neurones
Uses noradrenaline as the neurotransmitter	Uses acetylcholine as the neurotransmitter
Increases activity – prepares body for activity	Decreases activity – conserves energy
Most active at times of stress	Most active during sleep or relaxation
Effects include: • increases heart rate • dilates pupils • increases ventilation rate • reduces digestive activity • orgasm	Effects include: • decreases heart rate • constricts pupils • reduces ventilation rate • increases digestive activity • sexual arousal

Table 1 The differences between the sympathetic and parasympathetic systems.

Questions

1. Explain why the sensory neurones and somatic motor neurones are myelinated.

2. Explain why the control of pupil diameter in the eye is described as an autonomic response.

3. Suggest why the sympathetic system inhibits digestive activity.

4. List the responses that may prepare the body for an activity such as sprinting.

5. Explain why the relay neurones in the brain have multiple connections.

⑥ The brain

By the end of this topic, you should be able to demonstrate and apply your knowledge and understanding of:

* the structure of the human brain and the functions of its parts

The structure of the human brain

The human brain has four main parts:

* The **cerebrum**, which is the largest part and organises most of our higher thought processes, such as conscious thought and memory.
* The **cerebellum**, which coordinates movement and balance.
* The **hypothalamus** and **pituitary** complex, which organises homeostatic responses and controls various physiological processes.
* The **medulla oblongata**, which coordinates many of the autonomic responses.

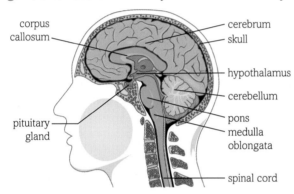

Figure 1 The main parts of the human brain.

> **KEY DEFINITIONS**
>
> **cerebellum:** region of the brain coordinating balance and fine control of movement.
> **cerebrum:** region of the brain dealing with the higher functions such as conscious thought; it is divided into two cerebral hemispheres.
> **hypothalamus:** the part of the brain that coordinates homeostatic responses.
> **medulla oblongata:** region of the brain that controls physiological processes.
> **pituitary gland:** endocrine gland at the base of the brain, below but attached to the hypothalamus; the anterior lobe secretes many hormones; the posterior lobe stores and releases hormones made in the hypothalamus.

The cerebrum

The cerebrum has two cerebral hemispheres, which are connected via major tracts of neurones called the **corpus callosum**. The outermost layer of the cerebrum consists of a thin layer of nerve cell bodies called the **cerebral cortex** (Figure 2a).

The cerebrum is much more highly developed in humans than in any other organism. It controls the 'higher brain' functions including:

* conscious thought
* conscious actions (including the ability to override some reflexes)
* emotional responses
* intelligence, reasoning, judgement and decision making
* factual memory.

The cerebral cortex is subdivided into areas responsible for specific activities and body regions:

* Sensory areas receive action potentials indirectly from the sensory receptors. The sizes of the regions allocated to receive input from different receptors are related to the sensitivity of the area that inputs are received from.
* Association areas compare sensory inputs with previous experience, interpret what the input means, and judge an appropriate response.
* Motor areas send action potentials to various effectors (muscles and glands). The sizes of the regions allocated to deal with different effectors are related to the complexity of the movements needed in the parts of the body, as shown in Figure 2(b). Motor areas on the left side of the brain control the effectors on the right side of the body and vice versa.

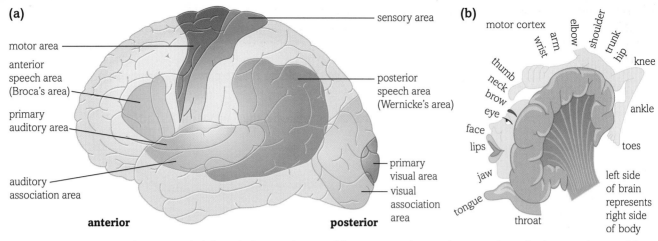

Figure 2 (a) The left cerebral cortex; (b) parts of the motor area showing there is a relationship between the size of the motor area and the complexity of the movements controlled.

> **DID YOU KNOW?**
> The human brain is wrinkled, because our cerebral cortex has become enlarged to enable the higher processes to occur. As the cerebral cortex is only a thin layer of cells, it has become enlarged by increasing its surface area – which is about 2.5 m^2 – and folding the surface to make it fit inside the head.

The cerebellum

The cerebellum contains over half of all of the neurones in the brain. It is involved with balance and fine coordination of movement. To do this it must receive information from many sensory receptors and process the information accurately. The sensory receptors that supply information to the cerebellum include the retina, the balance organs in the inner ear, and spindle fibres in the muscles, which give information about muscle length and the joints.

The conscious decision to contract voluntary muscles is initiated in the cerebral cortex. However, the cerebral cortex does not provide the complex signals required to coordinate complex movements. The cerebellum coordinates the fine control of muscular movements such as:

- maintaining body position and balance, such as when riding a bicycle
- judging the position of objects and limbs while moving about or playing sport
- tensioning muscles in order to use tools and play musical instruments effectively
- coordinating contraction and relaxation of antagonistic skeletal muscles when walking and running.

This control often requires learning. Once learnt, such activities may become second nature and involve much unconscious control. This sort of coordination requires complex nervous pathways. The nervous pathways are strengthened by practice. The complex activity becomes 'programmed' into the cerebellum, and neurones from the cerebellum conduct action potentials to the motor areas, so that motor output to the effectors can be finely controlled.

The cerebrum and cerebellum are connected by the pons.

The hypothalamus and pituitary complex

The hypothalamus controls homeostatic mechanisms in the body. It contains its own sensory receptors and acts by negative feedback to maintain a constant internal environment (see topic 5.1.2):

- Temperature regulation: the hypothalamus detects changes in core body temperature. However, it also receives sensory input from temperature receptors in the skin. It will initiate responses to temperature change that regulate body temperature within a narrow range. These responses may be mediated by the nervous system or by the hormonal system (via the pituitary gland).

- Osmoregulation: the hypothalamus contains osmoreceptors that monitor the water potential in the blood. When the water potential changes, the osmoregulatory centre initiates responses that bring about a reversal of this change. The responses are mediated by the hormonal system via the pituitary gland.

The pituitary gland acts in conjunction with the hypothalamus. The pituitary gland consists of two lobes:

- The posterior lobe is linked to the hypothalamus by specialised neurosecretory cells. Hormones such as ADH, which are manufactured in the hypothalamus, pass down the neurosecretory cells and are released into the blood from the pituitary gland (see topic 5.2.7).

- The anterior lobe produces its own hormones, which are released into the blood in response to **releasing factors** produced by the hypothalamus. These releasing factors are hormones that need to be transported only a short distance from the hypothalamus to the pituitary. Hormones from the anterior pituitary control a number of physiological processes in the body, including response to stress, growth, reproduction and lactation.

The medulla oblongata

The medulla oblongata controls the non-skeletal muscles (the cardiac muscles and involuntary smooth muscles) by sending action potentials out through the autonomic nervous system. The medulla oblongata contains centres for regulating several vital processes, including:

- the cardiac centre, which regulates heart rate
- the vasomotor centre, which regulates circulation and blood pressure
- the respiratory centre, which controls the rate and depth of breathing.

These centres receive sensory information and coordinate vital functions by negative feedback.

LEARNING TIP
Remember that brain activity relies on nervous conduction and synaptic transmission. This topic links closely with the topics in Chapter 5.3, and exam questions may link the topics together.

Questions

1 Suggest how damage to the cerebellum may affect the behaviour of a person.

2 Explain why damage to the left side of the brain may cause paralysis to the right side of the body.

3 What is the role of the corpus callosum?

4 List the areas of the brain that would be involved in driving a car.

5 List the areas of the brain that would be involved in responding to a question verbally.

6 Suggest why a blow to the back of the head can result in blurred vision.

(7) Reflex actions

By the end of this topic, you should be able to demonstrate and apply your knowledge and understanding of:

* reflex actions

What is a reflex action?

Reflex actions are responses to changes in the environment that do not involve any processing in the brain to coordinate the movement. The nervous pathway is as short as possible so that the reflex is rapid. Most reflex pathways consist of just three neurones:

sensory neurone → relay neurone → motor neurone

The brain may be informed that the reflex has happened, but is not involved in coordinating the response.

Reflex actions always have a survival value. A reflex may be used to get out of danger, to avoid damage to part of the body, or it may be used to maintain balance.

Reflex actions include the blinking reflex and the knee jerk reflex.

Blinking reflex

The blinking reflex causes temporary closure of the eyelids to protect the eyes from damage.

The nervous pathway for the blinking reflex passes through part of the brain – in other words, the reflex is a **cranial reflex**. However, the pathway is a direct pathway that does not involve any thought processes in the higher parts of the brain. Since the receptor and the effector are in the same place, this is called a **reflex arc**.

Blinking may be stimulated by sudden changes in the environment such as:

* a foreign object touching the eye (the corneal reflex)
* sudden bright light (the optical reflex)
* loud sounds
* sudden movements close to the eye.

Corneal reflex

This reflex is mediated by a sensory neurone from the cornea, which enters the **pons**. A synapse connects the sensory neurone to a relay neurone, which passes the action potential to the motor neurone. The motor neurone passes back out of the brain to the facial muscles, causing the eyelid to blink. This is a very short and direct pathway, so the corneal reflex is very rapid – it takes about 0.1 seconds. The corneal reflex usually causes both eyes to blink, even if only one cornea is affected.

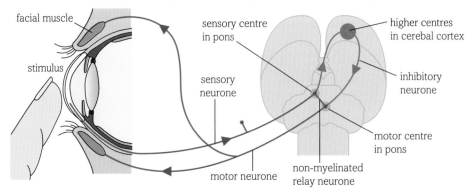

Figure 1 The corneal reflex, shown in blue. The red neurones indicate how the reflex can be overridden.

The sensory neurone involved in the corneal reflex also passes the action potential to myelinated neurones in the pons. These myelinated neurones carry the action potential to the sensory region in the cerebral cortex, to inform the higher centres of the brain that the stimulus has occurred. This allows the reflex to be overridden by conscious control. The higher parts of the brain (cerebral cortex) can send inhibitory signals to the motor centre in the pons. The myelinated neurones carrying impulses to and from the cerebral cortex transmit action potentials much more rapidly than the non-myelinated relay neurone in the pons. Therefore, the inhibitory action potentials can prevent the formation of an action potential in the motor neurone.

DID YOU KNOW?

Many reflex actions can be overridden. Overriding the corneal reflex is essential for people who wear contact lenses.

Optical reflex

This protects the light-sensitive cells of the retina from damage. The stimulus is detected by the retina and the reflex is mediated by the optical centre in the cerebral cortex. The **optical reflex** is a little slower than the corneal reflex.

LEARNING TIP

Use simple flow diagrams to help you recall the sequence of events in a reflex action or other more complex responses.

Knee jerk reflex

The **knee jerk reflex** is a **spinal reflex** – the nervous pathway passes through the spinal cord rather than through the brain.

The knee jerk reflex is involved in coordinated movement and balance. The muscle at the front of the thigh (quadriceps) contracts to straighten the leg. This muscle is attached to the lower leg via the patella tendon that connects the patella to the lower leg bones at the front of the knee. When the muscles at the front of the thigh are stretched, specialised stretch receptors called **muscle spindles** detect the increase in length of the muscle. If this stretching is unexpected, a reflex action causes contraction of the same muscle.

This is part of the mechanism that enables us to balance on two legs. Consider a situation where you are standing still. Under these conditions, the muscle in front of the thigh will stretch if the knee is bending or the body is starting to lean backwards. Contraction of the muscles straightens the knee or brings the body back above the legs. Such a response must be very rapid, so that the body can remain balanced.

The knee jerk reflex is unusual in that the nervous pathway consists of only two neurones:

sensory neurone → motor neurone

Therefore, there is one less synapse involved and the response is quicker.

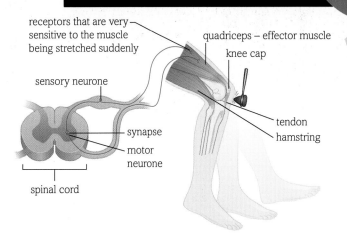

Figure 2 The knee jerk reflex.

As happens with other reflex actions, the higher parts of the brain are informed that the reflex is occurring. However, because there is no relay neurone, the brain cannot inhibit the reflex. Inhibition relies on rapid myelinated neurones carrying the inhibitory action potentials to the synapse before the motor neurone is stimulated. In the absence of a relay neurone, the motor neurone is stimulated directly by the sensory neurone and there is insufficient delay to enable inhibition. This is why doctors test your reflexes by tapping the tendon below the knee cap. It causes an immediate response that cannot be inhibited.

While we are walking or running, the knee must bend and will stimulate the muscle spindles. However, the complex pattern of nervous impulses coming from the cerebellum is able to inhibit the reflex contractions. As action potentials are sent to the muscles behind the thigh (hamstring), stimulating it to contract, inhibitory action potentials are sent to the synapse in the reflex arc to prevent the reflex contraction of the opposing muscle.

Questions

1 Explain the survival value of the following reflex actions:
   (a) an earthworm withdraws into its burrow when touched
   (b) we lift our foot quickly if we step on a sharp object.

2 Explain why a reflex action is so rapid.

3 When we pick up a hot object, a reflex action occurs to make us drop the object. However, if the object is important or valuable (such as our dinner on a hot plate), we are able to hold on to it long enough to put it down gently. Explain how this is achieved.

4 Why must the action potentials to override a reflex be carried by myelinated neurones?

5 What are the differences between the blinking reflex and the knee jerk reflex?

Coordinating responses

By the end of this topic, you should be able to demonstrate and apply your knowledge and understanding of:

* the coordination of responses by the nervous and endocrine systems

Stimulus and response

Mammals have complex sensory mechanisms that monitor changes in both the internal and external environment. These provide input to the brain, which must assimilate the inputs and coordinate a response which ensures survival. Even a relatively simple task such as standing up requires input from a range of receptors, including the eyes, the balance organs and muscle stretch receptors.

More complex activities and responses receive input from a wider range of receptors. Input may include information about blood glucose levels from the islets of Langerhans in the pancreas, or information from stretch receptors in the stomach or pain receptors in the skin and joints.

Responses may be short-term responses such as the homeostatic mechanisms of temperature control, or they may be longer-term responses such as the behaviours associated with reproduction.

The brain coordinates responses through output to the effectors. This output may include:

* action potentials in the somatic nervous system
* action potentials in the sympathetic and parasympathetic parts of the autonomic nervous system
* release of hormones via the hypothalamus and pituitary gland.

The 'fight or flight' response

Detecting a threat to survival stimulates the 'fight or flight' response. In mammals, this leads to a range of physiological changes that prepare the animal for activity. That activity may be running away, or it may be a direct challenge to the perceived threat.

Physiological changes

Table 1 lists physiological changes associated with the fight or flight response and the survival value of each change.

DID YOU KNOW?

Hairs standing on end make the mammal look bigger and therefore more threatening – this may serve to avoid a real fight and therefore enhance chances of survival.

Physiological change	Survival value
Pupils dilate	Allows more light to enter the eyes, making the retina more sensitive
Heart rate and blood pressure increase	Increases the rate of blood flow to deliver more oxygen and glucose to the muscles and to remove carbon dioxide and other toxins
Arterioles to the digestive system and skin are constricted, whilst those to the muscles and liver are dilated	Diverts blood flow away from the skin and digestive system and towards the muscles
Blood glucose levels increase	Supplies energy for muscular contraction
Metabolic rate increases	Converts glucose to useable forms of energy such as ATP
Erector pili muscles in the skin contract	Makes hairs stand up – which is a sign of aggression
Ventilation rate and depth increase	Increases gaseous exchange so that more oxygen enters the blood and supplies aerobic respiration
Endorphins (natural painkillers) are released in the brain	Wounds inflicted on the mammal do not prevent activity

Table 1 Physiological changes associated with the fight or flight response and the survival value of each change.

Coordination of the fight or flight response

Receptors that can detect an external threat include the eyes, ears and nose. Internal receptors may detect a threat, such as pain or a sudden increase or decrease in blood pressure. The cerebrum uses such sensory input to coordinate a suitable response:

1. Inputs feed into the sensory centres in the cerebrum.

2. The cerebrum passes signals to the association centres.

3. If a threat is recognised, the cerebrum stimulates the hypothalamus.

4. The hypothalamus increases activity in the sympathetic nervous system and stimulates the release of hormones from the anterior pituitary gland.

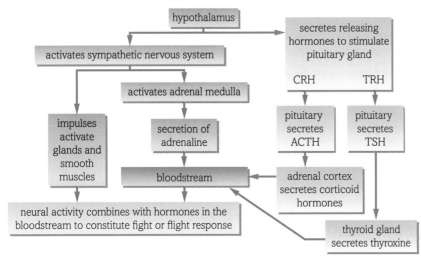

Figure 1 The action of the hypothalamus.

The role of the sympathetic nervous system

The autonomic nervous system controls many of the physiological mechanisms in Table 1. Increasing stimulation of the sympathetic nervous system will increase the activity of the effectors, as described in topic 5.5.5. However, nervous communication is used for rapid response rather than prolonged response. A fight or a flight from danger may need a prolonged response. This is achieved by the endocrine system.

The sympathetic nervous system stimulates the adrenal medulla. Adrenaline released from the adrenal medulla has a wide range of effects on cells, as described in topic 5.4.2.

The mechanism of adrenaline action

Adrenaline is a first messenger. It is an amino acid derivative and is, therefore, unable to enter the target cell; it must cause an effect inside the cell, without entering the cell itself (see Figure 2).

> **LEARNING TIP**
>
> This is an obvious topic to link back to membrane structure and how substances pass through membranes.

1. Adrenaline binds to the adrenaline receptor on the plasma membrane. This receptor is associated with a G protein on the inner surface of the plasma membrane, which is stimulated to activate the enzyme **adenyl cyclase**.

2. Adenyl cyclase converts ATP to **cyclic AMP (cAMP)**, which is the second messenger inside the cell.

3. cAMP causes an effect inside the cell by activating enzyme action. The precise effect depends upon the cell that the adrenaline has bound to.

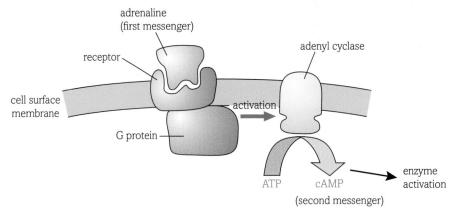

Figure 2 The mechanism of adrenaline action.

The release of hormones from the anterior pituitary

The hypothalamus secretes releasing hormones (also known as releasing factors) into the blood. These pass down a portal vessel to the pituitary gland and stimulate the release of **tropic hormones** from the anterior part of the pituitary gland. These stimulate activity in a variety of endocrine glands.

- **Corticotropin-releasing hormone** (CRH) from the hypothalamus causes the release of **adrenocorticotropic hormone** (ACTH). ACTH passes around the blood system and stimulates the adrenal cortex to release a number of different (corticosteroid) hormones. These include glucocorticoids such as cortisol (see Figure 3), which regulate the metabolism of carbohydrates. As a result, more glucose is released from glycogen stores. New glucose may also be produced from fat and protein stores.

- **Thyrotropin-releasing hormone** (TRH) causes the release of **thyroid-stimulating hormone** (TSH), which stimulates the thyroid gland to release more thyroid hormone (thyroxine). Thyroxine acts on nearly every cell of the body, increasing the metabolic rate and making the cells more sensitive to adrenaline.

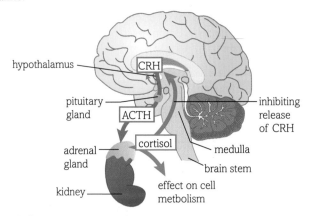

Figure 3 The release of cortisol.

Questions

1. The release of adrenaline is associated with stress. List three situations that would cause adrenaline to be released.

2. Suggest why digestive disorders are a sign of long-term stress.

3. Suggest how it is possible for an imagined threat to stimulate the fight or flight response.

4. Why does the parasympathetic system have little or no role in the fight or flight response?

⑨ Controlling heart rate

By the end of this topic, you should be able to demonstrate and apply your knowledge and understanding of:

* the effects of hormones and nervous mechanisms on heart rate

The human heart

The heart pumps blood around the circulatory system. This circulation has several important roles:

* Transport of oxygen and nutrients, such as glucose, fatty acids and amino acids to the tissues.

* Removal of waste products, such as carbon dioxide from the tissues to prevent accumulation that may become toxic.

* Transport of urea from the liver to the kidneys.

* Distribute heat around the body or deliver it to the skin to be radiated away.

The requirements of the cells and tissues vary according to their level of activity. When you are being physically active, your muscle cells need more oxygen and glucose so that they can respire more, releasing the energy for contraction. Your heart muscle cells also need more oxygen and fatty acids. All the muscles will also need to remove more carbon dioxide and heat.

It is essential that the circulatory system can adapt to meet the needs of the tissues. Part of this adaptation is controlling the activity of the heart. The heart action can be modified by:

* Raising or lowering the heart rate. This is the number of beats per minute.

* Altering the force of the contractions of the ventricular walls.

* Altering the stroke volume (volume of blood pumped per beat).

Heart rate

The rate at which the heart beats is affected by a number of factors.

Cardiac muscle in the heart is **myogenic**. This means that it can initiate its own beat at regular intervals. However, the atrial muscle has a higher myogenic rate than the ventricular muscle. The two pairs of chambers must contract in a coordinated fashion or the heart action will be ineffective. Therefore, a coordination mechanism is essential.

The heart contains its own pacemaker – the sinoatrial node (SAN). The SAN initiates waves of excitation that usually override the myogenic action of the cardiac muscle. The SAN is a region of tissue that can initiate an action potential, which travels as a wave of excitation over the atrial walls, through the AVN (atrio-ventricular node) and down the Purkyne fibres to the walls of the ventricles, causing them to contract.

The heart muscle also responds directly to the hormone adrenaline in the blood. This increases heart rate.

> **LEARNING TIP**
>
> The coordination of heart action is a topic covered in Book 1. Each heart beat takes the same length of time – here the frequency of stimulations is altered. Remember to say the frequency is altered, rather than the heart beats more quickly.

Control of heart rate by the cardiovascular centre

At rest, the heart rate is controlled by the SAN. This has a set frequency, varying from person to person, at which it initiates waves of excitation. The frequency of excitation is typically 60–80 per minute. However, the frequency of these excitation waves is altered by the output from the **cardiovascular centre** in the medulla oblongata.

Nerves from the cardiovascular centre in the medulla oblongata of the brain supply the SAN. These nerves are part of the autonomic nervous system. The nerves do not initiate a contraction, but can affect the frequency of contractions:

* Action potentials sent down a sympathetic nerve (the **accelerans nerve**) cause the release of the neurotransmitter noradrenaline at the SAN. This increases the heart rate.

* Action potentials sent down the **vagus nerve** release the neurotransmitter acetylcholine, which reduces the heart rate.

A range of environmental factors affect heart rate. Input from sensory receptors is fed to the cardiovascular centre in the medulla oblongata. Some inputs increase heart rate, others decrease it. The interaction of these inputs is coordinated by the cardiovascular centre to ensure that the output to the SAN is appropriate to the overall conditions.

Sensory input to the cardiovascular centre includes:

* Stretch receptors in the muscles detect movement of the limbs. These send impulses to the cardiovascular centre, informing it that extra oxygen may soon be needed. This leads to an increase in heart rate.

> **DID YOU KNOW?**
>
> The 'warming up' period for athletes includes stretching. This boosts heart activity in preparation for the race, ensuring that the muscles will have a good supply of oxygen.

- **Chemoreceptors** in the carotid arteries, the aorta and the brain monitor the pH of the blood. When we exercise, the muscles produce more carbon dioxide. Some of this reacts with the water in the blood plasma to produce a weak acid (carbonic acid). This reduces the pH of the blood, which will affect the transport of oxygen. The change in pH is detected by the chemoreceptors, which send action potentials to the cardiovascular centre. This will tend to increase the heart rate.

- The concentration of carbon dioxide in the blood. When we stop exercising, the concentration of carbon dioxide in the blood falls. This reduces the activity of the accelerator pathway. Therefore, the heart rate declines.

- Stretch receptors in the walls of the **carotid sinus** monitor blood pressure. The carotid sinus is a small swelling in the carotid artery. An increase in blood pressure, perhaps during vigorous exercise, is detected by these stretch receptors. If pressure rises too high, the stretch receptors send action potentials to the cardiovascular centre, leading to a reduction in heart rate.

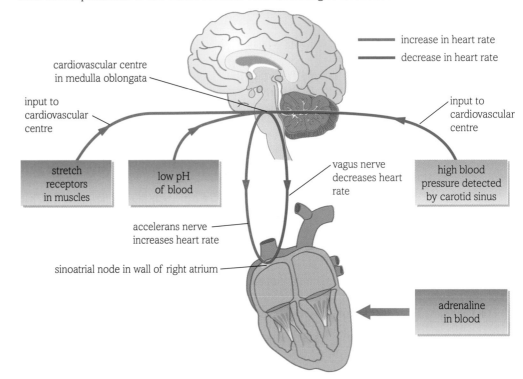

Figure 1 The role of the cardiovascular centre.

Artificial control of heart rate

If the mechanism controlling the heart rate fails, then an artificial pacemaker must be fitted. A pacemaker delivers an electrical impulse to the heart muscle. A pacemaker is implanted under the skin and fat on the chest (or sometimes within the chest cavity itself). An artificial pacemaker may be connected to the SAN or directly to the ventricle muscle.

Figure 2 An artificial pacemaker inserted under the skin.

Questions

1. Explain what is meant by the term *myogenic*.

2. Explain why the heart must be able to respond to changes in physical activity.

3. Accelerating the heart rate is likely to produce more carbon dioxide in the heart muscle. What effect might this have on the heart rate? What type of feedback is this?

4. Describe how sensory input is used to prevent the heart rate rising too high during intense exercise.

5. Suggest how stretch receptors convert the movement of the muscle to an action potential.

⑩ Muscle

By the end of this topic, you should be able to demonstrate and apply your knowledge and understanding of:

* the structure of mammalian muscle and the mechanism of muscular contraction

* the examination of stained sections or photomicrographs of skeletal muscle

KEY DEFINITIONS

cardiac muscle: muscle found in the heart walls.
involuntary muscle: smooth muscle that contracts without conscious control.
neuromuscular junction: the structure at which a nerve meets the muscle; it is similar in action to a synapse.
skeletal (striated) muscle: muscle under voluntary control.

Three types of muscle

Muscles are composed of cells arranged to form fibres. These fibres can contract to become shorter, which produces a force.

Contraction is achieved by interaction between two protein filaments (**actin** and **myosin**) in the muscle cells. Muscle cannot elongate without an **antagonist**. Therefore, muscles are usually arranged in opposing pairs, so that one contracts as the other elongates. In some cases, the antagonist may be elastic recoil or hydrostatic pressure in a chamber.

There are three types of muscle: **involuntary** (smooth), **cardiac**, and voluntary (**skeletal**, or **striated**) **muscle**.

Involuntary (smooth) muscle

Involuntary (smooth) muscle consists of individual cells, tapered at both ends (spindle shaped). At rest, each cell is about 500 μm long and 5 μm wide. Each cell contains a nucleus and bundles of actin and myosin.

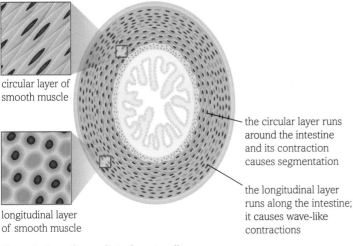

circular layer of smooth muscle

longitudinal layer of smooth muscle

the circular layer runs around the intestine and its contraction causes segmentation

the longitudinal layer runs along the intestine; it causes wave-like contractions

Figure 1 Smooth muscle in the gut wall.

This type of muscle contracts slowly and regularly. It does not tire quickly. It is controlled by the autonomic nervous system.

Involuntary muscle is found in the walls of tubular structures, such as the digestive system and blood vessels. The muscle is usually arranged in longitudinal and circular layers that oppose each other (see Figure 1).

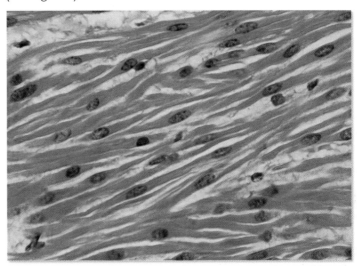

Figure 2 A light micrograph of smooth muscle.

Cardiac muscle

Cardiac muscle (Figure 3) forms the muscular part of the heart. The individual cells form long fibres, which branch to form cross-bridges between the fibres. These cross-bridges help to ensure that electrical stimulation spreads evenly over the walls of the chambers. When the muscle contracts, this arrangement also ensures that the contraction is a squeezing action rather than one-dimensional.

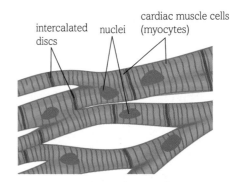

intercalated discs   nuclei   cardiac muscle cells (myocytes)

Figure 3 Cardiac muscle.

The cells are joined by **intercalated discs**. These are specialised cell surface membranes fused to produce gap junctions that allow free diffusion of ions between the cells. Action potentials pass easily and quickly along and between the cardiac muscle fibres.

Cardiac muscle contracts and relaxes continuously throughout life. It can contract powerfully and does not fatigue easily. Some muscle fibres in the heart (Purkyne fibres) are modified to carry electrical impulses. These coordinate the contraction of the chamber walls. Heart muscle is myogenic – it can initiate its own contraction. However, the rate of contraction is normally controlled by the SAN.

Cardiac muscle appears striated (striped) when viewed under the microscope (Figure 4).

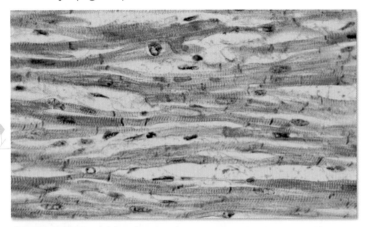

Figure 4 Micrograph of cardiac muscle.

> **LEARNING TIP**
> Remember that in skeletal muscle fibres, myofibrils and filaments all lie longitudinally in the muscle – this means the muscle can only contract in that one direction. Cardiac muscle has cross-bridges, which enable it to contract in a squeezing action.

Voluntary (skeletal, or striated) muscle

Skeletal muscle (Figure 5) occurs at the joints in the skeleton. Contraction causes movement of the skeleton by bending or straightening the joint. The muscles are arranged in pairs called antagonistic pairs. When one contracts, the other elongates.

The muscle cells form fibres of about 100 μm in diameter. Each fibre is multinucleate (contains many nuclei) and is surrounded by a membrane called the **sarcolemma**.

Muscle cell cytoplasm is known as **sarcoplasm**, and is specialised to contain many mitochondria and an extensive **sarcoplasmic reticulum** (specialised endoplasmic reticulum).

The contents of the fibres are arranged into a number of **myofibrils**, which are the contractile elements (see topic 5.5.11). These myofibrils are divided into a chain of subunits called **sarcomeres**. Sarcomeres contain the protein filaments actin and myosin.

Actin and myosin are arranged in a particular banded pattern, which gives the muscle a striped or striated appearance. Dark bands are known as the A bands and lighter bands are the I bands.

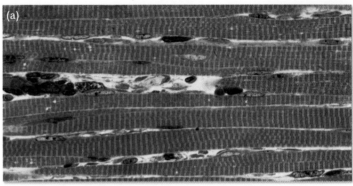

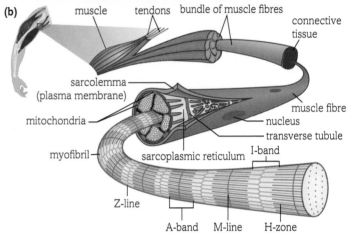

Figure 5 (a) Skeletal muscle (×540). (b) The structure of skeletal muscle.

Voluntary muscle contracts quickly and powerfully. It also fatigues quickly.

> **INVESTIGATION**
> View a prepared slide of skeletal muscle at low power and high power under the microscope (see Figure 5a). Note the large nuclei and the obvious striations across the fibres.

The neuromuscular junction

Skeletal muscle is under voluntary control. Its contractions are stimulated by the somatic nervous system (see topic 5.5.5). The junction between the nervous system and the muscle is called a **neuromuscular junction**. It has many similarities to a synapse.

Stimulation of contraction

1. Action potentials arriving at the end of the axon open calcium ion channels in the membrane. Calcium ions flood into the end of the axon.

2. Vesicles of acetylcholine move towards and fuse with the end membrane.

3. Acetylcholine molecules diffuse across the gap and fuse with receptors in the sarcolemma.

4. This opens sodium ion channels, which allow sodium ions to enter the muscle fibre, causing depolarisation of the sarcolemma.

5. A wave of depolarisation spreads along the sarcolemma and down transverse tubules into the muscle fibre.

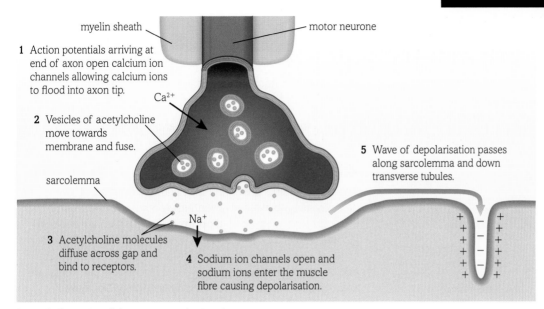

1 Action potentials arriving at end of axon open calcium ion channels allowing calcium ions to flood into axon tip.

Ca²⁺

2 Vesicles of acetylcholine move towards membrane and fuse.

sarcolemma

5 Wave of depolarisation passes along sarcolemma and down transverse tubules.

Na⁺

3 Acetylcholine molecules diffuse across gap and bind to receptors.

4 Sodium ion channels open and sodium ions enter the muscle fibre causing depolarisation.

Figure 6 The action of the neuromuscular junction.

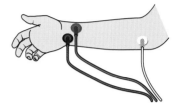

Figure 7 Placing EMG electrodes.

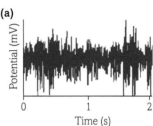

(a)

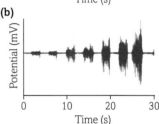

(b)

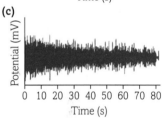

(c)

Figure 8 EMG traces showing (a) a trace from a single contraction, (b) a trace as ball is squeezed harder and harder, and (c) a trace showing how the amplitude decreases with fatigue.

The motor unit

Some motor neurones stimulate single muscle fibres. However, many motor neurones divide and connect to several muscle fibres. All these muscles fibres contract together, providing a stronger contraction. This is called a motor unit.

INVESTIGATION

Investigating muscle stimulation

The electrical activity of muscles can be investigated using an electromyograph (EMG). When a muscle is stimulated, the motor neurone creates action potentials in the muscle fibres. Electrodes applied to the surface of the skin detect the combined effects of these action potentials. A simple contraction of the muscle is seen as a series of apparently disorganised peaks on the trace. However, the amplitude of the EMG recording reflects the number and size of the motor units involved in the contraction – so a more powerful contraction is seen as higher amplitude.

Placing electrical leads on the forearm, as shown in Figure 7, allows activity in the forearm muscles to be recorded.

The recorded trace is called an electromyogram. Trace (a) in Figure 8 shows the disorganised peaks resulting from flexing a finger. Trace (b) shows how the amplitude of the trace increases as a rubber ball is squeezed with greater force. Trace (c) shows how the amplitude of the EMG trace decreases as a muscle fatigues.

Questions

1 Identify precisely where the three types of muscle can be found in the chest cavity.

2 What stretches out the muscle after contraction in:
(a) the walls of the heart
(b) the blood vessels
(c) the airways?

3 Draw a simple table to compare the structure and function of the three types of muscle.

4 Describe three ways in which a neuromuscular junction is similar to a synapse.

5.5 ⑪ Muscle contraction

By the end of this topic, you should be able to demonstrate and apply your knowledge and understanding of:

* the structure of mammalian muscle and the mechanism of muscular contraction

KEY DEFINITION

creatine phosphate: a compound in muscle that acts as a store of phosphates and can supply phosphates to make ATP rapidly.

Structure of the myofibril

Myofibrils are the contractile units of skeletal muscle and contain two types of protein filament:

* Thin filaments, which are aligned to make up the **light band**; these are held together by the **Z line**.
* Thick filaments, which make up the **dark band**.

The thick and thin filaments overlap, but in the middle of the dark band there is no overlap. This is called the **H zone**.

The distance between two Z lines is called a **sarcomere**. This is the functional unit of the muscle. At rest, a sarcomere is about 2.5 μm long.

The thick and thin filaments are surrounded by sarcoplasmic reticulum.

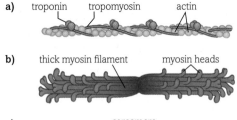

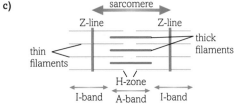

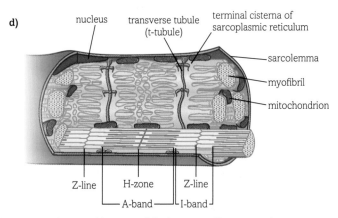

Figure 1 (a) The actin filament and (b) the myosin filament. (c) The arrangement of the filaments and (d) the filaments surrounded by sarcoplasmic reticulum.

Thin filaments

The thin filaments are actin. Each filament consists of two chains of actin subunits twisted around each other. Wound around the actin is a molecule of **tropomyosin** to which are attached globular molecules of **troponin**. Each troponin complex consists of three polypeptides: one binds to actin, one to tropomyosin and the third binds to calcium when it is available. Tropomyosin and troponin are part of the mechanism to control muscular contraction. At rest, these molecules cover binding sites to which the thick filaments can bind.

Thick filaments

Each thick filament consists of a bundle of myosin molecules. Each myosin molecule has two protruding heads, which stick out at each end of the molecule. These heads are mobile and can bind to the actin when the binding sites are exposed.

The sliding filament hypothesis

During contraction, the light band and the H zone get shorter. Therefore, the Z lines move closer together and the sarcomere gets shorter. This observation led to the sliding filament hypothesis. During contraction, the thick and thin filaments slide past one another.

The mechanism of contraction

The sliding action is caused by the movement of the myosin heads. When the muscle is stimulated, the tropomyosin is moved aside, exposing the binding sites on the actin (see Figure 2). The myosin heads attach to the actin and move, causing the actin to slide past the myosin.

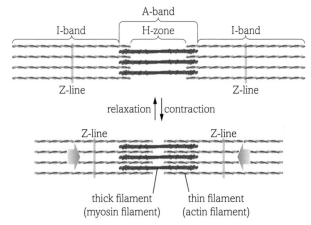

Figure 2 The filaments slide past one another during contraction.

106

Control of contraction

1. When the muscle is stimulated, the action potential passes along the sarcolemma and down the transverse tubules (t-tubules) into the muscle fibre.

2. The action potential is carried to the sarcoplasmic reticulum, which stores calcium ions, and causes the release of calcium ions into the sarcoplasm.

3. The calcium ions bind to the troponin, which alters the shape pulling the tropomyosin aside. This exposes the binding sites on the actin.

4. Myosin heads bind to the actin, forming cross-bridges between the filaments.

5. The myosin heads move, pulling the actin filament past the myosin filament.

6. The myosin heads detach from the actin and can bind again further up the actin filament.

Millions of cross-bridges can be formed between the actin and the myosin filaments. Once contraction has occurred, the calcium ions are rapidly pumped back into the sarcoplasmic reticulum, allowing the muscle to relax.

The role of ATP

ATP supplies the energy for contraction. Part of the myosin head acts as ATPase and can hydrolyse the ATP to ADP and inorganic phosphate (P_i), releasing energy (see Figure 3):

1. The myosin head attaches to the actin filament, forming a cross-bridge.

2. The myosin head moves (tilts backwards), causing the thin filament to slide past the myosin filament. This is the power stroke. During the power stroke, ADP and P_i are released from the myosin head.

3. After the power stroke, a new ATP molecule attaches to the myosin head, breaking the cross-bridge.

4. The myosin head then returns to its original position (swings forward again) as the ATP is hydrolysed, releasing the energy to make this movement occur. The myosin head can now make a new cross-bridge further along the actin filament.

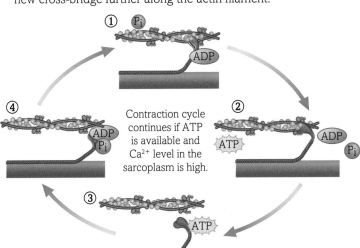

Contraction cycle continues if ATP is available and Ca^{2+} level in the sarcoplasm is high.

Figure 3 How ATP is used during contraction.

Maintaining the supply of ATP

As there are millions of myosin heads involved in muscle contraction, there is a huge requirement for ATP.

The ATP available in muscle tissue is only enough to support at most 1–2 seconds' worth of contraction. ATP must be regenerated very quickly in order to allow continued contraction. Three mechanisms are involved in maintaining the supply of ATP:

- Aerobic respiration in mitochondria. Muscle tissue contains a large number of mitochondria in which aerobic respiration can occur. The Bohr effect helps to release more oxygen from the haemoglobin in the blood. However, during intense activity, the rate at which ATP can be produced will be limited by the delivery of oxygen to the muscle tissue.

- Anaerobic respiration in the sarcoplasm of the muscle tissue. Anaerobic respiration can release a little more ATP from the respiratory substrates. However, it leads to the production of lactate (lactic acid), which is toxic. Anaerobic respiration can only last a few seconds before lactic acid build-up starts to cause fatigue.

- **Creatine phosphate** in the sarcoplasm acts as a reserve store of phosphate groups. The phosphate can be transferred from the creatine phosphate to ADP molecules, creating ATP molecules very rapidly. The enzyme creatine phosphotransferase is involved. The supply of creatine phosphate is sufficient to support muscular contraction for a further 2–4 seconds.

> **DID YOU KNOW?**
>
> For prolonged muscular contraction, activity must be limited to a level that can be maintained by the regeneration of ATP by aerobic respiration. This is why the efficiency of the cardiovascular and gaseous exchange systems is so important to athletes.

Questions

1. What happens when the thick and thin filaments overlap?

2. Which supply of ATP is most important for:
   (a) a marathon runner
   (b) a sprinter?

3. Describe the role of ATP during muscle contraction.

4. How is ATP used while the muscle is at rest?

5. A few hours after death, the muscles contract in rigor mortis. Suggest why this occurs.

IMPROVING FLAVOURS

Supermarkets today stock fruit and vegetables from around the world, ripe and ready to eat whatever the season. But have you ever thought about how this is managed?

MAKING FRUIT AND VEGETABLES TASTIER

Consider the dilemma that cantaloupe melons presented to plant breeders. To enjoy a cantaloupe's full flavour, you must pick and eat it at peak ripeness, before it goes too soft. Toward the later stages of a cantaloupe's development, a burst of the hormone ethylene causes the fruit to ripen and soften quite quickly. This speedy puberty made transporting cantaloupes from one country to another problematic: even on ice the melons turned to mush by the voyage's end. So plant breeders decreased the level of ethylene in cantaloupes intended for long-distance shipping by cross-pollinating only melons that naturally produced the lowest amounts of the hormone. Without a strong spurt of ethylene, the melons stay firm on the trip from field to produce aisle, but the chemical reactions that produce a ripe melon's aroma and taste never happen.

Breeders have had some success in overcoming this predicament. In the 1990s, Dominique Chambeyron – a plant breeder employed by the Dutch De Ruiter Seeds Group – managed to create a variety of small striped cantaloupe that retained both its firmness and flavour for weeks after it was harvested. Known as Melorange, this cultivar is grown in Central America and shipped to… chain stores in the USA between December and April, when it is too cold to grow melons locally…. It can take more than a decade to perfect a consistently impressive new cultivar. Breeders must cross-pollinate plants over and over, hoping that some of the offspring will inherit the right characteristics. And they generally have to wait for the plants to grow and produce ripe fruit to find out. Much of what they produce is way off the mark and entirely unusable.

Genetics has recently offered an alternative path. At Monsanto, Jeff Mills and his colleagues are able to predict the quality of a cantaloupe plant's eventual fruit before they ever put a single seed in the ground. First, Mills and his team pinpointed the melon genes underlying Melorange's unique combination of flavour and firmness. They can look for these genetic "markers" in cantaloupe seeds with help from a group of cooperative and largely autonomous robots.

A seed chipper shaves off a sliver of a seed for DNA analysis, leaving the rest of the kernel unharmed and suitable for sowing in a greenhouse or field. Another robot extracts the DNA from that tiny bit of seed and adds the necessary molecules and enzymes to chemically give fluorescent tags to the relevant genes, if they are there. Yet another machine amplifies the number of these glowing tags to measure the light they emit and determine whether a gene is present.

All this talk of DNA analysis sounds suspiciously like genetic engineering – the gene-editing technique that creates genetically modified organisms. But it is not. It is a completely non-GMO technology. In fact, that is one of the main reasons it is so attractive to researchers.

Source
- *Taken from* Building tastier fruits and veggies (no GMOs required). *Scientific American* **311**: 56–61 (July 2014)

Where else will I encounter these themes?

Let's start by considering the nature of the writing in the article.

1 Do you think this article is aimed at scientists, the general public, or people who are not scientists but who have an interest in science? Provide examples from the article to support your answer.

Now we will look at the biology in, or connected to, this article.

2 Describe the role of ethylene (also known as ethene) in cantaloupe melons.

3 Ethylene (ethene) is a small gaseous molecule. Suggest why this makes it a good signalling molecule in plants.

4 Explain the meaning of the term 'genetic marker'.

5 Seed germination commonly requires scarification (scratching) of the testa (the outer protective layer). This allows water to diffuse into the seed itself, where starch is stored; this helps to trigger germination.

  a What is diffusion?

  b By thinking about the conditions required for respiration to occur, suggest why water is required for germination.

  c Suggest why taking a shaving from the outside coat of the seed does not inhibit germination.

6 When ethylene (ethene) binds to glycoprotein receptors on plant cell membranes, it sets off a cascade pathway which turns two inhibitor genes off: ETR1 and CTR1. This allows expression of other genes which are transcribed and translated to produce pectinases, amylases, and enzymes that break down chlorophyll.

  a Explain the meaning of the term 'cascade pathway', using an example with which you are familiar.

  b Explain the difference between transcription and translation.

  c Describe the structure of glycoproteins and their location in the cell surface membrane.

  d By referring to pectinases and amylases, explain why fruit softens when it ripens.

  e Using what you know about enzyme action, suggest why fruit and vegetables are commonly chilled after harvest.

  f Suggest why fruit changes colour when it ripens.

7 The polymerase chain reaction can be used to amplify sections of DNA. Draw out a flow chart to explain how this reaction works.

8 Fruit tends to ripen in many species in response to stress. Suggest how such a response may have evolved, using the principles of natural selection to explain your answer.

9 Create a table which compares and contrasts genetic engineering with artificial selection. Think about mechanism, purpose and outcome to help you include a variety of ideas.

Activity

Write a storyboard and script for an information video which explains why the technique in the article does not involve genetic engineering. Your storyboard and script should explain to an anti-GMO activist how this technique mirrors the kind of artificial selection which humans have employed for many years. You should include some scientific content about artificial selection, how genetic engineering happens, and what the role of gene analysis is in selecting melons for breeding.

1. Which row correctly describes the responses of a plant to its environment? [1]

	Response	Stimulus	Type of response
A	A new root grows down into the soil	Water potential	Geotropism
B	A shoot grows upwards towards light	Gravity	Phototropism
C	A pollen tube grows towards the anther	Chemical	Chemotropism
D	A bean shoot curls around a stick	Touch	Thigmotropism

2. How do plant growth substances move around the plant? [1]
   A. Active transport
   B. Diffusion
   C. Mass flow in xylem and phloem
   D. All three of the above

3. Read the following statements about plant growth substances.
   (i) Cytokinins promote cell division.
   (ii) Abscisic acid closes stomata.
   (iii) Gibberellins promote leaf growth.
   (iv) Auxin promotes cell elongation.
   (v) Ethene prevents fruit ripening.
   Which statement(s) is/are true? [1]
   A. All statements
   B. (ii), (iii) and (v)
   C. (i), (ii) and (iv)
   D. All statements except (i) and (ii)

4. What happens when skeletal muscle contracts? [1]
   A. The H zone and the I band get shorter.
   B. The H zone, the I band and the A band get shorter.
   C. The H zone and the A band get shorter.
   D. The I band and the A band get shorter.

5. Figure 1 is a diagram of the human brain.

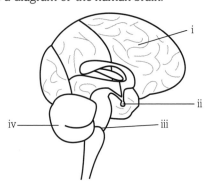

Figure 1

Which row correctly identifies the main parts of the brain? [1]

	i	ii	iii	iv
A	cerebellum	cerebrum	medulla oblongata	pituitary gland
B	cerebrum	pituitary gland	medulla oblongata	cerebellum
C	cerebellum	medulla oblongata	pituitary gland	cerebrum
D	cerebrum	cerebellum	medulla oblongata	pituitary gland

[Total: 5]

6. A student investigated the effect of plant growth substances (IAA and GA) on the elongation of pea stems. The student marked a 10 mm length of stem on each of four plants and applied growth substances to the stem. Figure 2 shows a graph of the student's results.

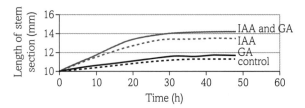

Figure 2

   (a) Describe the effects on growth of the stem for the two growth substances applied separately. [4]
   (b) The maximum growth of the control was an increase in length of 1.3 mm. The maximum growth after application of growth substances was 4.2 mm. Calculate the percentage increase in growth caused by the growth substances. [2]

(c) Suggest a suitable treatment for the control. [2]

(d) Suggest how the student could have improved this investigation. [5]

[Total: 13]

7. (a) A farmer who wants to create a thick stock-proof hedge will cut the hedge plants regularly. Explain how this makes the hedge thicker. [4]

(b) Plant growth substances have a range of commercial uses. Describe the use of three plant growth substances in commercial applications. [6]

[Total: 10]

8. Figure 3 is a drawing of the human brain.

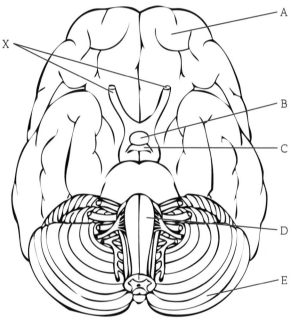

Figure 3

(a) Identify the parts of the brain labelled A–E. [5]

(b) The brain has been drawn from underneath. State one piece of evidence that confirms this. [1]

(c) The nerves labelled X are attached to two sensory organs in the head. Suggest which sensory organs they are attached to. [1]

(d) Many of the nerves emerging from part D of the brain belong to the autonomic nervous system. Describe the organisation of the autonomic nervous system. [7]

[Total: 14]

9. Table 1 compares the functions of the sympathetic and parasympathetic nervous systems.

(a) Complete the table. [4]

Sympathetic nervous system	Parasympathetic nervous system
Most active during _____	Most active during _____
Causes ventilation rate to _____	Causes ventilation rate to _____
_____ pupils	_____ pupils
_____ activity of digestive system	_____ activity of digestive system

(b) During exercise, the heart rate increases. Describe the mechanism that increases the heart rate. [6]

[Total: 10]

10. Figure 4 shows the protein filaments from a skeletal muscle during contraction.

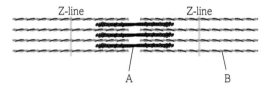

Figure 4

(a) Identify filaments A and B. [2]

(b) Name the length of myofibril between two Z lines. [1]

(c) Draw the same protein filaments as you would expect to see them in a relaxed muscle. [2]

(d) ATP is the energy source for muscle contraction. Most ATP is produced in aerobic respiration. Describe how ATP supply is maintained when insufficient oxygen is delivered to an active muscle. [3]

(e) Outline the role of calcium ions in muscle contraction. [3]

[Total: 11]

Communication, homoeostasis and energy

PHOTOSYNTHESIS

Introduction

Nearly all living organisms, the exception being chemosynthetic organisms such as those that live near thermal oceanic vents, depend on photosynthesis for their existence. Plants, as well as some algae and some bacteria, are autotrophic feeders. This means that they use sunlight energy to make organic molecules, containing chemical energy, from simple inorganic chemicals – water and carbon dioxide.

We eat plants, animals that have eaten plants, and products made using bacteria and fungi that have been grown on plant wastes. We also use many plant products, such as a variety of medicinal products, wood, paper, cotton and linen, as well as many substances synthesised from cellulose.

The evolution of photosynthesis led eventually to the presence of free oxygen in the Earth's atmosphere, which contributed to the evolution of many different types of living organisms.

In the early seventeenth century, a Flemish chemist and medic Jan Baptista van Helmont wondered how plants grew. In order to answer this question he grew a willow tree in a tub for five years. He discovered that the tree gained about 80 kilograms in mass, but that the mineral content of the soil reduced by only a few grams. He thought that the increase in the tree's mass came solely from the water he had added to the tub. In 1772, Joseph Priestley discovered that plants produce oxygen, and in 1779 Jan Ingenhousz found that plants need light to do this. In 1781, carbon dioxide gas was identified in the atmosphere, and in 1804 carbon dioxide was found to be the source of carbon compounds in plants.

In the late nineteenth century, Theodor Engelmann shone different colours of light onto a green alga and discovered that photosynthesis only occurs where chlorophyll is present, and that red and blue light are more effective than white light.

In 1939, Robert Hill showed that isolated chloroplasts suspended in aqueous solution can produce oxygen.

In this chapter, you will find out about the two stages of photosynthesis: the light-dependent stage and the light-independent stage.

All the maths you need

To unlock the puzzles of this chapter you need the following maths:

* Know units of measurement
* Know powers of 10
* Use division and multiplication, for example when making serial dilutions
* Calculate ratios, percentages and fractions
* Use an appropriate number of significant figures when handling data from investigations
* Calculate percentage error
* Find arithmetic means
* Calculate the rate of change from a graph
* Select and use a statistical test
* Calculate the volume of a cylinder

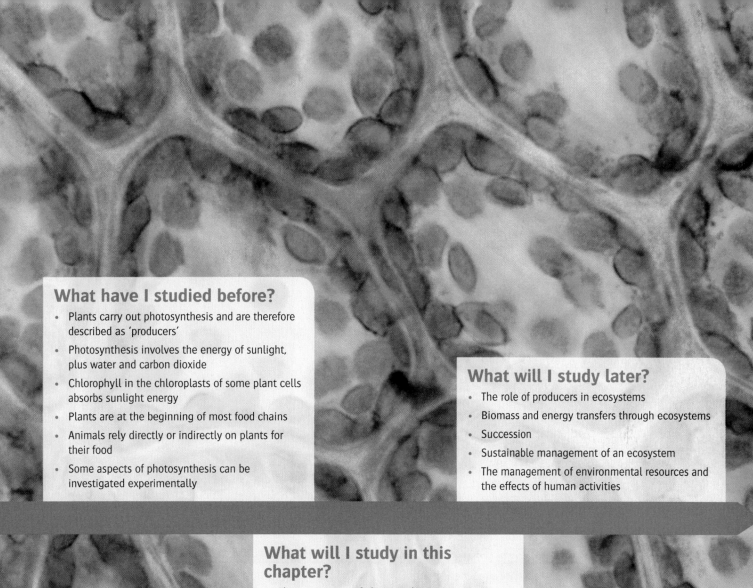

What have I studied before?

- Plants carry out photosynthesis and are therefore described as 'producers'
- Photosynthesis involves the energy of sunlight, plus water and carbon dioxide
- Chlorophyll in the chloroplasts of some plant cells absorbs sunlight energy
- Plants are at the beginning of most food chains
- Animals rely directly or indirectly on plants for their food
- Some aspects of photosynthesis can be investigated experimentally

What will I study later?

- The role of producers in ecosystems
- Biomass and energy transfers through ecosystems
- Succession
- Sustainable management of an ecosystem
- The management of environmental resources and the effects of human activities

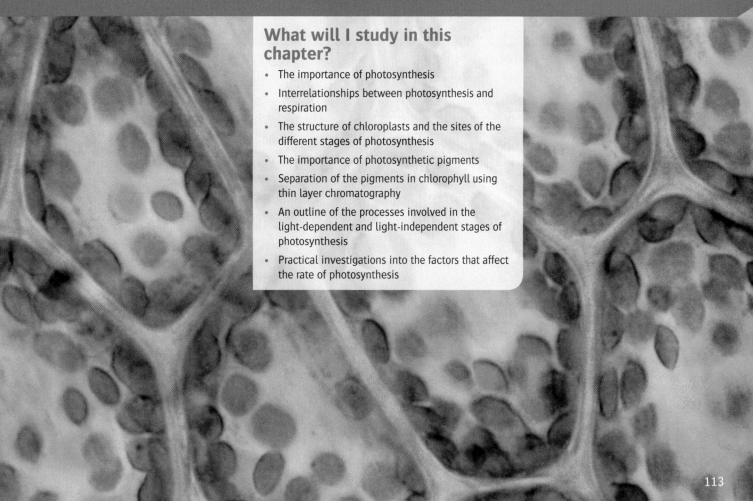

What will I study in this chapter?

- The importance of photosynthesis
- Interrelationships between photosynthesis and respiration
- The structure of chloroplasts and the sites of the different stages of photosynthesis
- The importance of photosynthetic pigments
- Separation of the pigments in chlorophyll using thin layer chromatography
- An outline of the processes involved in the light-dependent and light-independent stages of photosynthesis
- Practical investigations into the factors that affect the rate of photosynthesis

The interrelationship between photosynthesis and respiration

By the end of this topic, you should be able to demonstrate and apply your knowledge and understanding of:

* the interrelationship between the processes of photosynthesis and respiration

The importance of photosynthesis

Photosynthesis is a physiological process used by plants, algae and some types of bacteria to convert light energy from sunlight into chemical energy. Organisms can use this chemical energy to synthesise large organic molecules, which form the building blocks of living cells, from simple inorganic molecules such as water and carbon dioxide – this is **autotrophic nutrition**.

Organisms that photosynthesise are called 'photoautotrophs', because they use light as the energy source for autotrophic nutrition. These organisms are also described as 'producers', because they are at the beginning (first trophic level) of a food chain and provide energy and organic molecules to other, non-photosynthetic, organisms.

The general equation for photosynthesis is:

$$6CO_2 + 6H_2O + \text{energy from photons} \xrightarrow{\text{chlorophyll}} C_6H_{12}O_6 + 6O_2$$

A photon is a particle of light; each photon contains an amount (a quantum) of energy.

The main product of photosynthesis is monosaccharide sugar, which can be converted to disaccharides for transport (see Book 1, topic 3.3.1) and then to starch for storage.

Photosynthesis is an example of carbon fixation – the process by which carbon dioxide is converted into sugars. The carbon for synthesising all types of organic molecule is provided by carbon fixation.

Carbon fixation is endothermic, and so needs energy. Carbon fixation also needs electrons; the addition of electrons is a reduction reaction.

Carbon fixation helps regulate the concentration of carbon dioxide in the atmosphere and oceans.

Most forms of life on Earth rely directly or indirectly on photosynthesis.

DID YOU KNOW?

Around 2 billion years ago, cyanobacteria, such as that shown in Figure 1, carried out photosynthesis using carbon dioxide and water as raw materials. The breakdown of water to provide electrons and protons (hydrogen ions) generated free oxygen into the Earth's atmosphere. The oxygen in the atmosphere enabled the evolution of more complex life forms.

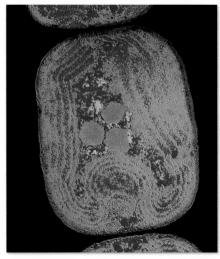

Figure 1 Coloured TEM of a single cell of the cyanobacterium *Pseudanabaena* sp. The photosynthetic lamellae, seen here in green, are the site of photosynthesis. The red circles are granules of polyphosphate (×24 000).

DID YOU KNOW?

It has been estimated that the average rate of energy capture by photosynthesis across the globe is around 1.3×10^{15} watts (1.3 petawatts), which is about six times greater than the present global consumption of energy by the human population. It has also been estimated that 3.5×10^{15} kg of carbon is fixed by plants each year on Earth.

Figure 2 Polarised light micrograph of *Spirogyra* sp. alga. This filamentous protoctist consists of cylindrical cells that connect end to end forming filaments. Inside each cell is a spiral chloroplast (×200).

Respiration

Plants and other organisms that photosynthesise also respire. During respiration, they oxidise the organic molecules that they have previously synthesised by photosynthesis and stored, releasing chemical energy.

Non-photosynthetic organisms such as fungi, animals, many protoctists and many types of bacteria are described as 'heterotrophs'. They obtain energy by digesting complex organic molecules of food to smaller molecules that they can use as respiratory substrates. They obtain energy from the products of digestion by respiration.

During respiration, glucose and other organic compounds are oxidised to produce carbon dioxide and water. Respiration releases chemical energy (it is exothermic) that can drive the organism's metabolism:

$$C_6H_{12}O_6 + 6O_2 \rightarrow 6H_2O + 6CO_2 + energy$$

How photosynthesis and respiration interrelate

Both photosynthesis and aerobic respiration are important in cycling carbon dioxide and oxygen in the atmosphere. The products of one process are the raw materials for the other process: aerobic respiration removes oxygen from the atmosphere and adds carbon dioxide, while photosynthesis does the opposite.

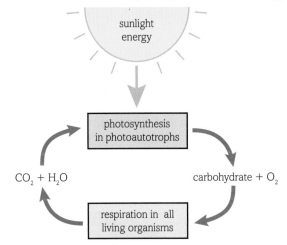

Figure 3 Photosynthesis and respiration cycle oxygen and carbon dioxide within the atmosphere.

Figure 4 Plants can be kept in a sealed container. The oxygen produced during photosynthesis is used for respiration; the carbon dioxide produced by respiration is used for photosynthesis. How is water balance maintained?

Compensation point

Plants respire *all* the time. However, they only photosynthesise during daylight. Plants often compete with each other for light. The intensity of light has to be sufficient to allow photosynthesis at a rate that replenishes the carbohydrate stores used up by respiration.

When photosynthesis and respiration proceed at the same rate, so that there is no net gain or loss of carbohydrate, the plant is at its **compensation point**. The time a plant takes to reach its compensation point is called the compensation period. The compensation period is different for different plant species. Shade plants can utilise light of lower intensity than sun plants can. When exposed to light after being in darkness, shade plants reach their compensation point sooner (they have a shorter compensation period) than sun plants, which require a higher light intensity to achieve their optimum rate of photosynthesis.

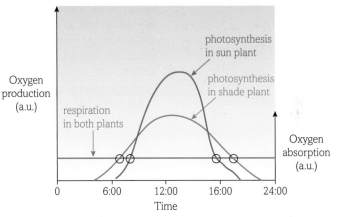

Figure 5 Graph showing rates of photosynthesis and respiration for sun and shade plants, on a sunny summer's day.

LEARNING TIP

Remember that all organisms, including plants, respire *all* the time — both during the night and during daylight.

Photoautotrophs (e.g. plants) photosynthesise and, because photosynthesis uses light energy, they can only photosynthesise during daylight.

Questions

1. Explain why most forms of life on Earth rely directly or indirectly on photosynthesis.

2. Explain why, as shown in Figure 5, the rate of respiration does not vary much throughout the 24-hour period, but the rate of photosynthesis does vary throughout the same period.

3. What is meant by compensation point?

4. In Figure 5, what is the morning compensation point for the:
   (a) shade plant?
   (b) sun plant?

5. Suggest which process – aerobic respiration or photosynthesis – evolved first on Earth. Give reasons for your answer.

2 Chloroplasts and photosynthetic pigments

By the end of this topic, you should be able to demonstrate and apply your knowledge and understanding of:

* the structure of a chloroplast and the sites of the two main stages of photosynthesis
* the importance of photosynthetic pigments in photosynthesis
* practical investigations using thin layer chromatography (TLC) to separate photosynthetic pigments

---

KEY DEFINITIONS

granum (plural: grana): inner part of chloroplasts made of stacks of thylakoid membranes, where the light-dependent stage of photosynthesis takes place.

photosynthetic pigment: pigment that absorbs specific wavelengths of light and traps the energy associated with the light; such pigments include chlorophylls a and b, carotene and xanthophyll.

photosystem: system of photosynthetic pigments found in thylakoids of chloroplasts; each photosystem contains about 300 molecules of chlorophyll that trap photons and pass their energy to a primary pigment reaction centre, a molecule of chlorophyll a, during the light-dependent stage of photosynthesis.

stroma: fluid-filled matrix of chloroplasts, where the light-independent stage of photosynthesis takes place.

thylakoid: flattened membrane-bound sac found inside chloroplasts; contains photosynthetic pigments/photosystems and is the site of the light-dependent stage of photosynthesis.

The structure of chloroplasts

You saw in Book 1, topic 2.1.4 that chloroplasts are the organelles within plant cells where photosynthesis takes place. Algae have chloroplasts, but photosynthetic bacteria do not.

Most plant chloroplasts are disc shaped and around 2–10 μm long. Each is surrounded by a double membrane, the envelope, with an **intermembrane space** of width 10–20 nm between the inner and outer membrane. The outer membrane is highly permeable.

There are two distinct regions, visible on electron micrographs, inside a chloroplast: the fluid-filled matrix called the **stroma**, and the **grana** that consists of stacks of thylakoid membranes.

(a)

inner membrane
outer membrane
stroma
granum
chloroplast envelope
intermembrane compartment
intergranal lamellae
thylakoids

(b)

granal thylakoid
intergranal lamella

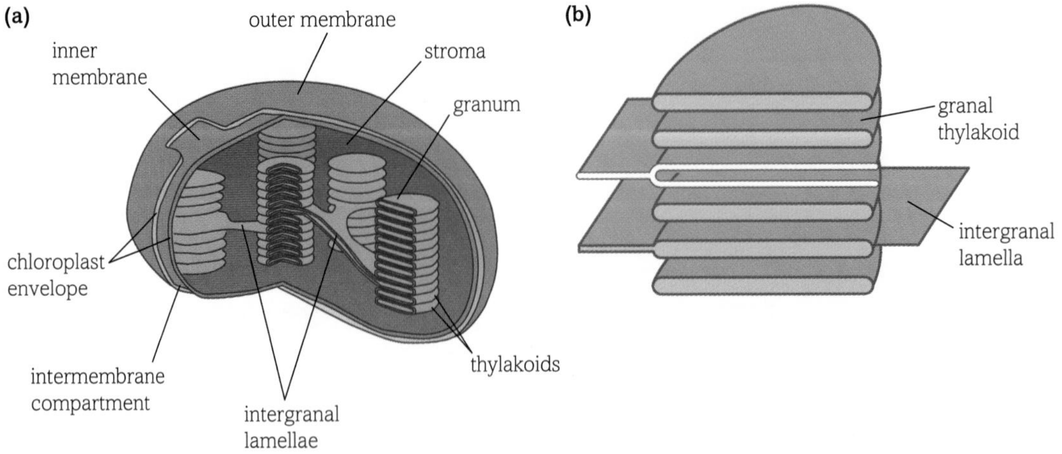

Figure 1 (a) Stereogram showing the internal structure of a whole chloroplast; (b) section of a single granum, showing its stack of thylakoids.

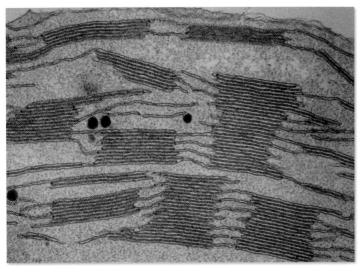

Figure 2 False colour TEM of stacks of grana in a chloroplast taken from a leaf of maize, *Zea mays*. The oil droplets act as a reserve of raw materials for the synthesis of new chloroplast membranes (×12 400).

Grana

The first stage of photosynthesis, the light-dependent stage (see topic 5.6.3), takes place in the grana.

Chloroplasts have three distinct membranes – outer, inner and thylakoid, giving three separate internal compartments – the intermembrane space, stroma and the thylakoid space. The thylakoids within a granum may be connected to thylakoids within another granum by intergranal lamellae (also known as intergranal thylakoids).

The thylakoid membrane of each chloroplast is less permeable and is folded into flattened disc-like sacs called **thylakoids** that form stacks. Each stack of thylakoids is called a granum. One granum may contain up to 100 thylakoids.

With many grana in every chloroplast and with many chloroplasts in each photosynthetic cell, there is a huge surface area for:

- the distribution of the **photosystems** that contain the **photosynthetic pigments** that trap sunlight energy
- the electron carriers and ATP synthase enzymes needed to convert that light energy into ATP.

Proteins embedded in the thylakoid membranes hold the photosystems in place.

The grana are surrounded by the stroma, so the products of the light-dependent stage can easily pass to the stroma to be used in the light-independent stage.

Stroma

The stroma is the fluid-filled matrix. It contains the enzymes needed to catalyse the reactions of the light-independent stage of photosynthesis (see topic 5.6.4), as well as starch grains, oil droplets, small ribosomes similar to those found in prokaryote cells (see Book 1, topic 2.1.7) and DNA.

The loop of DNA contains genes that code for some of the proteins needed for photosynthesis. These proteins are assembled at the chloroplast ribosomes.

> **LEARNING TIP**
>
> Make sure you refer to the matrix of chloroplasts as the *stroma*. Do not confuse that word with stoma (pore in leaf for gaseous exchange).

Photosynthetic pigments

Within the thylakoid membranes of each chloroplast are funnel-shaped structures called photosystems. These photosystems contain **photosynthetic pigments**. Each pigment absorbs light of a particular wavelength and reflects other wavelengths of light. Each pigment appears, to our eyes and brain, the colour of the wavelength of light it is reflecting.

The energy associated with the wavelengths of light captured is funnelled down to the primary pigment reaction centre, consisting of a type of chlorophyll, at the base of the photosystem (see Figure 3).

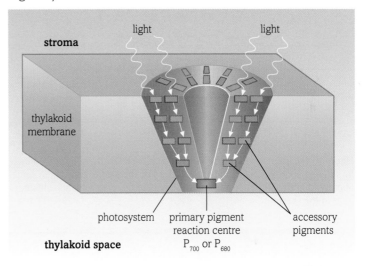

Figure 3 Diagram representing a photosystem consisting of a funnel-shaped light-harvesting cluster of photosynthetic pigments, held in place in a thylakoid membrane, by proteins. Only a few pigment molecules are shown. The primary pigment reaction centre is a molecule of chlorophyll a. The accessory pigments consist of molecules of chlorophyll b and carotenoids. There are two main types of photosystem, photosystem I and photosystem II.

Chlorophylls

Chlorophylls are a mixture of pigments. All have a similar molecular structure consisting of a porphyrin group, in which is a magnesium atom, and a long hydrocarbon chain.

Chlorophyll a

There are two forms of chlorophyll a – both of which appear blue-green. Both are situated at the centre of photosystems. Both absorb red light, but they have different absorption peaks:

- P_{680} is found in photosystem II (see topic 5.6.3) and its peak of absorption is light of wavelength 680 nm.
- P_{700} is found in photosystem I (see topic 5.6.3) and its peak of absorption is light of wavelength 700 nm.

Chlorophyll a also absorbs some blue light, of wavelength around 440 nm.

Chlorophyll b

Chlorophyll b absorbs light of wavelengths 400–500 nm and around 640 nm. It appears yellow green.

Accessory pigments

Carotenoids absorb blue light of wavelengths 400–500 nm. They reflect yellow and orange light.

Xanthophylls absorb blue and green light of wavelengths 375–550 nm. They reflect yellow light.

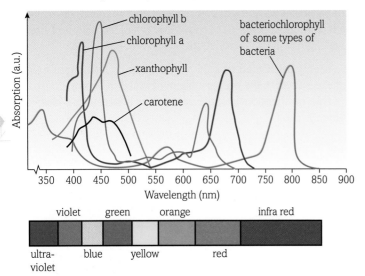

Figure 4 The absorption spectrum showing the wavelengths of light absorbed by different photosynthetic pigments. Notice that the pigment present in some photosynthetic bacteria can absorb light into both the ultraviolet and infrared parts of the spectrum.

Pigment	R_f value
carotene	0.91
phaeophytin	0.63–0.75
chlorophyll a	0.63
chlorophyll b	0.58
xanthophyll	0.32–0.53

Table 1 R_f values for the main photosynthetic pigments obtained using TLC with a solvent of 5 parts cyclohexane, 3 parts propanone and 2 parts petroleum ether.

Questions

1. In which part of the chloroplast does the light-dependent stage of photosynthesis take place?

2. In which part of the chloroplast does the light-independent stage take place?

3. Where exactly in the chloroplast are the photosystems situated?

4. Describe the structure and function of photosystems.

5. How does the structure of photosystem I differ from that of photosystem II?

6. Describe how the structure of chloroplasts enables then to carry out their functions.

7. Suggest why the R_f values of some pigments, as given in Table 1, show a range, rather than a single figure.

8. R_f values for the different pigments differ according to the solvent used. Suggest why this is the case.

INVESTIGATION

Separating photosynthetic pigments using thin layer chromatography (TLC) (adapted from SAPS protocol)

You can very easily separate the pigments in chlorophyll by scraping or mashing a green leaf. You can do this using two microscope slides. You can then spot the chlorophyll onto a slide coated with thin layer chromatography (TLC) material. When a solvent creeps up the slide the pigments separate out because they have different R_f values (see Table 1).

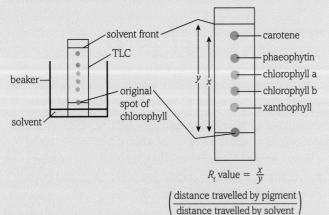

Figure 5 Separation of photosynthetic pigments using TLC.
Note: If using a highly flammable solvent such as ethanol, make sure there are no naked flames in the vicinity and no uncapped stock bottles. Avoid any leaves that are known to provoke an allergic reaction and wash hands after use.

(3) The light-dependent stage

By the end of this topic, you should be able to demonstrate and apply your knowledge and understanding of:

* the light-dependent stage of photosynthesis

The light-dependent stage of photosynthesis occurs in the grana (thylakoids) of chloroplasts and involves the photosystems. It involves the direct use of light energy. This stage consists of:

1. light harvesting at the photosystems (see topic 5.6.2)
2. **photolysis** of water
3. **photophosphorylation** – the production of ATP in the presence of light
4. the formation of reduced **NADP**.

Oxygen, the by-product of photosynthesis, is also produced in the light-dependent stage.

Two types of photosystem

Topic 5.6.2 described the structure and location of the photosystems and outlined the roles of the various pigments. There are two types of photosystem:

* In photosystem I (PSI), the pigment at the primary reaction centre is a type of chlorophyll a, which has a peak absorption of red light of wavelength 700 nm (P_{700}).
* In photosystem II (PSII), the pigment at the primary reaction centre is also a type of chlorophyll a, but this has a peak absorption of red light of wavelength 680 nm (P_{680}).

The role of water

In PSII, there is an enzyme that, *in the presence of light*, splits water molecules into protons (hydrogen ions), electrons and oxygen. The splitting of water in this way is called photolysis:

$$2H_2O \rightarrow 4H^+ + 4e^- + O_2$$

Some of the oxygen produced during photolysis is used by plant cells for aerobic respiration, but during periods of high light intensity the rate of photosynthesis is greater than the rate of respiration in the plant, so much of the oxygen by-product will diffuse out of the leaves, through stomata, into the surrounding atmosphere.

Water:

* is the source of protons (hydrogen ions) that will be used in photophosphorylation
* donates electrons to chlorophyll to replace those lost when light strikes chlorophyll
* is the source of the by-product, oxygen
* keeps plant cells turgid, enabling them to function.

Photophosphorylation

Photophosphorylation is the generation of ATP from ADP and inorganic phosphate, in the presence of light.

There are two types of photophosphorylation:

* Non-cyclic photophosphorylation involves PSI and PSII. It produces ATP, oxygen and reduced NADP.
* Cyclic photophosphorylation involves only PSI. It produces ATP but in smaller quantities than are made by non-cyclic photophosphorylation.

Both involve iron-containing proteins embedded in the thylakoid membranes that accept and donate electrons and form an electron transport system.

> **KEY DEFINITIONS**
>
> **electron carriers:** molecules that can accept one or more electrons and then donate those electrons to another carrier. Proteins embedded in thylakoid membranes are electron carriers, and form an electron transport chain or system. Ferredoxin, NAD and NADP are also electron carriers.
>
> **NADP:** nicotinamide adenine dinucleotide phosphate; a coenzyme and electron and hydrogen carrier.
>
> **photophosphorylation:** the generation of ATP from ADP and inorganic phosphate, in the presence of light.

Non-cyclic photophosphorylation

Figure 1 outlines what happens in non-cyclic photophosphorylation.

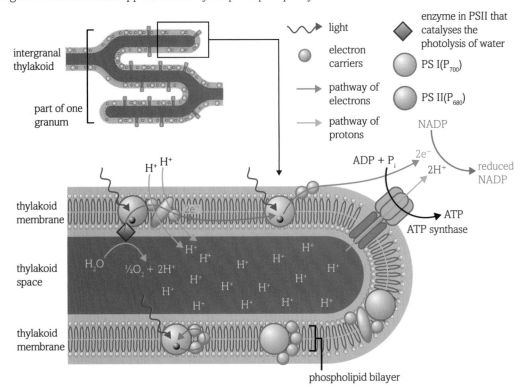

Figure 1 Diagram showing the process of photophosphorylation.

1. When a photon of light strikes PSII (P_{680}), its energy is channelled to the primary pigment reaction centre (see topic 5.6.2).

2. The light energy excites a pair of electrons inside the chlorophyll molecule.

3. The energised electrons escape from the chlorophyll molecule and are captured by an **electron carrier**, which is a protein with iron at its centre, embedded in the thylakoid membrane.

4. These electrons are replaced by electrons derived from photolysis.

5. When this iron ion combines with an electron it becomes reduced (Fe^{2+}). It can then donate the electron, becoming reoxidised (Fe^{3+}), to the next electron carrier in the chain.

6. As electrons are passed along a chain of electron carriers embedded in the thylakoid membrane, at each step some energy associated with the electrons is released.

7. This energy is used to *pump* protons across the thylakoid membrane into the thylakoid space.

8. Eventually the electrons are captured by another molecule of chlorophyll a in PSI. These electrons replace those lost from PSI due to excitation by light energy.

9. A protein–iron–sulfur complex called ferredoxin accepts the electrons from PSI and passes them to NADP in the stroma.

10. As protons accumulate in the thylakoid space, a proton gradient forms across the membrane.

11. Protons diffuse down their concentration gradient though special channels in the membrane associated with ATP synthase enzymes and, as they do so, the flow of protons causes ADP and inorganic phosphate to join, forming ATP.

12. As the protons pass through the channel they are accepted, along with electrons, by NADP which becomes reduced. The reduction of NADP is catalysed by the enzyme NADP reductase.

The light energy has been converted into chemical energy in the form of ATP by photophosphorylation. ATP and reduced NADP are now in the stroma ready for the light-independent stage of photosynthesis (see topic 5.6.4).

Use this mnemonic to help you remember whether electrons are lost or gained during reduction and oxidation:

   OIL RIG

Oxidation is loss. Reduction is gain.

Hydrogen ions (protons) are *pumped* across the thylakoid membrane into the thylakoid space. The energy comes from electrons.

Cyclic photophosphorylation

This uses only PSI (P_{700}). As light strikes PSI, a pair of electrons in the chlorophyll molecule at the reaction centre gain energy and become excited. They escape from the chlorophyll and pass to an electron carrier system and then pass back to PSI.

During the passage of electrons along the electron carriers, a small amount of ATP is generated. However, no photolysis of water occurs, so no protons or oxygen are produced. No reduced NADP is generated.

Chloroplasts in guard cells contain only PSI. They produce only ATP which actively brings potassium ions into the cells, lowering the water potential so that water follows by osmosis. This causes the guard cells to swell and opens the stoma.

Figure 2 summarises the main events of cyclic and non-cyclic photophosphorylation, in a diagram known as the Z scheme.

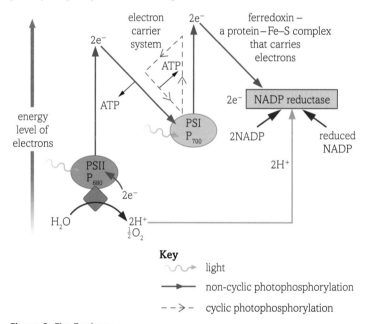

Figure 2 The Z scheme.

You may wonder why PSII is involved before PSI. The numbers refer to the order in which these photosystems were discovered and identified.

Red algae and brown algae live in water, at a level deeper than where green algae live. Green algae nearer the surface absorb much of the red and blue light, and green light passes through them. The red and brown algae have pigments, such as phycoerythrin and fucoxanthin respectively, that can absorb green light.

Figure 3 The red alga *Rhodymenia palmata* on rocks under water off the Devon coast.

Questions

1. Outline the role of water in photosynthesis.

2. Suggest why a lack of iron in soil may reduce growth in plants.

3. What is photolysis and where exactly does it take place?

4. If chloroplasts are isolated from plant cells, suspended in a solution of blue DCPIP and then illuminated, the DCPIP solution becomes reduced and loses its blue colour. Suggest how the DCPIP solution has become reduced.

5. Complete the table to compare cyclic and non-cyclic photophosphorylation.

	Cyclic photophosphorylation	Non-cyclic photophosphorylation
Photosystems involved		
Does photolysis occur?		
Fate of electrons released from chlorophyll		
Products		

(4) The light-independent stage

By the end of this topic, you should be able to demonstrate and apply your knowledge and understanding of:

* the fixation of carbon dioxide and the light-independent stage of photosynthesis
* the uses of triose phosphate (TP)

KEY DEFINITIONS

Calvin cycle: metabolic pathway of the light-independent stage of photosynthesis, occurring (in eukaryotic cells) in the stroma of chloroplasts where carbon dioxide is fixed, with the products of the light-dependent stage, to make organic compounds. The Calvin cycle also occurs in many photoautotrophic bacteria.

glycerate-3-phosphate (GP): an intermediate compound in the Calvin cycle.

ribulose bisphosphate (RuBP): a five-carbon compound present in chloroplasts; a carbon dioxide acceptor.

triose phosphate (TP): a three-carbon compound, and the product of the Calvin cycle; can be used to make other larger organic molecules.

Figure 1 Melvin Calvin in his photosynthesis research laboratory.

The light-independent stage of photosynthesis takes place in the stroma of chloroplasts. Although it does not directly use light energy, it uses the products of the light-dependent stage. If the plant is not illuminated, the light-independent stage soon ceases, because ATP and hydrogen are not available to reduce the carbon dioxide and synthesise large complex organic molecules.

The role of carbon dioxide

Carbon dioxide is the source of carbon for the production of all organic molecules found in all the carbon-based life forms on Earth. These organic molecules may be used as structures (e.g. cell membranes, antigens, enzymes, muscle proteins, cellulose cell walls) or act as energy stores (starch and glycogen).

Carbon dioxide in air enters the leaf through the stomata and diffuses through the spongy mesophyll layer to the palisade layer, into the palisade cells, through their thin cellulose cell walls, and then through the chloroplast envelope into the stroma. The fixation of carbon dioxide in the stroma maintains a concentration gradient that aids this diffusion. Carbon dioxide that is a by-product of respiration in plant cells may also be used for this stage of photosynthesis.

The series of reactions whereby carbon dioxide is converted to organic molecules is called the **Calvin cycle**.

DID YOU KNOW?

Between 1946 and 1953, the American biochemist Melvin Calvin and his associates, Andrew Benson and James Bassham, at the University of California, Berkeley, worked out the sequence of reactions in the light-independent stage. They grew the alga *Chlorella* in a solution containing carbon dioxide labelled with radioactive carbon, ^{14}C. At different time intervals they dropped some of the algae into boiling ethanol, to stop any reactions, and then identified which compounds were present. This was not an easy task. In 1961, Calvin won the Nobel Prize for this work.

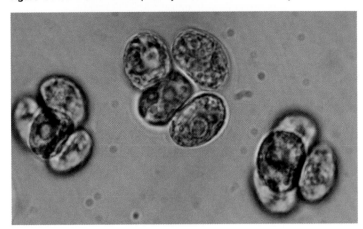

Figure 2 Light micrograph of *Chlorella vulgaris,* a unicellular photosynthetic alga that lives in freshwater ponds and lakes (×1900).

The Calvin cycle

1. Carbon dioxide combines with a carbon dioxide acceptor, a five-carbon compound called **ribulose bisphosphate** (**RuBP**). This reaction is catalysed by the enzyme **RuBisCO** (ribulose bisphosphate carboxylase-oxygenase).

2. RuBP, by accepting the carboxyl (COO^-) group, becomes carboxylated, forming an unstable intermediate six-carbon compound that immediately breaks down.

3. The product of this reaction is two molecules of a three-carbon compound, **GP** (**glycerate-3-phosphate**). The carbon dioxide has now been fixed.

4. GP is then reduced, using hydrogens from the reduced NADP made during the light-dependent stage, to **triose phosphate** (**TP**). Energy from ATP, also made during the light-dependent stage, is used at this stage at the rate of two molecules of ATP for every molecule of carbon dioxide fixed during stage 3.

5. In 10 of every 12 TP molecules, the atoms are rearranged to regenerate six molecules of RuBP. This process requires phosphate groups. Chloroplasts contain only low levels of RuBP, as it is continually being converted to GP, but is also continually being regenerated. The remaining two of the 12 molecules of TP are the product.

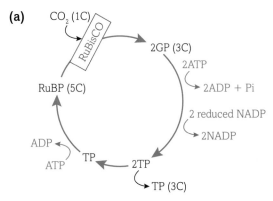

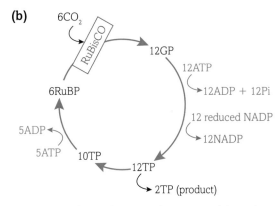

Figure 3 (a) An outline of the Calvin cycle. (b) Six turns of the cycle are needed for the formation of two molecules of TP that can then be used to make one molecule of glucose.

The Calvin cycle only runs during daylight

As already mentioned, the products of the light-dependent stage, namely ATP and reduced NADP, are continuously needed for the Calvin cycle to run.

During the light-dependent stage, hydrogen ions are pumped from the stroma into the thylakoid spaces, so the concentration of free protons in the stroma falls, raising the pH to around 8, which is optimum for the enzyme RuBisCO. RuBisCO is also activated by the presence of extra ATP in the stroma.

In daylight, the concentration of magnesium ions increases in the stroma. These ions attach to the active site of RuBisCO, acting as cofactors (Book 1, topic 2.4.2) to activate it.

The ferredoxin that is reduced by electrons from PSI (topic 5.6.3) activates enzymes involved in the reactions of the Calvin cycle.

RuBisCO is possibly the most abundant enzyme on Earth. It is found in plant chloroplasts (it is the most abundant protein in leaves), in photosynthetic algae and in many photosynthetic bacteria. Autotrophs and heterotrophs depend upon it.

LEARNING TIP

Many stages of the Calvin cycle have been omitted here for simplicity. This is as much biochemical detail as you need to know, but be aware that this metabolic pathway has several steps, each catalysed by a different enzyme. Hence the Calvin cycle is affected by factors that affect enzyme action.

The uses of triose phosphate (TP)

Some TP molecules are used to synthesise organic compounds, for example:

- Some glucose is converted to sucrose, some to starch and some to cellulose.

- Some TP is used to synthesise amino acids, fatty acids and glycerol.

The rest of the TP is recycled to regenerate the supply of RuBP (see Figure 3). Five molecules of the three-carbon compound TP interact to form three moleclues of the five-carbon compound RuBP.

Questions

1 Suggest why there are always only low levels of RuBP in the stroma of chloroplasts.

2 What is the product of the Calvin cycle?

3 Suggest how a lack of nitrate ions in the soil can affect the levels of RuBisCO in plants.

4 Which of the following contain nitrogen: NADP, ATP, GP, RuBP, electron carriers in the electron transport chain, hexose sugars, chlorophyll?

5 Explain why the illumination of chloroplasts leads to optimum conditions for the enzyme RuBisCO involved in the Calvin cycle.

6 Explain why RuBisCO is important to heterotrophs.

5 Factors affecting photosynthesis

By the end of this topic, you should be able to demonstrate and apply your knowledge and understanding of:

* factors affecting photosynthesis

Limiting factors

The factors that affect the rate of the complex process of photosynthesis operate simultaneously. These factors include the raw materials – carbon dioxide and water – as well as the energy source – **light intensity** – plus availability of chlorophyll, electron carriers and the relevant enzymes. Other factors, including temperature and turgidity of the cells, are also important.

At any given moment, the rate of a metabolic process that depends on a number of factors is limited by the factor that is present at its least favourable (lowest) level.

Light intensity

Light provides the energy to power the first stage of photosynthesis and produce ATP and reduced NADP needed for the next stage. Light also causes stomata to open so that gaseous exchange (Book 1, topic 2.6.5) can occur. When stomata are open **transpiration** also occurs, and this leads to uptake of water and its delivery to leaves (Book 1, topic 3.3.4).

At a constant favourable temperature and constant suitable carbon dioxide concentration, light intensity is the limiting factor. When light intensity is low, the rate of photosynthesis is low. As light intensity increases, the rate of photosynthesis increases.

At a certain point, even when light intensity increases, the rate of photosynthesis does not increase, as shown in Figure 1. Now a factor other than light intensity is limiting the process.

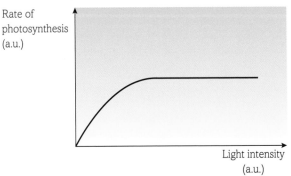

Figure 1 The rate of photosynthesis, at constant temperature and carbon dioxide concentration, with increasing light intensity.

Figure 2 shows that, where curve A levels out, carbon dioxide concentration limits the process. Where curve C plateaus, temperature is the limiting factor, as shown by curve D.

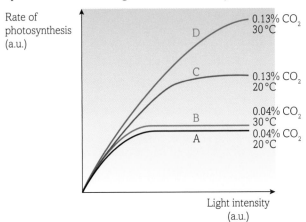

Figure 2 The effect of increasing light intensity on the rate of photosynthesis at two temperatures and two different carbon dioxide concentrations.

The effect of changing the light intensity on the Calvin cycle

Figure 3 shows what happens when there is little or no light:

1. GP cannot be reduced to TP.

2. TP levels fall and GP accumulates.

3. If TP levels fall, RuBP cannot be regenerated.

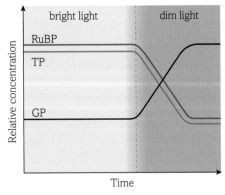

Figure 3 The effect of reduced light levels on the relative concentrations of GP, TP and RuBP in the Calvin cycle.

Carbon dioxide concentration

The levels of carbon dioxide in the atmosphere and in aquatic habitats are high enough that carbon dioxide is not usually a limiting factor.

The effect of changing the carbon dioxide concentration on the Calvin cycle

If the concentration of carbon dioxide falls below 0.01%:

1. RuBP cannot accept it, and accumulates.

2. GP cannot be made.

3. Therefore, TP cannot be made.

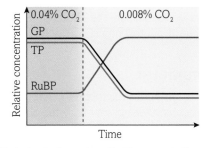

Figure 4 The effect of reducing carbon dioxide concentrations on the levels of GP, TP and RuBP in the Calvin cycle.

Temperature

The Calvin cycle involves many enzyme-catalysed reactions and therefore is sensitive to temperature.

The effects of changing temperature on the Calvin cycle are:

- From low temperatures to temperatures of 25–30 °C, if plants have enough water and carbon dioxide and a sufficient light intensity, the rate of photosynthesis increases as temperature increases.

- At temperatures above 30 °C, for most plants, growth rates may reduce due to photorespiration: oxygen competes with carbon dioxide for the enzyme RuBisCO's active site. This reduces the amount of carbon dioxide being accepted by RuBP and subsequently reduces the quantity of GP and therefore of TP being produced, whilst initially causing an accumulation of RuBP. However, due to lack of TP, RuBP cannot be regenerated.

- At temperatures above 45 °C, enzymes involved in photosynthesis may be denatured. This would reduce the concentrations of GP and TP, and eventually of RuBP as it could not be regenerated due to lack of TP.

Water stress

If a plant has access to sufficient water in the soil, then the transpiration stream (topic 3.3.5) has a cooling effect on the plant. The water passing up the xylem to leaves also keeps plant cells turgid so they can function. Turgid guard cells keep the stomata open for gaseous exchange.

If not enough water is available to the plant (**water stress**):

1. the roots are unable to take up enough water to replace that lost via transpiration

2. cells lose water and become plasmolysed

3. plant roots produce abscisic acid that, when translocated to leaves, causes stomata to close, reducing gaseous exchange

4. tissues become flaccid and leaves wilt

5. the rate of photosynthesis greatly reduces.

LEARNING TIP

Remember that *cells* become plasmolysed, but *tissues* become flaccid.

Questions

1. Look at curves A and B on the graph in Figure 2. Which factor is limiting the rate of photosynthesis when curve A plateaus? Give reasons for your answer.

2. Look at curves C and D on the graph in Figure 2. Which factor is limiting the rate of photosynthesis when curve C plateaus? Give reasons for your answer.

3. Look at Figure 3. Describe how and explain why the levels of RuBP, TP and GP change when the light intensity falls below a level sufficient for the light-dependent stage of photosynthesis to occur.

4. Look at Figure 4. Describe how and explain why the levels of RuBP, TP and GP change when the carbon dioxide concentration falls below 0.01%.

5. Describe how plants respond to water stress. Explain how this response affects their ability to photosynthesise.

6. Some scientists are working on genetically modifying plants to produce RuBisCO that would not be competitively inhibited by oxygen. Some parts of the enzyme are coded for by a gene in chloroplasts and the rest of the molecule is coded for by a gene in the cell nucleus.

   (a) Suggest why modification of the enzyme is thought to be needed.

   (b) Which part of the enzyme would need to be altered?

Factors affecting photosynthesis: practical investigations

By the end of this topic, you should be able to demonstrate and apply your knowledge and understanding of:

* practical investigations into factors affecting the rate of photosynthesis

Measuring the rate of photosynthesis

There are many ways to measure the rate of photosynthesis, including the rate of uptake of raw materials, such as carbon dioxide, or the rate of production of the by-product, oxygen. In each case, to measure the rate we need to calculate the quantity taken up or produced per unit time.

In school laboratories, the rate of photosynthesis is often found by measuring the volume of oxygen produced per minute by an aquatic plant. There are limitations with this method because:

* some of the oxygen produced by the plant will be used for its respiration
* there may be some dissolved nitrogen in the gas collected.

However, the same apparatus can be adapted and used to measure the effects of light intensity, temperature or carbon dioxide availability on the rate of photosynthesis.

Setting up and using a photosynthometer

Figure 1 shows a **photosynthometer**, also known as an Audus microburette. It is set up so that it is air tight and there are no air bubbles in the capillary tubing. Gas given off by the plant, over a known period of time, collects in the flared end of the capillary tube. As the experimenter manipulates the syringe, the gas bubble can be moved into the part of the capillary tube against the scale and its length measured. If the radius of the capillary tube bore is known, then this length can be converted to volume.

$$\text{Volume of gas collected} = \text{length of bubble} \times \pi r^2$$

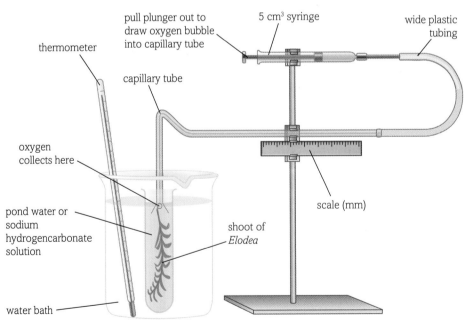

Figure 1 A photosynthometer – the apparatus to measure the rate of photosynthesis under various conditions.

If the same apparatus is used throughout the investigation, the radius of the tube bore is constant so comparison can be made using just the bubble lengths.

INVESTIGATION 1

Investigating the effect of light intensity on the rate of photosynthesis

Equipment
- photosynthometer, as shown in Figure 1
- piece of aquatic plant
- boiling tube
- beaker
- thermometer
- strong (at least 1000 lumens), cool and full spectrum light source
- metre ruler
- 0.2 M sodium hydrogencarbonate solution
- timer

This experiment should ideally be carried out in a darkened room, so that the only light received by the plant is from the light source. Avoid skin and eye contact with the sodium hydrogencarbonate solution.

Before setting up the apparatus, state your prediction and use your biological knowledge to justify it. You should also state the independent variable (IV) and the dependent variable (DV) and indicate the control variables.

1. Remove the plunger from the syringe and allow a gentle stream of tap water into the barrel of the syringe, until the whole barrel and plastic tube are full of water.
2. Replace the syringe plunger and gently push water out of the flared end of the tube until there are no air bubbles in the water of the capillary tube.
3. Cut a 7 cm length of well-illuminated pondweed and make sure that bubbles of gas are emerging from the cut stem. Place this piece of pondweed, cut end upwards, into a boiling tube containing some of the water in which it has been kept and add two drops of the hydrogencarbonate solution to it.
4. Stand the boiling tube in a beaker of water at around 25 °C (use the thermometer to check). Have some cold water ready to add, if the temperature begins to rise during the investigation.
5. Place the light source as close to the beaker as possible. Measure and record the distance, d, between the light source and the plant. Light intensity follows the inverse square law and is given by the formula:

$$I = \frac{1}{d^2}$$

   Alternatively, you may use a light meter to measure the light intensity.
6. Leave the apparatus with the capillary tube positioned, so that it is *not* collecting gas given off by the plant, for 5–10 minutes to allow the plant to acclimatise to that light intensity.
7. Position the capillary tube over the cut end of the plant stem and after a known period of time (e.g. 5 minutes) gently pull the syringe plunger and bring the oxygen bubble into the tube against the scale. Read the length of the bubble.
8. Repeat steps 6 and 7 twice more.
9. Move the light source further from the plant. Measure and record the distance, or measure the light intensity using a light meter. Repeat steps 6 and 7.
10. Continue the investigation at different light intensities. Tabulate your data and plot a graph of rate of photosynthesis against light intensity.
11. Interpret your findings and explain the shape of your graph. Discuss the limitations of this method and indicate how the method could be improved to overcome these limitations.

Using the photosynthometer to investigate other factors that affect the rate of photosynthesis

Investigation 1 can be adapted to investigate other factors such as wavelength of light (using coloured filters), temperature and carbon dioxide concentration.

Before you try the investigations you need to:

- make and justify a prediction

- state the IV and DV

- state the variables you will need to control, why you need to control them and how you will control them

- write a plan and ask your teacher to check it.

INVESTIGATION 2

An alternative method to investigate factors that affect the rate of photosynthesis
Equipment
- salad cress seedlings
- 10 cm³ syringe
- 25 cm³ beakers
- lamp with low energy bulb
- 0.2 M sodium hydrogencarbonate solution
- drinking straw
- fine forceps
- timer

1. Use a drinking straw to cut several leaf discs from cress **cotyledons**.
2. Place six or eight discs into a 10 cm³ syringe and half fill the syringe with dilute sodium hydrogencarbonate solution.

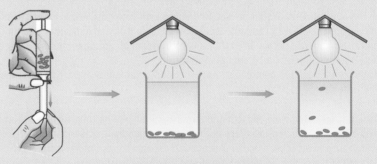

Figure 2 Using leaf discs to measure the rate of photosynthesis.

Figure 3 Cress cotyledons.

3. Hold the syringe upright, place your finger over the end of the syringe and gently pull on the plunger. This causes air to leave the spongy mesophyll layer of the leaf discs. The air is replaced by sodium hydrogencarbonate solution that enters the discs, making them more dense. They sink to the bottom of the syringe.
4. Once all the discs have sunk, transfer the contents of the syringe to a small beaker. Alternatively, keep the discs in the syringe, place the upside-down syringe in a boiling tube and position the tube underneath a light source. Use a light meter to monitor the light intensity.
5. Place the beaker of discs directly under a lamp and measure the time taken for one leaf disc to float to the top. As the discs carry out photosynthesis, oxygen produced collects in the spongy mesophyll and displaces the sodium hydrogencarbonate solution, making the discs less dense and more buoyant.
   The rate of photosynthesis can be expressed as the reciprocal of the time taken [$1/t$ (s^{-1})].
6. Repeat steps 3–5 twice more and find the mean rate of photosynthesis.
7. Plan how you could adapt this method to investigate the effect of carbon dioxide concentration, temperature or wavelength of light on the rate of photosynthesis.
8. If you have suitable sensors and data logging devices, you could adapt either of the methods to measure the changes of pH as an aquatic plant carries out photosynthesis. If the rate of photosynthesis is greater than the plant's rate of respiration, then carbon dioxide will be removed from the water, raising the pH.

In 1883 a German botanist, T.W. Engelmann, realised that it would be difficult to detect the site of photosynthesis in plant cells packed with chloroplasts. Instead he mounted a filament of the alga *Spirogyra* on a microscope slide, along with *Pseudomonas* bacteria that had been deprived of oxygen. Engelmann had previously discovered that *Pseudomonas* bacteria move in the presence of oxygen and cluster together where the concentration of oxygen is highest.

Engelmann kept the slide in the dark to immobilise the bacteria and then placed it under a microscope and illuminated it. He knew that the evolution of oxygen would indicate that the filamentous alga was carrying out photosynthesis. The motile bacteria moved to the immediate vicinity of the spiral chloroplast. Engelmann therefore deduced that photosynthesis occurs in chloroplasts.

This investigation was ingenious and used one organism to investigate the biochemistry of another organism.

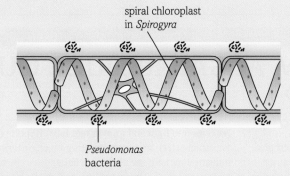

spiral chloroplast
in *Spirogyra*

Pseudomonas
bacteria

Figure 4 Motile bacteria move closer to chloroplasts.

The units for light intensity are lux – spelled with a lower case 'l'.

Questions

1 Explain why the leaf discs float after they have been illuminated for a few minutes.

2 Explain why, in each investigation, replicates should be carried out.

3 Predict, with reasons, what would happen in investigation 2 if the leaf discs were illuminated by red light, then blue light and then green light.

4 State, with reasons, at which stage of photosynthesis each of the following factors would have most effect on the rate of photosynthesis.
   (a) light wavelength
   (b) temperature
   (c) carbon dioxide concentration.

5 Outline the route by which air moves out of the cress cotyledon leaf disc in investigation 2 in this topic.

THE FUTURE OF FUEL

Algae can be turned into fuel via anaerobic digestion by bacteria to produce methane, or via transesterification where algal lipids are reacted with alcohols to produce biodiesel and glycerol. Whole algae can be heated under pressure to change their lipids to a range of fuel products. This information comes from an article in The Society of Biology magazine, *The Biologist*.

COULD BIOFUEL FROM ALGAE REDUCE OUR DEPENDENCE ON OIL?

Could algae, that green stuff that clogs ponds and rots on beaches making a pong, save the world? This diverse group of simple autotrophic organisms are being used to make advanced biofuels. These have a rich energy content and use minimal land space. They may help reduce carbon emissions and break our dependence on oil.

Algae are autotrophs and they harness the sun's light energy to turn simple inorganic molecules to energy rich hydrocarbons, such as lipids.

Biofuels are at present made from the oils of corn and soya beans but these crops take up land space; they are often the cause of deforestation leading to an increase of carbon emissions and can drive up the price of food.

Some species of microalgae can convert up to 60% of their biomass to oils, compared to 2–3% in soya beans. They grow faster than crop plants because they do not produce stems and leaves. At the University of Bath Professor Rod Scott, a plant molecular biologist, is carrying out research to develop strains of microalgae to be used solely for biofuel production. He says 'To provide 50% of the USA's fuel with corn oil, 846% of the available crop area would be needed – clearly impossible. That falls to just 2.5% with microalgae.'

Pyrolysis involves heating algae to 500–700 °C without oxygen; it produces biochar (charcoal) and other fuels. Hydrothermal liquefaction puts algae under high pressure at 250–350 °C with water, to make biocrude and hydrogen gas. This process mimics the natural production of crude oil but takes a day. Naturally produced crude oil forms when algaenan, a tough hydrocarbon polymer within algal cell walls, turns to oil when buried beneath the seabed under extreme heat and pressure, but this takes 30 million years! The challenge now is to replicate this process in real time and on a vast scale to meet the global demand for oil. At the moment humans use over 90 million barrels (14 billion litres) of oil every day. At the moment the cost of making algal biofuel is more than the cost of piping oil from the ground.

Algal cells produce more hydrocarbon if they are starved of nitrogen. Cells to produce the hydrocarbons must be separated from the growing cells. If cells are too densely packed in the medium then each cell receives less light and growth rate falls. Reports on biofuels say biology, via strain selection, plus translating bioengineering to the necessary industrial agricultural scale are challenges to be met for algal biofuel to be successful. If genetic modification is also used, risk assessments will have to be done to meet the regulatory requirements. Other factors have to be considered, for example seaweed beds on the ocean floor help prevent coastal erosion and beached seaweed cannot just be removed as this will interfere with coastal ecosystems. China is well ahead and grows 10 million tonnes a year of seaweed in shallow waters.

As with all challenges to meet the needs of a growing population, there will not be one single solution but algae are likely to be part of the mix and will provide a range of products gradually replacing oil. Meanwhile algal biotechnologists will continue to burn the midnight oil in the hope of making a valuable biofuel for the future.

Figure 1 Researcher at the Laboratory of Bioenergetics and Biotechnology in the Cadarache Research Centre, France, culturing microalgae for producing biofuel (and other products).

Source
● *Adapted from* Tom Ireland: Algal biofuel – in bloom or dead in the water? *The Biologist* **Vol 61 No 1 Feb/Mar 2014 p.20–23**

Where else will I encounter these themes?

DID YOU KNOW?

14 billion litres/90 million barrels of oil are used per day globally.
25% of global carbon emissions are caused by transport.
98% of fuel for transport is made from oil.
60% of algae such as eustigmatophytes can be turned into lipid (hydrocarbon).
0.2% of road fuel is predicted to come from algae by 2020.
50 m is the length to which brown seaweed such as giant kelp, *Macrocyctis pyrifera,* can grow.

DID YOU KNOW?

To provide biofuel for a lane of cars requires a strip of land the length of that lane and 8 km wide.

Let's start by considering the nature of the writing in the article.

1. What features of this article indicate it is aimed at readers with some knowledge of biology?
2. Were terms explained? Was the article clearly written? Give examples to explain your answer.

Here are some questions relating to the content of the article.

3. Calculate how many litres of oil are in one barrel of oil.
4. To which kingdom do algae belong?
5. Explain why the present use of land to grow crop plants such as soya beans and corn to make biofuel:
    a. drives up the price of food
    b. increases carbon emissions more than the burning of fossil fuels.
6. What do you think are the implications for the global population if food prices increase?
7. Suggest explanations for each of the following statements.
    a. Algae produce more hydrocarbons if they are starved of a source of nitrogen.
    b. Growing algal cells and hydrocarbon-producing algal cells must be separated when being cultured.
    c. The growth rate of algae falls if cells are too densely packed.
8. Give a biological explanation for why algae have a greater percentage of their biomass as hydrocarbons (lipids) than do crop plants such as soya and corn, presently used for biofuel.

> Think about the consequences of clearing forest to grow soy and corn.

> Think about the uneven distribution of wealth on Earth.

Activity

Use information:
- from the Did you know? boxes
- from the article here
- from the Internet and other journals
to create an infographic that represents the data in an interesting pictorial format.

Practice questions

1. What is at the primary reaction centre of a photosystem? [1]
   A. Chlorophyll a
   B. Chlorophyll b
   C. Xanthophyll
   D. Carotene

2. Read the following statements. [1]
   (i) There are two forms of chlorophyll a and both absorb red light.
   (ii) Plants are photoautotrophs and humans are heterotrophs.
   (iii) Photolysis is the generation of ATP in the presence of light.
   Which statement(s) is/are true?
   A. (i) only
   B. (i) and (ii) only
   C. (i) and (iii) only
   D. (i), (ii) and (iii)

3. Which is *not* a role of water in plants? [1]
   A. Source of protons
   B. Electron acceptor
   C. Source of oxygen
   D. Keep cells turgid

4. What is the product of the Calvin cycle? [1]
   A. ATP
   B. GP
   C. TP
   D. NAD

5. Read the following statements. [1]
   (i) RuBisCO is a carbon dioxide acceptor.
   (ii) If light intensity falls very low, the light-independent stage cannot continue.
   (iii) Chloroplasts contain small amounts of RuBP, because it is regenerated from GP.
   Which statement(s) is/are true?
   A. (i) only
   B. (ii) only
   C. (i) and (iii) only
   D. (ii) and (iii) only

   [Total: 5]

6. Hydrogencarbonate indicator solution is sensitive to small changes in pH. It is red when neutral, yellow at pH 6 and purple/red at pH slightly above 7.

   Figure 1 shows the results of an investigation where pieces of *Elodea*, an aquatic plant, were placed in hydrogencarbonate indicator solution. One piece was illuminated for 30 minutes.

   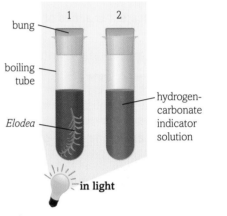

   bung
   boiling tube
   Elodea
   hydrogen-carbonate indicator solution
   in light   **in dark**

   Figure 1

   (a) Consider the colour changes in this investigation:
   (i) Explain the colour change of the hydrogencarbonate indicator solution in tube 1. [3]
   (ii) Explain the colour change of the hydrogencarbonate indicator solution in tube 3. [3]
   (iii) What is the purpose of tubes 2 and 4? [1]
   (iv) Predict, with reasons, the colour change that would have occurred in tube 1 if two small water snails were also present. [2]

   (b) A photosynthesising plant was exposed to radioactively labelled carbon dioxide, $^{14}CO_2$. In which order would the labelled (radioactive) carbon appear in the following compounds: starch, GP, glucose, TP?

   Explain your answer. [4]

   (c) In some glasshouses where tomato plants are grown, the atmosphere is enriched with carbon dioxide. Explain how this might increase the yield from the tomato plants. [7]

   [Total: 20]

7. Figure 2 represents some of the steps in the Calvin cycle.

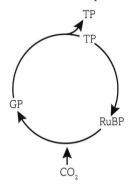

Figure 2

(a) Indicate where in the cycle the enzyme RuBisCO acts. [1]

(b) How many carbon atoms are in one molecule of each of the following compounds?
   (i) RuBP [1]
   (ii) GP [1]
   (iii) TP [1]

(c) Indicate where in the cycle reduced NADP is used. [1]

(d) How many molecules of carbon dioxide must enter the Calvin cycle for one molecule of hexose sugar to be produced? [1]

(e) State the name of the part of chloroplasts where the Calvin cycle occurs. [1]

(f) Suggest why chloroplasts contain only low quantities of ribulose bisphosphate. [1]

(g) Most sugar in plants is transported from leaves to other areas, such as roots, in the form of sucrose.
   (i) Which two hexose sugars does the plant have to make before it can make molecules of sucrose? [1]
   (ii) What type of reaction takes place to join these two hexose sugars? [1]

(h) Figure 3 shows the relative amounts of RuBP and TP production in the Calvin cycle before and after the light source was switched off.

Explain the change in the relative amounts of RUBP and TP after the light source was switched off. [3]

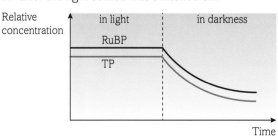

Figure 3

(i) Explain the roles of ATP and RuBP in the light-independent stage of photosynthesis. [3]

[Total: 16]

8. An investigation into photosynthesis and respiration in a leaf was carried out. The net uptake of carbon dioxide by the leaf in bright light and the mass of carbon dioxide released in the dark were measured at different temperatures.

The results are shown in Table 1.

Temperature (°C)	10	20	30	40	50
Net uptake of carbon dioxide in bright light ($mg\,g^{-1}\,h^{-1}$)	2.5	3.6	2.4	0.0	0.0
Release of carbon dioxide in the dark ($mg\,g^{-1}\,h^{-1}$)	0.8	1.6	3.2	5.8	0.0
True rate of photosynthesis ($mg\,g^{-1}\,h^{-1}$)					

Table 1

(a) Assuming that the rate of release of carbon dioxide is the same in daylight as in the dark, calculate the true rate of photosynthesis at each temperature. [2]

(b) Calculate the temperature coefficient, Q_{10}, for respiration between 20 °C and 30 °C. Express your answer to 1 d.p. [2]

(c) Calculate the temperature coefficient, Q_{10}, for photosynthesis over the same temperature range. Express your answer to the nearest whole number. [2]

(d) Comment on the Q_{10} values for respiration and photosynthesis at this temperature range. [4]

Figure 4 shows the result of an investigation where light of different wavelengths was shone onto a length of a filamentous photosynthetic alga whilst under a microscope. Oxygen-sensitive motile bacteria that cluster in areas of high oxygen concentration were added to the slide.

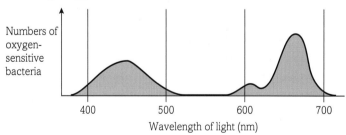

Figure 4

(e) Explain the results (the distribution of the bacteria around the alga) of this investigation. [6]

[Total: 16]

RESPIRATION

Introduction

All living organisms need energy to do work, such as movement, active transport, bulk transport, synthesis of molecules, conduction of nerve impulses and replication of DNA before cells divide. For most, respiration is the process that releases chemical energy from food, enabling them to carry out such work.

Life first evolved on Earth around 3500 million years ago. There was no free oxygen in the atmosphere. The first life forms, prokaryotic organisms, used various metabolic pathways to obtain energy from chemicals in their environment. About 2000 million years ago, some prokaryotes evolved photosynthesis, some using hydrogen sulfide and some using water as the source of hydrogen. The involvement of water released free oxygen into the atmosphere. However, with only some types of prokaryote producing oxygen, and with no photosynthetic algae or plants yet evolved, the accumulation of oxygen in the atmosphere, to between 10 and 20%, took a long time.

About 541 million years ago, at the start of the Cambrian period, the presence of oxygen in the atmosphere contributed to the great increase in the diversity of life forms, most of which relied on oxygen for respiration to release energy from respiratory substrates. Some types of organism could utilise chemical reactions to obtain energy; most organisms could also respire anaerobically, but not many types could rely solely on anaerobic respiration for survival.

Since the start of the Cambrian period, the concentration of oxygen in the Earth's atmosphere has fluctuated between 15 and 35%, and is now around 20%. At its peak about 300 million years ago, the concentration of oxygen may have contributed to the large size of arthropods and amphibians alive at that time.

Since Joseph Priestley and Antoine Lavoisier discovered oxygen in the atmosphere and its relevance to living organisms, many other scientists have contributed to our knowledge of the metabolic reactions involved, some of which are outlined in this chapter.

All the maths you need

To unlock the puzzles of this chapter you need the following maths:

- Know units of measurement
- Know powers of 10
- Use division and multiplication, for example when making serial dilutions
- Calculate ratios, percentages and fractions
- Use an appropriate number of significant figures when handling data from investigations
- Calculate percentage error
- Find arithmetic means
- Calculate the rate of change from a graph
- Select and use a statistical test
- Calculate the volume of a cylinder

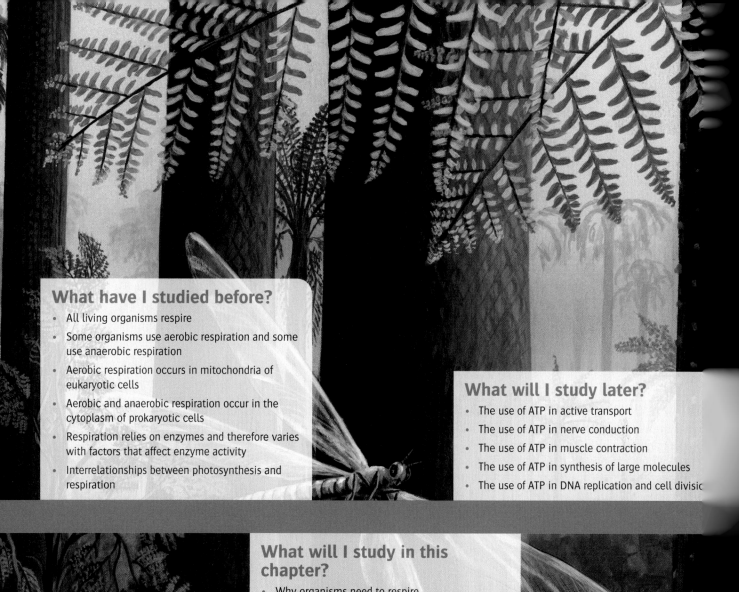

What have I studied before?

- All living organisms respire
- Some organisms use aerobic respiration and some use anaerobic respiration
- Aerobic respiration occurs in mitochondria of eukaryotic cells
- Aerobic and anaerobic respiration occur in the cytoplasm of prokaryotic cells
- Respiration relies on enzymes and therefore varies with factors that affect enzyme activity
- Interrelationships between photosynthesis and respiration

What will I study later?

- The use of ATP in active transport
- The use of ATP in nerve conduction
- The use of ATP in muscle contraction
- The use of ATP in synthesis of large molecules
- The use of ATP in DNA replication and cell division

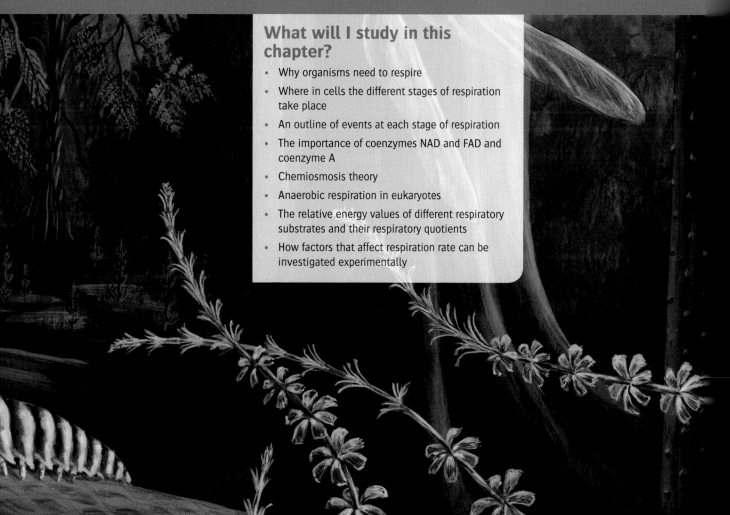

What will I study in this chapter?

- Why organisms need to respire
- Where in cells the different stages of respiration take place
- An outline of events at each stage of respiration
- The importance of coenzymes NAD and FAD and coenzyme A
- Chemiosmosis theory
- Anaerobic respiration in eukaryotes
- The relative energy values of different respiratory substrates and their respiratory quotients
- How factors that affect respiration rate can be investigated experimentally

By the end of this topic, you should be able to demonstrate and apply your knowledge and understanding of:

* the need for cellular respiration

Why do living organisms need to respire?

Respiration is the process that occurs in living cells and releases the energy stored in organic molecules such as glucose. The energy is immediately used to synthesise molecules of **ATP** from **ADP** and **inorganic phosphate** (**P_i**). ATP in cells can be hydrolysed to release energy needed to drive biological processes. Microorganisms (both eukaryotic microbes such as yeast, and prokaryotes such as bacteria), plants, animals, fungi and protoctists all respire to obtain energy.

Why do living organisms need energy?

Energy is the capacity to do work. The energy that is stored in complex organic molecules – e.g. fats, carbohydrates and proteins – is potential energy. It is also chemical energy, converted from light energy during the process of photosynthesis (topic 5.6.1). When this energy is released from organic molecules, via respiration, it can be used to make ATP to drive biological processes such as:

* active transport (Book 1, topic 2.5.4)

* endocytosis (Book 1, topic 2.5.4)

* exocytosis, including secretion of large molecules from cells (Book 1, topics 2.1.6 and 2.5.4)

* synthesis of large molecules such as proteins, e.g. collagen, enzymes and antibodies (Book 1, topic 2.3.3)

* DNA replication (Book 1, topic 2.3.2)

* cell division (Book 1, topics 2.6.2 and 2.6.3)

* movement – such as the movement of bacterial flagella, of eukaryotic cilia and undulipodia, and motor proteins that walk along cytoskeleton threads in cells, moving organelles (Book 1, topic 2.1.5)

* activation of chemicals – glucose is phosphorylated at the beginning of respiration so that it becomes more reactive and able to be broken down to release more energy (topic 5.7.2).

All the chemical reactions that take place within living cells are known collectively as metabolism or metabolic reactions.

* Anabolic reactions are metabolic reactions where large molecules are synthesised from smaller molecules.

* Catabolic reactions are metabolic reactions involving the hydrolysis of large molecules to smaller ones.

Within living cells atoms, ions and molecules have kinetic energy, and this allows them to move; for example, when molecules diffuse down a concentration gradient, moving from one place to another, they use their kinetic energy to do so.

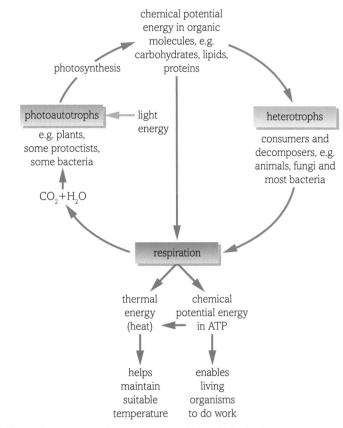

Figure 1 Energy transfer between and within living organisms.

> **LEARNING TIP**
>
> Respiration *releases* energy from respiratory substrates such as glucose, but it does *not* create or make energy. However, some of the released energy can be used to make ATP.

The role of ATP

ATP is the standard intermediary between energy-releasing and energy-consuming metabolic reactions in both eukaryotic and prokaryotic cells.

Figure 2 shows the structure of an ATP molecule; it is a phosphorylated nucleotide. Each molecule of ATP consists of adenosine, which is the nitrogenous base adenine plus the five-carbon sugar ribose, and three phosphate (phosphoryl) groups.

Respiration 5.7

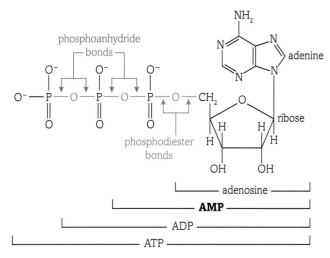

Figure 2 The structure of ATP.

ATP is relatively stable (it does not break down to ADP and P$_i$) when in solution in cells, but is readily hydrolysed by enzyme catalysis. However, whilst in solution, it can easily be moved from place to place within a cell.

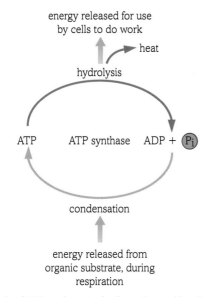

Figure 3 The role of ATP synthase in the formation and breakdown of ATP.

The energy-releasing hydrolysis of ATP is coupled with an energy-consuming metabolic reaction. ATP is the immediate energy source for this metabolic reaction. When ATP is hydrolysed to ADP and P$_i$, a small quantity of energy is released for use in the cells. Cells can therefore obtain the energy they need for a process in small manageable amounts that will not cause damage or be wasteful. ATP is referred to as the universal energy currency, as it occurs in all living cells and is a source of energy that can be used by cells in small amounts.

Some energy is released from the hydrolysis of ATP as heat. The release of heat, both in respiration and during ATP hydrolysis, may appear to be inefficient and wasteful. Heat, however, helps keep living organisms 'warm' and enables their enzyme-catalysed reactions to proceed at or near their optimum rate (Book 1, topic 2.4.4).

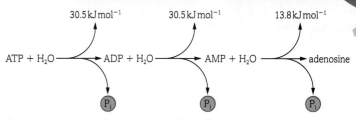

Figure 4 The chemical energy released from the hydrolysis of ATP.

DID YOU KNOW?

In your body you have, at any one time, about 5 g of ATP. However, you may use between 36–50 kg each day. This is possible because the ATP molecules are continually being hydrolysed and then resynthesised. At rest, a person consumes and continually regenerates ATP at the rate of about 1.5 kg per hour. When active this increases.

Questions

1 ATP is a nucleotide derivative. Do you think it is derived from a DNA nucleotide or from an RNA nucleotide? Give reasons for your answer.

2 For each of the following, state whether it is an anabolic or catabolic reaction:
(a) synthesis of the protein microtubules that form the spindle for mitosis or meiosis
(b) digestion of a pathogen inside a phagosome of a macrophage
(c) formation of insulin in cells of the pancreas.

3 Explain why ATP is known as the universal energy currency.

4 Suggest why ATP, being stable while in solution in cells but readily breaking down to ADP and P$_i$ by the action of an enzyme, is useful to living organisms.

5 Briefly outline the energy conversions needed for:
(a) fireflies to produce light energy from the food they eat
(b) electric eels to produce electricity from the food they eat.

Figure 5 Male firefly beetle. These animals use an enzyme, luciferase, to oxidise a chemical in a chamber in their abdomen and produce bioluminescence – flashes of light to attract a mate. They can control the flashes by regulating the amount of oxygen entering the chamber.

Figure 6 An electric eel, *Electrophorus electricus*, a fish native to the Amazon River, Brazil. This fish can produce powerful electric shocks.

137

By the end of this topic, you should be able to demonstrate and apply your knowledge and understanding of:

* the process and site of glycolysis

glycolysis: first stage of respiration; a 10-stage metabolic pathway that converts glucose to pyruvate.

Glycolysis

Glycolysis is a biochemical pathway that occurs in the cytoplasm of all living organisms that respire, including many prokaryotes.

The pathway involves a sequence of 10 reactions, each catalysed by a different enzyme, some with the help of the **coenzyme**, NAD.

You only need to know the pathway in outline, so we will consider the three main stages:

1. Phosphorylation of glucose to **hexose bisphosphate**.

2. Splitting each hexose bisphosphate molecule into two **triose phosphate** molecules.

3. **Oxidation** of triose phosphate to **pyruvate**.

Figure 1 summarises the process of glycolysis.

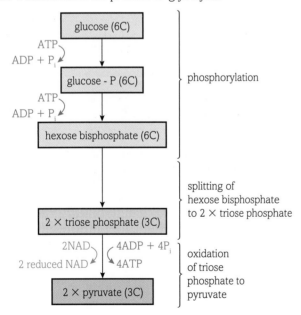

Figure 1 A summary of glycolysis.

NAD

Enzymes that catalyse oxidation and **reduction** reactions need the help of coenzymes (a type of cofactor – Book 1, topic 2.4.2) that accept the hydrogen atoms removed during oxidation.

NAD (nicotinamide adenine dinucleotide) is a non-protein molecule that helps dehydrogenase enzymes to carry out oxidation reactions. NAD oxidises substrate molecules during glycolysis, the link reaction and the Krebs cycle.

NAD is synthesised in living cells from nicotinamide (vitamin B_3), the five-carbon sugar ribose, the nucleotide base adenine and two phosphoryl groups. Figure 2 represents its molecular structure. The nicotinamide ring can accept two hydrogen atoms, becoming reduced NAD.

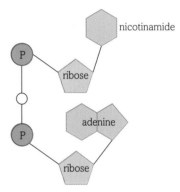

Figure 2 The molecular structure of the coenzyme NAD.

Reduced NAD carries the protons and electrons to the cristae of mitochondria (topic 5.7.3) and delivers them to be used in oxidative phosphorylation (topic 5.7.5) for the generation of ATP from ADP and P_i.

When reduced NAD gives up the protons and electrons that it accepted during one of the first three stages of respiration, it becomes oxidised and can be reused to oxidise more substrate, in the process becoming reduced again.

You need to remember that the coenzyme NAD is involved in respiration, and the coenzyme NADP is involved in photosynthesis. To help you, remember that photosynthesis begins with *P* and uses NAD*P*.

The three main stages of glycolysis

Phosphorylation

Glucose is a hexose sugar, which means it contains six carbon atoms. Its molecules are stable and need to be activated before they can be split into two three-carbon compounds.

1. One molecule of ATP is hydrolysed and the released phosphoryl group is added to glucose to make hexose monophosphate.

2. Another molecule of ATP is hydrolysed and the phosphoryl group added to the hexose phosphate to form a molecule of hexose bisphosphate. This sugar has one phosphate group at carbon atom number one and another at carbon atom number six.

The **energy** from the hydrolysed ATP molecules activates the hexose sugar and prevents it from being transported out of the cell.

Splitting the hexose bisphosphate

Each molecule of hexose bisphosphate is split into two three-carbon molecules, triose phosphate, each with a phosphate group attached.

Oxidation of triose phosphate to pyruvate

Although this process is anaerobic it involves oxidation, because it involves the removal of hydrogen atoms from substrate molecules:

1. Dehydrogenase enzymes, aided by the coenzyme NAD, remove hydrogens from triose phosphate.
2. The two molecules of NAD accept the hydrogen atoms (protons and electrons) and become reduced.
3. At this stage of glycolysis, two molecules of NAD are reduced for every molecule of glucose undergoing this process. Also at this stage, four molecules of ATP are made for every two triose phosphate molecules undergoing oxidation.

The products of glycolysis

From each molecule of glucose, at the end of glycolysis there are:

- two molecules of ATP; four have been made, but two were used to 'kick start' the process, so the net gain is two molecules of ATP
- two molecules of reduced NAD
- two molecules of pyruvate.

The stages of respiration

Respiration of glucose has four stages (Figure 3):

1. glycolysis
2. the link reaction
3. the Krebs cycle
4. oxidative phosphorylation.

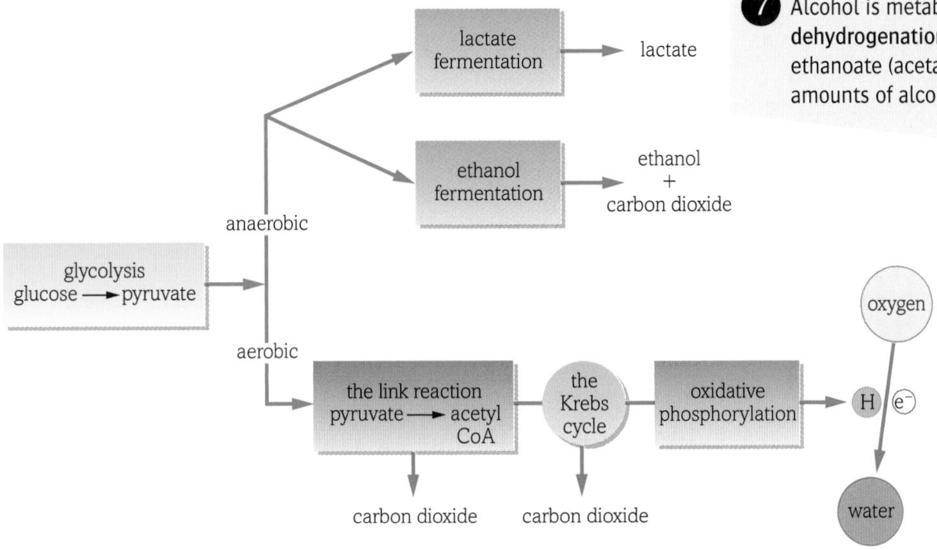

The last three stages only take place under aerobic conditions. Under **aerobic** conditions, the pyruvate molecules from glycolysis are actively transported into the mitochondria for the link reaction (topic 5.7.4).

In the absence of oxygen (i.e. **anaerobic** conditions) pyruvate is converted, in the cytoplasm, to lactate or ethanol (topic 5.7.6). In the process, the reduced NAD molecules are reoxidised so that glycolysis can continue to run, generating two molecules of ATP for every glucose molecule metabolised.

> **DID YOU KNOW?**
>
> The dietary deficiency disease pellagra (the symptoms are diarrhoea, dermatitis and dementia) is caused by lack of nicotinamide (vitamin B_3) in the diet. Humans can synthesise nicotinamide from the amino acid tryptophan. Corn does not contain tryptophan, but does contain nicotinamide in a form that has to be treated with an alkaline substance before it can be absorbed from the intestine. Mexican Indians are thought to have domesticated the corn plant, and their diet has always contained a lot of corn. They soak the corn in limewater (calcium hydroxide solution) before using it to make tortillas, and they do not suffer from pellagra. However, in the southern United States of America in the early part of the twentieth century, pellagra was endemic as the diet consisted mainly of corn products but no limewater was used in their preparation.

Questions

1. Suggest why living organisms have low levels of NAD in their cells, although they use many molecules of it throughout every day.

2. Explain why NAD is described as a nucleotide derivative.

3. Explain how glycolysis involves oxidation, although it is an anaerobic process.

4. Read the information in the Did You Know? box. Explain why the Mexican Indians did not suffer pellagra, while people in the southern parts of the USA did, despite having a similar diet.

5. Describe the role of NAD during glycolysis.

6. Explain why the net gain of ATP is two molecules per molecule of glucose undergoing glycolysis, although four molecules of ATP are made.

7. Alcohol is metabolised in the liver. It is oxidised, by **dehydrogenation**, to ethanal. Ethanal is then oxidised to ethanoate (acetate). Suggest why people who regularly drink large amounts of alcohol may be deficient in NAD.

Figure 3 The stages of respiration.

(3) The structure of the mitochondrion

By the end of this topic, you should be able to demonstrate and apply your knowledge and understanding of:

* the structure of the mitochondrion

> **KEY DEFINITIONS**
> **cristae:** inner highly-folded mitochondrial membrane.
> **mitochondrial matrix:** fluid-filled inner part of mitochondria.

Mitochondrial structure

Mitochondria are organelles that are present in all types of eukaryotic cells. Although mitochondria were first identified in animal cells in 1840 and in plant cells in 1900, their ultrastructure was not worked out until the 1950s, after extensive studies using electron microscopes.

Figures 1 and 2 show the structure of a mitochondrion.

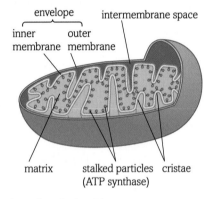

Figure 1 The structure of a mitochondrion.

Mitochondria may be rod-shaped, thread-like or spherical, with diameters of 0.5–1.0 μm and lengths of 2–5 μm, but occasionally up to 10 μm.

All mitochondria have an inner and an outer phospholipid membrane making up the **envelope**. The outer membrane is smooth, and the inner membrane is folded into **cristae** (singular: crista), giving it a large surface area.

Embedded in the inner membrane are proteins that transport electrons, and protein channels associated with ATP synthase enzyme that allow protons to diffuse through them.

Between the inner and outer mitochondrial membranes of the envelope, is an **intermembrane space**.

The **mitochondrial matrix**, enclosed by the inner membrane, is semi-rigid and gel-like; it contains mitochondrial ribosomes, looped mitochondrial DNA and enzymes for the link reaction and Krebs cycle (topic 5.7.4).

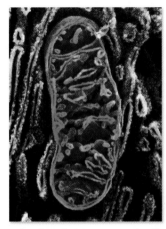

Figure 2 TEM of a mitochondrion from an intestinal cell (×54 000).

How the structure of mitochondria enables them to carry out their functions

The matrix

The matrix is where the link reaction and the Krebs cycle take place.

It contains:

* enzymes that catalyse the stages of these reactions
* molecules of the coenzymes NAD (nicotinamide adenine dinucleotide) and FAD (flavine adenine dinucleotide)
* oxaloacetate – the four-carbon compound that accepts the acetyl group from the link reaction
* mitochondrial DNA – some of which codes for mitochondrial enzymes and other proteins
* mitochondrial ribosomes, structurally similar to prokaryotic ribosomes (Book 1, topic 2.1.7), where these proteins are assembled.

The outer membrane

The phospholipid composition of the outer membrane is similar to that of membranes around other organelles in eukaryotic cells. It contains proteins, some of which form channels or carriers that allow the passage of molecules, such as pyruvate, into the mitochondrion.

The inner membrane

The lipid composition of the inner membrane differs from that of the outer membrane. This lipid bilayer is less permeable to small ions such as hydrogen ions (protons) than is the outer membrane.

The folds, cristae, in the inner membrane give a large surface area for the electron carriers and ATP synthase enzymes embedded in them.

The electron carriers are protein complexes arranged in **electron transport chains**. Electron transport chains are involved in the final stage of aerobic respiration, oxidative phosphorylation (topic 5.7.5).

The intermembrane space

The intermembrane space between the outer and inner layers of the mitochondrial envelope is also involved in oxidative phosphorylation.

The inner membrane is in close contact with the mitochondrial matrix, so the molecules of reduced NAD and FAD can easily deliver hydrogens to the electron transport chain.

The electron transport chain

Figure 3 summarises events that take place along the electron transport chain.

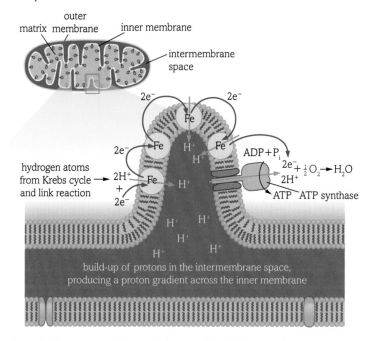

Figure 3 The structure of a part of the mitochondrial inner and outer membranes with the intermembrane space between them.

Each electron carrier protein contains a cofactor (Book 1, topic 2.4.2) – a non-protein haem group that contains an iron ion.

The iron ion can accept and donate electrons, because it can become reduced (Fe^{2+}) by gaining an electron and then become oxidised (Fe^{3+}) when donating the electron to the next electron carrier. Electron carrier proteins are oxido-reductase enzymes.

The electron carriers also have a coenzyme that, using energy released from the electrons, pumps protons from the matrix to the intermembrane space. Protons accumulate in the intermembrane space and a proton gradient forms across the membrane. The proton gradient can produce a flow of protons through the channels in the ATP synthase enzymes to make ATP. You will learn more about this in topic 5.7.5, and have already learnt about the synthesis of ATP by photophosphorylation in topic 5.6.3.

The ATP synthase enzymes

ATP synthase enzymes are large and protrude from the inner membrane into the matrix. Protons can pass through them.

Figure 4 represents the structure of ATP synthase.

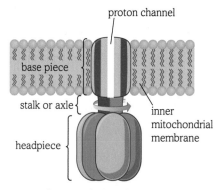

Figure 4 The structure of ATP synthase and its associated proton channel.

> **LEARNING TIP**
>
> The outer layer of the mitochondrion should always be described as an envelope, because it consists of two membranes.

> **DID YOU KNOW?**
>
> All the mitochondria in your cells come from your mother. This is because although sperm have mitochondria just above the tail, only the head or nucleus enters the egg cell. The egg cell contains mitochondria and these divide each time the zygote, and subsequent embryo cells, divide. Mitochondrial DNA (mtDNA) can be used for ancestry tracing, along the maternal line. In 1987, an article in the science journal *Nature* showed that all people on Earth today are descended from an African population of humans, supporting the theory that modern humans evolved in Africa around 200 000 years ago and then spread to other regions of the world.

Questions

1. Suggest why the synaptic knobs of neurones contain many mitochondria.

2. Suggest, with reasons, how the structure of a mitochondrion in a skin cell may differ from that of a mitochondrion in a muscle cell.

3. Explain the origin of the mitochondria in a new cell, produced by mitosis or meiosis.

4. Which stages of respiration take place in the mitochondrial matrix?

5. Which stage of respiration takes place at the cristae?

6. In what ways is the structure of a mitochondrion similar to that of a chloroplast?

7. In what ways is the structure of a mitochondrion different from that of a chloroplast?

By the end of this topic, you should be able to demonstrate and apply your knowledge and understanding of:

* the link reaction and its site in the cell

* the process and site of the Krebs cycle

* the importance of coenzymes in cellular respiration

KEY DEFINITIONS

decarboxylation: removal of a carboxyl group from a substrate molecule.
dehydrogenation: removal of hydrogen atoms from a substrate molecule.
substrate-level phosphorylation: production of ATP from ADP and P_i during glycolysis and the Krebs cycle.

Pyruvate

Pyruvate produced during glycolysis (topic 5.7.2) is transported across the outer and inner mitochondrial membranes via a specific pyruvate-H^+ **symport**, a transport protein that transports two ions or molecules in the same direction, and into the matrix. Then:

1. pyruvate is converted to a two-carbon acetyl group during the link reaction

2. the acetyl group is oxidised during the Krebs cycle.

The link reaction

The link reaction occurs in the mitochondrial matrix.

Pyruvate is **decarboxylated** and **dehydrogenated**, catalysed by a large multi-enzyme complex, pyruvate dehydrogenase, which catalyses the sequence of reactions that occur during the link reaction.

No ATP is produced during this reaction.

1. The carboxyl group is removed and is the origin of some of the carbon dioxide produced during respiration.

2. This decarboxylation of pyruvate together with dehydrogenation produces an acetyl group.

3. The acetyl group combines with coenzyme A (CoA) to become acetyl CoA.

4. The coenzyme NAD becomes reduced.

The equation summarises the link reaction for two molecules of pyruvate derived from one molecule of glucose:

$$2 \text{ pyruvate} + 2NAD + 2CoA \rightarrow 2CO_2 + 2 \text{ reduced NAD} + 2 \text{ acetyl CoA}$$

Coenzyme A accepts the acetyl group and, in the form of acetyl CoA, carries the acetyl group on to the Krebs cycle.

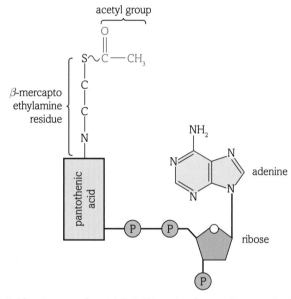

Figure 1 The structure of acetyl CoA. This molecule contains pantothenic acid, which is vitamin B_5. You do not need to learn this structure.

The Krebs cycle

Like the link reaction, the Krebs cycle takes place in the mitochondrial matrix.

The Krebs cycle is a series of enzyme-catalysed reactions that oxidise the acetate from the link reaction to two molecules of carbon dioxide, while conserving energy by reducing the coenzymes NAD and FAD (flavine adenine dinucleotide).

These reduced coenzymes then carry the hydrogen atoms to the electron transport chain on the cristae, where they will be involved in the production of many more ATP molecules (topic 5.7.5).

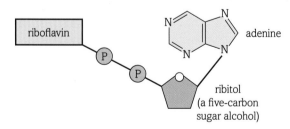

Figure 2 The structure of FAD. FAD contains a molecule of riboflavin (vitamin B_2). You do not need to learn this structure.

Figure 3 summarises some of the stages of the Krebs cycle.

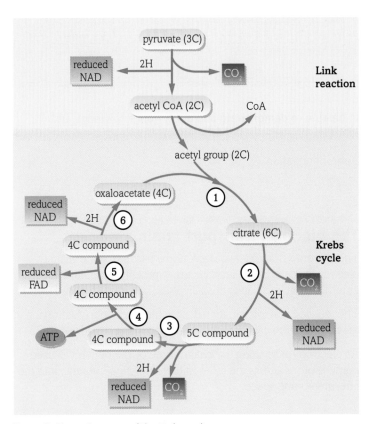

Figure 3 The main stages of the Krebs cycle.

1. The acetyl group released from acetyl CoA combines with a four-carbon compound, oxaloacetate, to form a six-carbon compound, citrate.

2. Citrate is decarboxylated and dehydrogenated, producing a five-carbon compound, one molecule of carbon dioxide and one molecule of reduced NAD.

3. This five-carbon compound is further decarboxylated and dehydrogenated, producing a four-carbon compound, one molecule of carbon dioxide and one molecule of reduced NAD.

4. This four-carbon compound combines temporarily with, and is then released from, coenzyme A. At this stage, **substrate-level phosphorylation** takes place, producing one molecule of ATP.

5. The four-carbon compound is dehydrogenated, producing a different four-carbon compound and a molecule of reduced FAD.

6. Rearrangement of the atoms in the four-carbon molecule, catalysed by an isomerase enzyme, followed by further dehydrogenation, regenerate a molecule of oxaloacetate, so the cycle can continue.

For every molecule of glucose there are two turns of the Krebs cycle.

The products of the link reaction and the Krebs cycle

Table 1 shows the products of the link reaction and the Krebs cycle for one molecule of glucose.

Product per molecule of glucose	The link reaction	The Krebs cycle
Reduced NAD	2	6
Reduced FAD	0	2
Carbon dioxide	2	4
ATP	0	2

Table 1 Products of the link reaction and the Krebs cycle for one molecule of glucose.

Although oxygen is not directly used in the link reaction and Krebs cycle, these stages will not occur in the absence of oxygen, so they are aerobic.

By the end of the Krebs cycle, the production of carbon dioxide from glucose is completed.

Other substrates besides glucose can be respired aerobically:

- Fatty acids are broken down to many molecules of acetate that enter the Krebs cycle via acetyl CoA.

- Glycerol may be converted to pyruvate and enter the Krebs cycle via the link reaction.

- Amino acids may be deaminated (the amino group [NH$_2$] is removed) and the rest of the molecule can enter the Krebs cycle directly or be changed to pyruvate or acetyl CoA.

LEARNING TIP

The link reaction is the only stage of respiration that does not produce any ATP.

DID YOU KNOW?

The coenzymes FAD and NAD are both derived from B vitamins. A dietary deficiency of the B vitamins can affect metabolism.

Questions

1. Suggest why living organisms have only small amounts of oxaloacetate in their cells.

2. Where, precisely, is the oxaloacetate in the cells of living organisms?

3. Explain why each stage of the Krebs cycle has to be catalysed by a different enzyme.

4. The link reaction is a metabolic pathway that involves five steps. Each step is catalysed by a different enzyme that is within a large multi-enzyme complex. Suggest how a multi-enzyme complex increases the efficiency of catalysing the steps in such a metabolic pathway.

5. Explain why fatty acids and amino acids can only be respired aerobically.

6. Describe how each of the following coenzymes is important in the link reaction and the Krebs cycle: NAD, FAD and CoA.

(5) Oxidative phosphorylation and the chemiosmotic theory

By the end of this topic, you should be able to demonstrate and apply your knowledge and understanding of:

* the process and site of oxidative phosphorylation
* the chemiosmotic theory

KEY DEFINITIONS

chemiosmosis: flow of protons, down their concentration gradient, across a membrane, through a channel associated with ATP synthase.
oxidative phosphorylation: the formation of ATP using energy released in the electron transport chain and in the presence of oxygen. It is the last stage in aerobic respiration.

The final stage of aerobic respiration

The final stage of aerobic respiration is **oxidative phosphorylation** – the production of ATP in the presence of oxygen. Oxidative phosphorylation takes place in mitochondria. It involves electron carrier proteins, arranged in chains called the electron transport chains, embedded in the inner mitochondrial membranes (the cristae), and a process called **chemiosmosis** (see Figure 1). The folded cristae give a large surface area for the electron carrier proteins and the ATP synthase enzymes (topic 5.7.3 gives details of mitochondrial structure).

1. Reduced NAD and reduced FAD are reoxidised when they deliver their hydrogen atoms to the electron transport chain.

2. The hydrogen atoms released from the reduced coenzymes split into protons and electrons.

3. The protons go into solution in the mitochondrial matrix.

The electron transport chain

The electrons from the hydrogen atoms pass along the chain of electron carriers. Each electron carrier protein has an iron ion at its core. The iron ions can gain an electron, becoming reduced (Fe^{2+}).

The reduced iron ion can then donate the electron to the iron ion in the next electron carrier in the chain, becoming reoxidised to Fe^{3+}.

As electrons pass along the chain, some of their energy is used to pump protons across the inner mitochondrial membrane, into the intermembrane space.

The proton gradient and chemiosmosis

As protons accumulate in the intermembrane space, a proton gradient forms across the membrane.

Proton gradients generate a chemiosmotic potential that is also known as a proton motive force, pmf. They are a source of potential energy.

ATP is made using the energy of the proton motive force.

Protons cannot easily diffuse through the lipid bilayer of the mitochondrial membranes, as the outer membrane has a low degree of permeability to protons and the inner membrane is

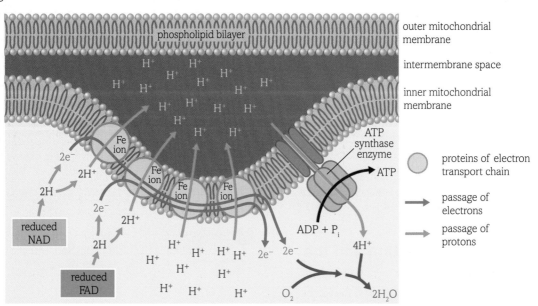

Figure 1 The electron transport chain and oxidative phosphorylation by chemiosmosis.

impermeable to protons. Protons can, however, diffuse through the protein channels associated with ATP synthase enzymes that are in the inner membrane. As protons diffuse down their concentration gradient through these channels, the flow of protons causes a conformational (shape) change in the ATP synthase enzyme that allows ADP and P_i to combine, forming ATP. This flow of protons is known as chemiosmosis. It is coupled to the formation of ATP.

The formation of ATP in this way, in the presence of oxygen, is oxidative phosphorylation.

Oxygen is the final electron acceptor. It combines with electrons coming off the electron transport chain and with protons, diffusing down the ATP synthase channel, forming water. The equation below summarises this reaction.

$$4H^+ + 4e^- + O_2 \rightarrow 2H_2O$$

DID YOU KNOW?

Peter Mitchell proposed the chemiosmotic theory in 1961. Mitchell realised that the accumulation of protons on one side of a membrane and the movement of those protons across the membrane, down the electrochemical gradient, could provide the energy to form ATP from ADP and P_i. Previously, scientists thought a high-energy intermediate compound provided the mechanism behind oxidative phosphorylation. Mitchell's theory was not readily accepted. However, by 1978, so much experimental evidence supported the idea of chemiosmosis that Mitchell's theory became widely accepted. Mitchell was awarded the Nobel Prize for Chemistry. Chemiosmosis is also responsible for the formation of ATP during photophosphorylation during the light-dependent stage of photosynthesis in chloroplasts.

DID YOU KNOW?

Some bacteria use ATP synthase but 'in reverse' to power the movement of their flagella. They use ATP to produce a proton gradient and the energy associated with that gradient causes a flow of protons that rotates the flagella.

LEARNING TIP

The transport of protons across the cristae into the intermembrane space is an active process but the energy comes from the electrons, *not* from ATP, so this transport is described as 'protons being *pumped* across the membrane'.

How much ATP is made during oxidative phosphorylation?

Name of molecule produced	Stage of respiration		
	Glycolysis	The link reaction	The Krebs cycle
Reduced NAD	2	2	6
Reduced FAD	0	0	2

Table 1 The number of reduced coenzymes formed during respiration.

The reduced coenzymes provide both protons and electrons to the electron transport chain.

The protons and electrons from the 10 molecules of reduced NAD can theoretically produce 25 molecules of ATP. The protons and electrons from the two molecules of reduced FAD can theoretically produce three molecules of ATP. Oxidative phosphorylation may therefore produce 28 molecules of ATP per molecule of glucose.

The total ATP tally per molecule of glucose during aerobic respiration

Stage of respiration	Net gain of ATP per molecule of glucose
Glycolysis	2
The link reaction	0
The Krebs cycle	2
Oxidative phosphorylation	28
Total	32

Table 2 ATP yield per molecule of glucose at each stage of respiration.

This theoretical yield is rarely achieved, and the actual yield may be closer to 30 molecules of ATP per molecule of glucose, or even less, because:

- some ATP is used to actively transport pyruvate into the mitochondria

- some ATP is used in a shuttle system that transports reduced NAD, made during glycolysis, into mitochondria

- some protons may leak out through the outer mitochondrial membrane.

Questions

1. Explain why the proton gradient across the inner mitochondrial membrane is a source of potential energy.

2. Explain why oxygen is described as the final electron acceptor in aerobic respiration.

3. Describe the pathway taken by a molecule of oxygen from a red blood cell in a capillary to the matrix of a mitochondrion in a respiring cell.

4. Explain why the pH measured within the intermembrane space of mitochondria is lower than that of both the mitochondrial matrix and the cytoplasm.

5. Suggest why the coenzymes NAD and FAD are also sometimes called 'hydrogen carriers'.

6. State precisely where oxidative phosphorylation takes place.

7. What is meant by chemiosmotic theory?

8. Explain why the formation of water at the end of oxidative phosphorylation helps maintain the proton gradient across the inner mitochondrial membrane.

6 Anaerobic respiration in eukaryotes

By the end of this topic, you should be able to demonstrate and apply your knowledge and understanding of:

* the process of anaerobic respiration in eukaryotes

Respiration in the absence of oxygen

If oxygen is absent:

1. Oxygen cannot act as the final electron acceptor at the end of oxidative phosphorylation. Protons diffusing through channels associated with ATP synthase are not able to combine with electrons and oxygen to form water.

2. The concentration of protons increases in the matrix and reduces the proton gradient across the inner mitochondrial membrane.

3. Oxidative phosphorylation ceases.

4. Reduced NAD and reduced FAD are not able to unload their hydrogen atoms and cannot be reoxidised.

5. The Krebs cycle stops, as does the link reaction.

For the organism to survive these adverse conditions, glycolysis can take place, but the reduced NAD generated during the oxidation of triose phosphate to pyruvate has to be reoxidised so that glycolysis can continue. These reduced coenzyme molecules cannot be reoxidised at the electron transport chain, so another metabolic pathway must operate to reoxidise them.

Reduced NAD has to be reoxidised

Eukaryotic cells have two metabolic pathways to reoxidise the reduced NAD:

* Fungi, such as yeast, and plants use the **ethanol** fermentation pathway.
* Mammals use the **lactate** fermentation pathway.

Both take place in the cytoplasm of cells.

The ethanol fermentation pathway

Figure 1 shows the fate of pyruvate under anaerobic conditions in yeast cells.

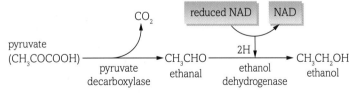

Figure 1 The ethanol fermentation pathway.

This is also known as alcoholic fermentation:

1. Each molecule of pyruvate produced during glycolysis is decarboxylated and converted to **ethanal**. This stage in the pathway is catalysed by pyruvate decarboxylase, which has a coenzyme, thiamine diphosphate, bound to it.

2. The ethanal accepts hydrogen atoms from reduced NAD, becoming reduced to ethanol. The enzyme ethanol dehydrogenase catalyses the reaction.

3. In the process, the reduced NAD is re-oxidised and made available to accept more hydrogen atoms from triose phosphate, thus allowing glycolysis to continue.

> **DID YOU KNOW?**
> Yeast is a facultative anaerobe. It can live without oxygen, but if oxygen is present it will respire aerobically. However, under anaerobic conditions, when the accumulation of ethanol reaches about 15%, it will kill the yeast cells.

The lactate fermentation pathway

Lactate fermentation occurs in mammalian muscle tissue during vigorous activity, such as when running fast to escape a predator, when the demand for ATP for muscle contraction is high and there is an oxygen deficit.

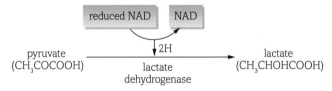

Figure 2 The lactate fermentation pathway.

1. Pyruvate, produced during glycolysis, accepts hydrogen atoms from the reduced NAD, also made during glycolysis. The enzyme lactate dehydrogenase catalyses the reaction. There are two outcomes:
   * Pyruvate is reduced to lactate.
   * The reduced NAD becomes reoxidised.

2. The reoxidised NAD can accept more hydrogen atoms from triose phosphate during glycolysis, and glycolysis can continue to produce enough ATP to sustain muscle contraction for a short period.

The fate of lactate

The lactate produced in the muscle tissue is carried away from the muscles, in the blood, to the liver. When more oxygen is available, the lactate may be either:

* converted to pyruvate, which may enter the Krebs cycle via the link reaction
* recycled to glucose and glycogen.

If lactate were not removed from the muscle tissues, the pH would be lowered and this would inhibit the action of many of the enzymes involved in glycolysis and muscle contraction.

The ATP yield from anaerobic respiration

Neither ethanol fermentation nor lactate fermentation produces any ATP. However, because this allows glycolysis to continue, the net gain of two molecules of ATP per molecule of glucose is still obtained.

Because the glucose is only partly broken down, many more molecules can undergo glycolysis per minute, and therefore the overall yield of ATP is quite large. However, for each molecule of glucose, the yield of ATP via anaerobic respiration is about 1/15 of that produced during aerobic respiration.

> **LEARNING TIP**
> You may have noticed that the enzyme pyruvate decarboxylase found in yeast has a coenzyme, thiamine diphosphate, bound to it. Thiamine is vitamin B_1. Do not confuse it with thymine, a nucleotide base.

> **DID YOU KNOW?**
> Fast-twitch muscle fibres have few if any mitochondria and use glycolysis to power their short-duration contractions. They fatigue easily and appear pale in colour due to the lack of electron transport proteins. They also lack myoglobin, a protein that stores oxygen in some muscles.
> Slow-twitch muscle fibres are dark red, contain many mitochondria and are slow to fatigue. They operate aerobically for endurance exercise. The breast meat of chickens is pale as they have mainly fast twitch fibres and only fly occasionally – just enough to try to escape from a predator. Ducks and geese often fly long distances and their breast meat is very dark, due to the high amount of slow twitch muscle fibres.

> **DID YOU KNOW?**
> Thiamine is present in meat and whole grains. People who eat a diet low in meat and high in polished rice may suffer from the deficiency disease beriberi. This disease was recognised as early as 4600 years ago in China.

Figure 3 Red jungle fowl – this bird only flies to escape danger and the flight is of very short duration, more like a jump with some wing flapping.

Questions

1 Complete the table to compare the ethanol and lactate fermentation pathways.

	Lactate fermentation	Ethanol fermentation
Hydrogen acceptor		
Is carbon dioxide produced?		
Is ATP produced?		
End products		
Enzymes involved		
Is NAD reoxidised?		
Site of pathway		

2 Compare the ATP yield per molecule of glucose made during anaerobic respiration with the ATP yield per molecule of glucose made during aerobic respiration.

3 Glycolysis is described as 'an ancient metabolic pathway'. It occurs in the cytoplasm of all living organisms that respire, including some bacteria that have inhabited the Earth for billions of years. Suggest, with reasons, whether glycolysis evolved before or at the same time that aerobic respiration evolved.

4 Explain why the breast meat of geese is much darker than that of chickens.

5 Suggest why muscle contraction cannot be sustained for more than a few seconds using ATP stores within the muscle.

6 Suggest why non-fortified wine is unlikely to have an alcohol content above 15%.

Figure 4 Migrating geese fly very long distances – a great feat of endurance.

By the end of this topic, you should be able to demonstrate and apply your knowledge and understanding of:

* practical investigations into respiration rates in yeast, under aerobic and anaerobic conditions

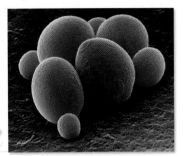

Figure 1 Coloured SEM of yeast *Saccharomyces cerevisiae*. This single-celled fungus reproduces asexually. A daughter cell is budding off from the mother cell at lower left (×2000).

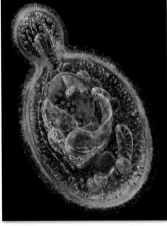

Figure 2 Computer artwork of a section through a budding yeast, *Candida albicans*. A daughter cell is budding off from the parent cell. A mitochondrion is being moved into the daughter cell along cytoskeleton threads. The DNA will replicate and the nucleus (purple) will divide, so that the daughter cell will receive the same genetic information that is in the mother cell.

Yeast, *Saccharomyces cerevisiae*, is a **facultative anaerobe**. It is a single-celled fungus and is eukaryotic. Its cells contain mitochondria. Yeast cells may reproduce asexually by dividing by mitosis.

If oxygen is available, then yeast cells respire aerobically using glycolysis, the link reaction, the Krebs cycle and oxidative phosphorylation, producing many molecules of ATP per molecule of glucose.

If oxygen is lacking, then the yeast cells respire anaerobically using glycolysis and the ethanol pathway. This produces only a few molecules of ATP per molecule of glucose.

For yeast cells to divide they require ATP and the rate of reproduction depends on the amount of ATP available. We would expect yeast to have a faster rate of reproduction under aerobic conditions.

Yeast can also oxidise ethanol under aerobic conditions. The alcohol content of cider is around 6% and so not enough to kill the yeast. However, with anaerobic respiration this may increase, whereas it will decrease in the aerobic flasks.

INVESTIGATION 1

The rate of reproduction of yeast cells under aerobic and anaerobic conditions

Equipment: microscope; haemocytometer; conical flasks of different sizes (e.g.: 50 cm³, 100 cm³, 150 cm³, 250 cm³ and 500 cm³); 50 cm³ measuring cylinder; muslin, cheesecloth or tights; pipettes; elastic bands; culture of brewer's yeast; dry cider

The conical flasks have to be clean and sterilised — they can be heated in a microwave oven for 5 minutes on high setting. The cider should be uncontaminated.

In the presence of oxygen, yeast cells can oxidise the ethanol and any sugar in the cider, and will not be killed by the alcohol. They will respire aerobically and produce more ATP that can be used for cell division. Under anaerobic conditions the ethanol cannot be oxidised and will eventually kill the yeast. Therefore, where the respiration is aerobic, more yeast cells should be present per cm³ at the end of a week.

In the small flasks, the cider is deep and with a small surface area for oxygen absorption and a large diffusion distance for oxygen to each the yeast cells. The converse is true for the large flasks.

Steps 1–4 may be done for you by the lab technicians, as only one of each size of flask is needed per class. You may carry out steps 5–9.

1. Pour 50 cm³ of cider into each of the conical flasks.
2. Using a clean pipette, add one drop of yeast suspension to each conical flask. Make sure, before taking the yeast suspension from its flask, that the flask is swirled to mix its contents.
3. Place four layers of muslin, cheesecloth or tights material over the mouth of each conical flask and secure with an elastic band. This allows oxygen to enter the flask, but keeps out dust and contaminants.
4. Leave the flask in a warm place for about a week.
5. Mix the contents of each flask thoroughly by swirling and, using a clean pipette, withdraw some of the flask contents and place a drop onto a haemocytometer slide with its coverslip in place (see Investigation 2 on how to use a haemocytometer).
6. Count the number of yeast cells in the centre square and each corner square. In each square where cells are counted, some cells may be lying on the boundaries. In this case only those cells touching the north (top) and west (left) side of the square are counted — as well as cells within the boundaries of that square.
7. Each of the five squares where you counted contained 16 smaller squares. So you counted cells in 80 very small squares, having a total volume of 0.02 mm³. If you multiply the cell count you obtained by 50 000, you will find the number of yeast cells per cm³.
8. Carry out three counts for each size of flask. Calculate the mean number of yeast cells per cm³ for each flask.
9. Tabulate and graph your data. What conclusions can you draw?

INVESTIGATION 2

Using a haemocytometer

Haemocytometers are so named because they were first used for counting red blood cells. Each one is a special thick slide with bevelled edges and grooves. If the grooves form an H shape, there will be two etched grids, so that two counts can be made from the same sample. The coverslips used are also much thicker than normal coverslips. Both slide and coverslip should be clean.

1. Breathe onto the underside of the coverslip to moisten it.
2. Slide the coverslip horizontally onto the slide and carefully press down with the index fingers whilst pushing with the thumbs.
3. When the coverslip is correctly in position, you will see six rainbow patterns (Newton's rings). The depth of the central chamber is now 0.1 mm.
4. Place the pipette tip at the entrance to the groove and allow liquid to fill the chamber. Leave for five minutes before counting.
5. Place the haemocytometer slide on the microscope stage with one of the grids over the stage aperture. Focus, using total ×40 magnification.
6. Now focus using total ×100 magnification. The central portion of the grid will now fill the field of view.
7. Count cells in the central and four corner squares.

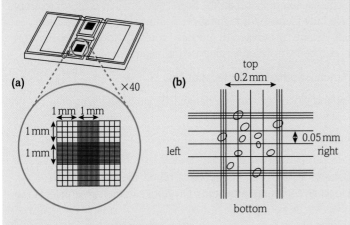

Figure 3 Using a haemocytometer.

Other ways to compare aerobic and anaerobic respiration in yeast

The rate of respiration can be measured by measuring the rate of evolution of carbon dioxide. As carbon dioxide dissolves in the culture medium it lowers the pH, and this can be measured using a pH meter.

LEARNING TIP

If each square in the haemocytometer grid is 0.2 mm × 0.2 mm and has a depth of liquid above it of 0.1 mm, then the volume of liquid over one small square is $0.2 \times 0.2 \times 0.1 \text{ mm}^3 = 0.004 \text{ mm}^3$.

You counted 5 such squares, making the volume $5 \times 0.004 = 0.02 \text{ mm}^3$. There are $50 \times 0.02 \text{ mm}^3$ in 1 mm^3 and 1000 mm^3 in 1 cm^3.

Questions

1. Suggest how glucose enters the yeast cells.

2. One limitation of using a haemocytometer to do a total cell count, as described in this topic, is that dead cells will also be counted. Discuss whether this will make a difference to this investigation.

3. Yeast, and other single-celled organisms growing in liquid media, produce heat as a by-product of their respiration. What are the implications of this during industrial fermentation processes involving large vessels of microorganisms growing and respiring in liquid media?

4. In an investigation similar to the one described in this topic, instead of cider, 5% glucose solution was used for yeast to grow in under aerobic and anaerobic conditions. The accumulation of ethanol only occurred when conditions were anaerobic and its effect on the yeast was much slower. Because the number of cells produced was much greater, samples were diluted by 1 in 10 before cell counts were made. If the mean cell count for a diluted sample was found to be 50 cells per 0.02 mm^3, calculate the cell count per cm^3 of the undiluted sample.

5. Outline how you could use the yeast cell count method to investigate the effect of temperature on the rate of respiration, and therefore on the rate of growth, of yeast at different temperatures.

Energy values of different respiratory substrates

By the end of this topic, you should be able to demonstrate and apply your knowledge and understanding of:

* the difference in relative energy values of carbohydrates, lipids and proteins as respiratory substrates

* the use and interpretation of the respiratory quotient (RQ)

KEY DEFINITION

respiratory substrate: an organic substance that can be oxidised by respiration, releasing energy to make molecules of ATP.

Respiratory substrates

Besides carbohydrates, lipids and proteins can also provide **respiratory substrates**. They can be oxidised in the presence of oxygen to produce molecules of ATP, carbon dioxide and water. They each have different relative energy values.

Carbohydrates

The monosaccharide glucose is the chief respiratory substrate. Some mammalian cells, for example brain cells and red blood cells, can use only glucose for respiration. Animals and some bacteria store carbohydrate as glycogen, which can be hydrolysed to glucose for respiration. Plant cells store carbohydrate as starch, and this can also be hydrolysed to glucose for respiration:

* Disaccharides can be digested to monosaccharides for respiration.

* Monosaccharides such as fructose and galactose can be changed, by isomerase enzymes, to glucose for respiration.

Lipids

Lipids are important respiratory substrates for a number of types of tissue, including muscle. Triglycerides are hydrolysed by lipase to glycerol and fatty acids. Glycerol can then be converted to triose phosphate and respired.

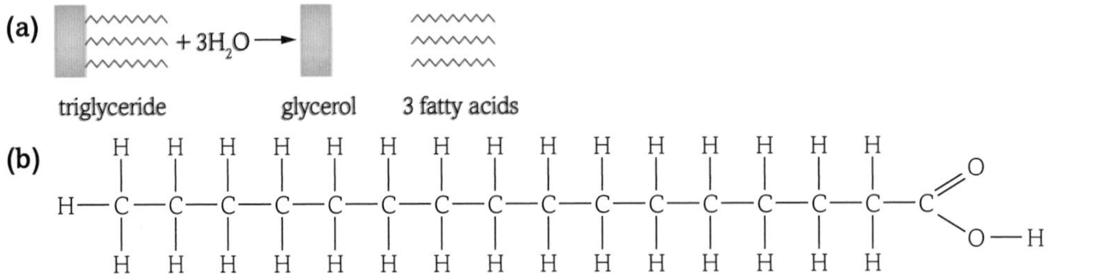

(a) triglyceride $+ 3H_2O \rightarrow$ glycerol   3 fatty acids

(b) palmitic acid

Figure 1 (a) Hydrolysis of a triglyceride molecule to glycerol and three fatty acids; (b) a molecule of the fatty acid palmitic acid.

Fatty acids are long-chain hydrocarbons (Book 1, topic 2.2.6) with a carboxylic acid group. Hence in each molecule there are many carbon atoms, many hydrogen atoms and very few oxygen atoms. These molecules are a source of many protons for oxidative phosphorylation, and so fats produce much more ATP than an equivalent mass of carbohydrate.

1. With the aid of some energy from the hydrolysis of one molecule of ATP to AMP, each fatty acid is combined with coenzyme A.

2. The fatty acid–CoA complex is transported into the mitochondrial matrix, where it is broken down into two-carbon acetyl groups, each attached to CoA.

3. This beta-oxidation pathway generates reduced NAD and reduced FAD.

4. The acetyl groups are released from CoA and enter the Krebs cycle by combining with the four-carbon oxaloacetate.

For every acetyl group oxidised in the Krebs cycle, three molecules of reduced NAD, one molecule of reduced FAD and one molecule of ATP, by substrate-level phosphorylation, are made.

Proteins

Excess amino acids, released after the digestion of proteins, are deaminated in the liver. Deamination of an amino acid involves removal of the amino group and its subsequent conversion to urea that is removed via the kidney (topic 5.2.5). The rest of the amino acid molecule, a keto acid, enters the respiratory pathway as pyruvate, acetyl CoA or a Krebs cycle acid such as oxaloacetic acid.

During fasting, starvation or prolonged exercise, when insufficient glucose or lipid are available for respiration, protein from muscle can be hydrolysed to amino acids which are then respired. These amino acids may be converted to pyruvate or acetate and enter the Krebs cycle.

Figure 2 shows how various amino acids enter the Krebs cycle for respiration.

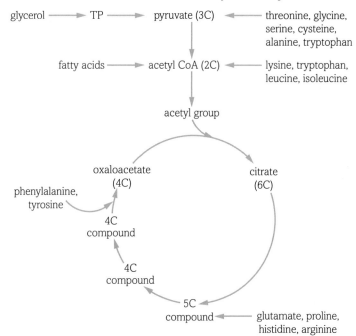

Figure 2 How amino acids and the products of fat digestion enter the Krebs cycle.

Energy values of different respiratory substrates

Topic 5.7.5 shows how most of the ATP produced during aerobic respiration is made during oxidative phosphorylation. The greater the availability of protons for chemiosmosis, the more ATP can be produced. Therefore the more hydrogen atoms there are in a molecule of respiratory substrate, the more ATP can be generated per molecule of substrate.

Respiratory substrate	Mean energy value (kJ g^{-1})
Carbohydrate	15.8
Lipid	39.4
Protein	17.0

Table 1 The mean energy values per gram of different respiratory substrates.

As the protons (hydrogen ions) ultimately combine with oxygen atoms to form water, the greater the proportion of hydrogen atoms in a molecule, the more oxygen will be needed for its respiration.

Respiratory quotient

The respiratory quotient, RQ, for different respiratory substrates can be calculated using the formula:

$$RQ = \frac{CO_2 \text{ produced}}{O_2 \text{ consumed}}$$

As it is a ratio, there are no units.

Respiratory substrate	Calculation of RQ value	RQ value(s)
Glucose	$C_6H_{12}O_6 + 6O_2 \rightarrow 6CO_2 + 6H_2O$	$6/6 = 1$
Fatty acids	Palmitic acid $C_{15}H_{31}COOH + 23O_2 \rightarrow 16CO_2 + 16H_2O$	$16/23 = 0.7$
Amino acids	The keto acid derived from the deamination of glycine has a formula CH_3COOH $CH_3COOH + 5O_2 \rightarrow 4CO_2 + 4H_2O$	$4/5 = 0.8$ values of 0.8–0.9 for the various amino acids

Table 2 How RQ values for different respiratory substrates are calculated.

If the RQ value is greater than 1, this indicates that some anaerobic respiration is taking place, because it shows that more carbon dioxide is being produced than oxygen is being consumed.

Questions

1. Explain why a diet high in fat is also high in energy.

2. Explain why palmitic acid, a large molecule, can pass into the matrix of the mitochondria.

3. Explain why children whose diet does not contain sufficient fat or carbohydrate may suffer muscle wastage.

4. Camel humps contain stored fat. One kilogram of fat, when respired, produces 1 kg metabolic water. Thus these humps provide energy and water for the camels. Explain why, when fat is respired, it produces more metabolic water than respiration of an equivalent quantity of stored carbohydrate.

5. Is the lipid in camel humps respired aerobically or anaerobically? Explain your answer.

6. In an investigation into respiration by woodlice, the RQ value was found to be 0.8. Suggest which type(s) of respiratory substrate were being used.

7. In an investigation into respiration by germinating seeds, the RQ value was found to be 2.3. What may be deduced from this finding?

9 Practical investigations into factors affecting the rate of respiration

By the end of this topic, you should be able to demonstrate and apply your knowledge and understanding of:

* practical investigations into the effect of factors such as temperature, substrate concentration and different respiratory substrates on the rate of respiration

Using a respirometer

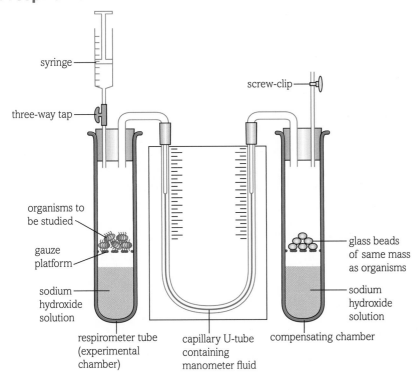

Figure 1 A simple respirometer.

> **KEY DEFINITION**
>
> **respirometer:** apparatus used to measure the rate of respiration of living organisms by measuring the rate for exchange of oxygen and carbon dioxide.

The principle

Organisms that are respiring aerobically absorb oxygen and give out carbon dioxide.

If the carbon dioxide produced is absorbed by sodium hydroxide solution or solid soda lime, then the only volume change within the **respirometer** is due to the volume of oxygen absorbed by the organisms.

If oxygen is absorbed from the tube containing the organisms, then that tube has a reduced volume of air in it, exerting less pressure than the greater volume of air in the other tube. As a result, the coloured liquid in the manometer tube rises up towards the respirometer tube.

If the original level of liquid in the manometer tube is marked and the radius of the bore in the capillary tube is known, the volume of oxygen absorbed during a specific period can be calculated.

To reset the apparatus, the syringe is depressed to inject air into the system and reset the liquid in the manometer tube back to its original position. This also allows a reading of the volume of oxygen absorbed by noting the change in level of the syringe plunger, as measured from the graduated scale on the syringe barrel.

Setting up the apparatus

1. After placing the coloured liquid, for example methylene blue solution that has one drop of detergent added to it, into the manometer tube, the apparatus is connected with the taps open. This enables the air in the apparatus to connect with the atmosphere.

2. The mass of living organisms (e.g. woodlice) to be used should be found.

3. With the taps still open the whole set-up, with the living organisms in place, is placed in a water bath for at least 10 minutes until it reaches the temperature of the water bath.

4. The syringe plunger should be near the top of the scale on the syringe barrel and its level noted.

5. The levels of coloured liquid in the manometer tubes can be marked with a felt tip pen or chinagraph pencil.

6. The taps are closed and the apparatus left in the water bath for a specific period, such as 10 minutes.

7. The change in level of manometer liquid can be measured, and the syringe barrel depressed to reset the apparatus. This also enables you to measure the volume of oxygen absorbed.

8. You can then calculate the volume of oxygen absorbed per minute per gram of living organism.

LEARNING TIP

Remember that the apparatus is left for about 10 minutes open to the atmosphere so that whilst it is reaching the required temperature, the organisms' respiration is not affecting the fluid in the manometer.

DID YOU KNOW?

Respirometers can also be used to measure rates of respiration in plants or other photosynthetic organisms such as some algae. However, the tubes containing them must be wrapped in foil or black paper to exclude light and prevent photosynthesis.

Figure 2 A woodlouse, *Porcellio scaber*. Woodlice are small terrestrial crustaceans belonging to the order isopods. They have a segmented, oval, rather flattened body and live under stones or bark, or in decaying vegetation, where it is moist. They are detritivores.

Measuring the effect of temperature

The effect of temperature on the rate of respiration can be investigated using the respirometer. Three readings should be taken at each temperature. In between each temperature reading, the apparatus and organisms should be allowed to adjust to the new temperature.

Suitable living organisms can be blowfly maggots, woodlice, yeast in glucose suspension or soaked pea seeds that are beginning to germinate. Animal specimens should only be used over a narrow range of temperatures, such as from 10 to 40 °C. For more extremes of temperature, fungal material should be used.

You can predict what you expect to happen, based on your knowledge of the process of respiration and the fact that there are many enzyme-catalysed stages involved.

Investigating the effect of substrate concentration on the rate of respiration in yeast

The respirometer may be used so that a suspension of yeast, with differing concentrations of glucose solution, is placed in one of the tubes. If the sodium hydroxide solution is omitted, the evolution of carbon dioxide during a specific time period can be measured.

① Using a pipette, place
10 drops of sugar solution
and 10 drops of yeast
suspension into each tube.
Top up with distilled water.
For the control, use 10 drops
of distilled water in place of
the sugar solution.

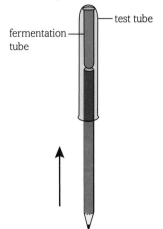

INVESTIGATION

The effect of different respiratory substrates on the rate of respiration in yeast

Respiration is a series of enzyme-catalysed reactions. All enzymes have specificity for a particular substrate. Yeast may be able to produce **isomerase enzymes** to change some types of monosaccharide to glucose so that glycolysis can take place. Yeast may also be able to produce enzymes to hydrolyse disaccharides to monosaccharides.

The respirometer, without sodium hydroxide solution, could be used to measure evolution of carbon dioxide within a specified period.

Alternatively, the carbon dioxide produced during the anaerobic respiration of yeast can be collected in fermentation tubes, as shown in Figure 3. The liquid in the fermentation tube is a solution of a particular sugar plus two drops of yeast suspension.

Figure 4 shows the structural formulae of some monosaccharides.

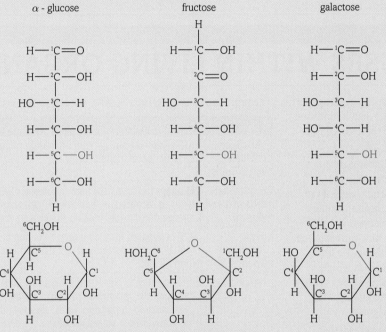

Figure 4 The structural formulae of glucose, fructose and galactose.

② Push the fluid-filled
fermentation tube
to the end of a
test tube and invert.

③ Measure the height
of the fluid in the
fermentation tube.

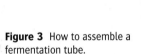

Figure 3 How to assemble a fermentation tube.

If you set up a series of fermentation tubes as shown, with all conditions the same, except the type of sugar being used, and measure the height of the carbon dioxide bubble after a set period of time, you can see how efficiently yeast can respire the different substrates. You could also try disaccharide sugars, such as maltose, sucrose and lactose.

Record your results.

Evaluate this method and suggest how you could use a pH meter and data logger to compare the efficiency with which yeast can use the different respiratory substrates.

LEARNING TIP

Avoid skin contact with methylene blue or sodium hydroxide. Wear eye protection, and gloves if possible.

Questions

1. Explain why, when oxygen is absorbed from the air in the respirometer tube that contains living organisms, the liquid in the manometer tube moves towards the respirometer tube.

2. What is the purpose, in a respirometer, of the other tube that does not contain any living organisms?

3. What is the purpose of the sodium hydroxide in the respirometer tubes?

4. Explain how the respiration rate of a living organism can be calculated from the data obtained using this apparatus.

5. Suggest why only plant or fungal organisms should be used to investigate the effect of temperatures above 40 °C on the rate of respiration.

6. Suggest an explanation for the results you obtained in the investigation of how yeast respires different sugars.

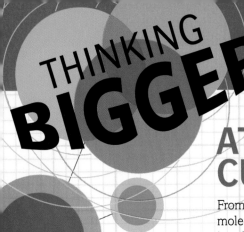

THINKING BIGGER

ATP – MORE THAN JUST ENERGY CURRENCY

From its discovery in 1929 as a source of energy in muscle tissue, through to the confirmation of its molecular structure in 1945, ATP was perceived in terms of it being an immediate source of energy in cells. However, scientists have since discovered that it has many other uses within living organisms.

ATP HAS OTHER USES WITHIN LIVING ORGANISMS

Biology students learn about the role of ATP (adenosine triphosphate) as the universal energy currency in living cells that, when hydrolysed, releases energy to drive biological reactions. However it also has an important role, as a signalling molecule, outside of cells.

In 1959 a scientist at the Cambridge Physiological laboratory published data that seemed to indicate sensory nerves released molecules of ATP. Another scientist who read the article carried out many investigations and showed that when other neurotransmitters were blocked, signals reached muscle tissue, from nerves, via ATP. He proposed that ATP acted with neurotransmitters as a cotransmitter, carrying information from motor nerves to muscle tissue. He also proposed that, as well as the already recognised synapses which release acetylcholine, dopamine or adrenaline, some synapses release purine derivatives such as ATP from the pre-synaptic knobs of the motor neurones. He described such synapses as purigenic.

Although many scientists were sceptical and did not believe that such a widely occurring molecule could act as a neurotransmitter, evidence for purigenic synapses was demonstrated by scientists from universities in London and Melbourne, Australia. By the early 1990s, specific receptors for ATP were identified in smooth muscle tissue. The presence of such receptors implied that ATP could act as a signalling molecule.

Information from the Human Genome Project identified genes that code for ATP receptors. These receptors are found on many different cell types and have also been shown to be targeted by various drugs, thereby modulating ATP signalling.

Enzymes called ectoATPases sit on the external surface of cells and can turn ATP to ADP and then to AMP and then to adenosine. Each of these breakdown products may exert an effect on the cell's metabolism. Sometimes:

- ATP and adenosine work together in the brain stem to regulate breathing, heart rhythm and gastrointestinal action.
- ATP and adenosine work antagonistically.
- ATP acts with other neurotransmitters, such as acetylcholine or noradrenaline, as a cotransmitter.

The table shows some effects of ATP on various tissues.

Tissue	Effect of ATP
Muscle cells of blood vessel walls	ATP released from sympathetic nerve endings causes constriction of blood vessels.
Endothelial cells of blood vessels	Changes in blood flow lead to release of ATP, which binds to receptors on endothelial cells causing them to release nitric oxide, making the vessels dilate.
Blood vessels at wounds	ATP is released from damaged cells and broken down to ADP that binds to receptors on platelets, which then form a clot.
Arteries	After surgery to clear a blocked artery, ATP released from damaged tissue binds to receptors on endothelial muscle cells, causing them to multiply, resulting in narrowing of the artery. This is called restenosis.
Nerve cells in retina of eye	Influences the information received from rods and cones and helps convey information to the visual cortex of the brain. Also involved in embryonic eye development.
Ear	Involved in development of cochlea and in subsequent working of the hair cells
Taste buds	Conveys information to taste centres in the brain
Skin	ATP receptors in sensory nerve endings in skin trigger the sensation of pain.
Immune system	ATP released from injured tissue causes inflammation and helps T cells kill infected cells.
Bone	Activation of ATP receptors stimulates bone development and represses bone-destroying cells.

Table 1 Some effects of ATP on various tissues.

Reference
- Baljit S Khakh and Geoffrey Burnstock, 'The Double Life of ATP', *Scientific American* 301, 84–92 (2009) (http://www.ucl.ac.uk/ani/GB's%20PDF%20 file%20copies/CV1412(Sci%20Am).pdf http://www.ucl.ac.uk/ani/GB's PDF file copies/CV1412(Sci Am).pdf)

Where else will I encounter these themes?

1. What features of the writing in this article indicate that it may be from a science magazine that is aimed at an audience with quite a lot of scientific knowledge?

Now let us look at the biological concepts underlying the information in this article.

2. Where, and during which process/processes, in eukaryotic cells, is ATP made?

3. Suggest why synapses that release ATP are termed 'purigenic' synapses.

4. a. By what process are neurotransmitter molecules released from the pre-synaptic knob of a neurone at a synapse?

   b. By what process do neurotransmitter molecules pass from the pre-synaptic membrane to the post-synaptic membrane of a synapse?

5. Explain how receptor proteins on the receiver membranes of a synapse are specific to particular neurotransmitters.

6. Suggest why ATP modulation could play a role in combating osteoporosis.

DID YOU KNOW?

ATP receptors have been found in plants, amoebae and green algae. Many reports suggest that ATP exerts potent effects on invertebrates and lower vertebrates. This suggests that the role of ATP as a signalling molecule arose early in the evolution of life on Earth and has a widespread influence on cell behaviour.

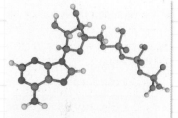

Figure 1 Molecular model of ATP.

Activity

Scientists have found that injections of ATP into the skin of humans cause pain. This pain is triggered by activation of ATP receptors on the sensory nerve endings in the skin.

Neuropathic pain is a different sort of pain and is triggered by damage to nerves and activation of ATP receptors on special immune cells within the spinal cord, called microglial cells. These microglia, when triggered in this way, release chemicals that irritate nerve fibres, leading to chronic pain. Several pharmaceutical companies are trying to develop drugs that target the receptors involved in chronic pain, in an attempt to alleviate such pain in patients.

Drugs that target other ATP receptors in the gut may help treat irritable bowel syndrome, IBS, by controlling the passage of food through the gut.

Suppression of inflammation and blood clotting may also be helpful in combating certain cardiovascular patholologies.

Find out, using the Internet, more about the role of ATP modulation in one of the above topics – chronic pain, IBS or cardiovascular disease – and write a short article, one side of A4 maximum, for an A level biology textbook.

Practice questions

1. Read the following statements.

    (i) FAD, NAD and coenzyme A all contain adenine.

    (ii) ATP is a phosphorylated nucleotide.

    (iii) FAD, NAD, coenzyme A and ATP are all derived from B vitamins.

   Which statement(s) is/are true? [1]

   A. (i), (ii) and (iii)

   B. (i) and (ii) only

   C. (i) and (iii) only

   D. (ii) and (iii) only

2. Which row correctly shows the net gain of ATP per molecule of glucose at each stage of aerobic respiration? [1]

	Glycolysis	Link reaction	Krebs cycle	Oxidative phosphorylation
A	4	0	1	32
B	2	2	2	28
C	2	0	2	28
D	4	0	1	32

3. What are the products of glycolysis? [1]

   A. Acetyl coenzyme A, ATP and reduced NAD

   B. Pyruvate, ATP and reduced NAD

   C. Pyruvate, ATP and reduced FAD

   D. Triose phosphate, ATP and reduced NAD

4. Read the following statements.

    (i) There is one turn of the Krebs cycle for each molecule of glucose.

    (ii) The Krebs cycle takes place in the mitochondrial matrix.

    (iii) During the Krebs cycle, dehydrogenation, decarboxylation and substrate-level phosphorylation all occur.

   Which statement(s) is/are true? [1]

   A. (i) and (ii) only

   B. (ii) and (iii) only

   C. (i) and (iii) only

   D. (i), (ii) and (iii)

5. Which correctly shows the products of anaerobic respiration in mammalian muscle tissue? [1]

   A. Ethanol + carbon dioxide + NAD

   B. Ethanol + carbon dioxide + reduced NAD

   C. Lactate + carbon dioxide + NAD

   D. Lactate + NAD

   [Total: 5]

6. Figure 1 outlines some of the stages in aerobic respiration.

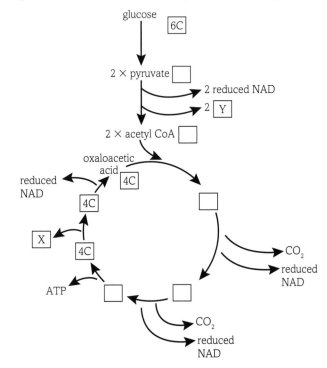

Figure 1

   (a) Fill in the empty boxes to show the number of carbon atoms present in the compounds at those stages of respiration. [5]

   (b) What is represented by each of the letters X and Y? [2]

   (c) Where, in a eukaryotic cell, does the conversion of glucose to pyruvate take place? [1]

   (d) (i) Where, in a eukaryotic cell, does the Krebs cycle take place? [1]

       (ii) Explain how the acetyl group is carried onto the Krebs cycle. [3]

   (e) Describe the fate of the reduced NAD produced during these stages of aerobic respiration. [5]

   [Total: 17]

7. Figure 2 shows the relationship between various metabolic processes.

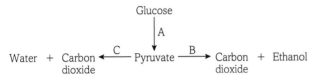

Figure 2

(a) (i) Identify the three metabolic processes, represented by the letters A, B and C. [3]

 (ii) In which stage, A, B or C, is ATP utilised? [1]

 (iii) Name one type of organism in which these processes occur. [1]

(b) In an investigation, mammalian liver cells were homogenised in a blender and centrifuged to isolate mitochondria. The following samples were then set up at 35 °C:

 1 homogenised uncentrifuged liver cells

 2 homogenised cells with cytoplasm, but no mitochondria

 3 mitochondria only

Each sample was treated in four different ways:

A. with glucose

B. with pyruvate

C. with glucose and cyanide

D. with pyruvate and cyanide

Cyanide inhibits enzymes involved in electron transfer. It inhibits oxidative phosphorylation.

Table 1 summarises the results of this investigation.

 (i) Explain why mitochondria produce carbon dioxide when they are incubated with pyruvate, but do not produce carbon dioxide when incubated with glucose. [3]

 (ii) Explain why lactate is not produced by mitochondria incubated with pyruvate. [2]

 (iii) Explain why, when cyanide is present, lactate is produced but not carbon dioxide. [3]

[Total: 13]

8. (a) What is meant by the term 'respiratory quotient, RQ'? [1]

(b) Figure 3 shows the RQ of two seeds whilst germinating over a period of 10 days. The percentage dry mass of stored organic compounds that could be used as respiratory substrate is also shown.

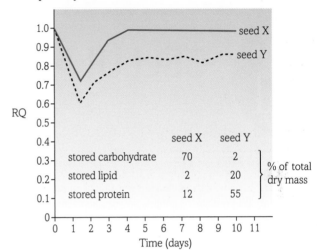

Figure 3

 (i) Suggest, with reasons, which organic compound is being used as a respiratory substrate during the first day of germination, for both seeds. [3]

 (ii) Describe and explain how the respiration of seed X and seed Y differ during days 4–10 of germination. [5]

(c) Describe how you could determine the RQ of germinating seeds in a laboratory investigation. You may use annotated diagrams to illustrate your answer. [8]

(d) Under what circumstances may the RQ of a respiring organism be greater than 1.0? [1]

[Total: 18]

	Complete homogenised cells		Cytoplasm only		Mitochondria only	
	CO_2 produced	lactate produced	CO_2 produced	lactate produced	CO_2 produced	lactate produced
Glucose	yes	yes	no	yes	no	no
Pyruvate	yes	yes	no	yes	yes	no
Glucose and cyanide	no	yes	no	yes	no	no
Pyruvate and cyanide	no	yes	no	yes	no	no

Table 1

CHAPTER
6.1

CELLULAR CONTROL

Introduction

In 1974, Sydney Brenner began his studies of a simple multicellular eukaryotic organism, *Caenorhabditis elegans*. This transparent, soil-dwelling, bacteria-eating nematode (roundworm) that lives in temperate regions was the first multicellular organism to have its genome sequenced; it has roughly the same number of genes as humans, indicating that in humans the interactions between genes and their products is crucial in the development of a much more complex organism. *C. elegans* has also had its connectome – the neuronal linkage pattern, worked out.

It is a small simple organism with bilateral symmetry. Because it is transparent, the migration of its cells produced during development can be followed. It was found that the worms have a fixed number of cells. Once cell division has ceased at the end of the larval period, growth results only from an increase in size of the cells. Most *C. elegans* are female hermaphrodites that have 959 cells. About 1 in 1000 are males that have 1031 cells. During development, more cells than the final number are produced and the excess cells are removed via apoptosis (programmed cell death).

This simple worm has a mouth, gut, reproductive organs, four bands of muscles, a neuronal network and an outer cuticle. It has become a model organism for much research into development, as it is easy to keep and breed, it can be frozen and is viable when thawed. It has provided much information about apoptosis and genes that regulate the process. Thirty-five per cent of its genes are conserved (still present) in humans and it has been used for research into nicotine addiction, ageing and sleep. Some of these worms have also spent time in the Space Station, for research into the effect of zero gravity on muscles. Three teams of scientists have won the Nobel Prize (two for Physiology or Medicine and one for Chemistry) for research using this humble nematode. It has been a key player in the study of evolutionary development, a recent branch of biology also known as evo-devo.

In this chapter, you will find out about the regulatory mechanisms that control gene expression; and about the importance of certain genes, apoptosis and mitosis in controlling the development of body plans and cell differentiation during development of organisms.

All the maths you need

To unlock the puzzles of this chapter you need the following maths:

- Know units of measurement

What have I studied before?

- Chromosomes are made of DNA and contain genes
- The genetic material in the nucleus of an organism's cells contains the blueprint (coded instructions) for the development of that organism
- The basic structure of DNA is the same in all living organisms
- Different organisms have some genes in common and some genes that are different
- Mutations can cause genetic differences
- During development of an organism from embryo to adult, cells become differentiated and specialised to carry out particular functions

What will I study later?

- How the interrelations between genes and environment contribute to phenotypic variation
- Patterns of inheritance
- The factors that affect the evolution of a species
- The principles and ethical considerations of artificial selection
- DNA sequencing
- The principles and uses of DNA profiling
- The polymerase chain reaction
- Principles and applications of genetic engineering
- The principles and potential for gene therapy in medicine
- Cloning and biotechnology

What will I study in this chapter?

- The way in which various types of gene mutation can affect protein structure and function
- How some mutations are neutral or beneficial
- How gene expression is controlled in cells
- How certain genes control the development of body plans in different living organisms
- The importance of mitosis and apoptosis in controlling the development of body form

Gene mutations

By the end of this topic, you should be able to demonstrate and apply your knowledge and understanding of:

* types of gene mutations and their possible effects on protein production and function

A mutation is a random change to the genetic material. Some mutations involve changes to the structure or number of chromosomes. A gene mutation is a change to the DNA.

Mutations may occur spontaneously during DNA replication before cell division. Certain chemicals, such as tar in tobacco smoke, and ionising radiation such as UV light, X-rays and gamma rays, may be **mutagenic**.

Types of gene mutation

The structure of the DNA molecule makes it stable and fairly resistant to corruption of the genetic information stored within it. Errors may occur, however, during the replication of a DNA molecule.

Mutations associated with mitotic division are somatic mutations and are not passed to offspring. They may be associated with the development of cancerous tumours.

Mutations associated with meiosis and gamete formation may be inherited by offspring.

Gene mutations may affect protein production and function.

There are two main classes of DNA mutation:

* Point mutation: one base pair replaces (is *substituted* for) another.

* Insertion or deletion (indel) mutation: one or more nucleotides are inserted or deleted from a length of DNA. These may cause a **frameshift.**

First position	Second position				Third position
	T	C	A	G	
T	Phe	Ser	Tyr	Cys	T
	Phe	Ser	Tyr	Cys	C
	Leu	Ser	STOP	STOP	A
	Leu	Ser	STOP	Trp	G
C	Leu	Pro	His	Arg	T
	Leu	Pro	His	Arg	C
	Leu	Pro	Gln	Arg	A
	Leu	Pro	Gln	Arg	G
A	Ile	Thr	Asn	Ser	T
	Ile	Thr	Asn	Ser	C
	Ile	Thr	Lys	Arg	A
	Met	Thr	Lys	Arg	G
G	Val	Ala	Asp	Gly	T
	Val	Ala	Asp	Gly	C
	Val	Ala	Glu	Gly	A
	Val	Ala	Glu	Gly	G

Key:
- Asp Aspartic acid
- Glu Glutamic acid
- His Histidine
- Ile Isoleucine
- Arg Arginine
- Thr Threonine
- Ser Serine
- Lys Lysine
- Gly Glycine
- Asn Asparagine
- Gln Glutamine
- Trp Tryptophan
- Tyr Tyrosine
- Ala Alanine
- Cys Cysteine
- Phe Phenylalanine
- Leu Leucine
- Met Methionine
- Pro Proline
- Val Valine

Figure 1 The DNA coding strand base triplets and the corresponding amino acids. Three base triplets do not code for an amino acid and act as stop codes.

Point mutations

The genetic code consists of nucleotide base triplets within the DNA. During **transcription** of a gene, this code is copied to a length of **mRNA** as codons, complementary to the base triplets on the template strand of the length of DNA (see Book 1, topic 2.3.3). The sequence of codons on the mRNA is therefore a copy of the sequence of base triplets on the gene (coding strand of the DNA).

There are three types of point mutation:

* silent

* missense

* nonsense.

Silent mutations

Figure 1 shows that all amino acids involved in protein synthesis, apart from methionine, have more than one base triplet code. This reduces the effect of point mutations, as they do not always cause a change to the sequence of amino acids in a protein. This is often called the 'redundancy' or 'degeneracy' of the genetic code.

A point mutation involving a change to the base triplet, where that triplet still codes for the same amino acid, is a **silent mutation**. The primary structure of the protein, and therefore the secondary and tertiary structure, is not altered (see Figure 2).

normal	ATG CAG CAG CAG TTT TTA CGC AAT CCC	DNA
	Met | Gln | Gln | Gln | Phe | Leu | Arg | Asn | Pro	polypeptide
silent mutation	ATG CAG CAG CAG TTT **TTG** CGC AAT CCC	DNA
	Met | Gln | Gln | Gln | Phe | **Leu** | Arg | Asn | Pro	polypeptide

Figure 2 A silent mutation does not alter the amino acid sequence of the polypeptide.

Missense mutations

A change to the base triplet sequence that leads to a change in the amino acid sequence in a protein is a **missense mutation** (see Figure 3).

Within a gene, such a point mutation may have a significant effect on the protein produced. The alteration to the primary structure leads to a change to the tertiary structure of the protein, altering its shape and preventing it from carrying out its usual function.

normal	ATG CAG CAG CAG TTT TTA CGC AAT CCC	DNA
	Met | Gln | Gln | Gln | Phe | Leu | Arg | Asn | Pro	polypeptide
missense mutation	ATG CAG CAG CAG TTT **TCA** CGC AAT CCC	DNA
	Met | Gln | Gln | Gln | Phe | **Ser** | Arg | Asn | Pro	polypeptide

Figure 3 How a missense mutation may alter a polypeptide.

Sickle cell anaemia results from a missense mutation on the sixth base triplet of the gene for the β-polypeptide chains of haemoglobin: the amino acid valine, instead of glutamic acid, is inserted at this position. This results in deoxygenated haemoglobin crystallising within erythrocytes, causing them to become sickle shaped, blocking capillaries and depriving tissues of oxygen.

Nonsense mutations

A point mutation may alter a base triplet, so that it becomes a termination (stop) triplet. This particularly disruptive point mutation results in a truncated protein that will not function (see Figure 4). This abnormal protein will most likely be degraded within the cell. The genetic disease Duchenne muscular dystrophy is the result of a **nonsense mutation**.

normal	ATG CAG CAG CAG TTT TTA CGC AAT CCC	DNA
	Met | Gln | Gln | Gln | Phe | Leu | Arg | Asn | Pro	polypeptide
nonsense mutation	ATG CAG CAG CAG TTT **TAA** CGC AAT CCC	DNA
	Met | Gln | Gln | Gln | Phe | **Stop**	polypeptide

Figure 4 A nonsense mutation severely disrupts the structure of the protein.

Indel mutations

Both *in*sertions and *del*etions cause a frameshift.

Insertions and deletions

If nucleotide base pairs, *not in multiples of three*, are inserted in the gene or deleted from the gene, because the code is non-overlapping and read in groups of three bases, all the subsequent base triplets are altered. This is a frameshift. When the mRNA from such a mutated gene is translated, the amino acid sequence after the frameshift is severely disrupted. The primary sequence of the protein, and subsequently the tertiary structure, is much altered. Consequently, the protein cannot carry out its normal function. If the protein is very abnormal, it will be rapidly degraded within the cell.

normal	ATG CAG CAG CAG TTT TTA CGC AAT CCC	DNA
	Met | Gln | Gln | Gln | Phe | Leu | Arg | Asn | Pro	polypeptide
frameshift mutation	ATG CAG CAG CAG TTT **TAC GCA ATC** CC-	DNA
	Met | Gln | Gln | Gln | Phe | **Tyr** | **Ala** | **Ile**	polypeptide

Figure 5 How a frameshift may affect the structure of a polypeptide.

Some forms of thalassaemia, a haemoglobin disorder, result from frameshifts due to deletions of nucleotide bases.

Insertions or deletions of a triplet of base pairs result in the addition or loss of an amino acid, and not in a frameshift.

Expanding triple nucleotide repeats

Some genes contain a repeating triplet such as -CAG CAG CAG-. In an expanding triple nucleotide repeat, the number of CAG triplets increases at meiosis and again from generation to generation.

Huntington disease results from an expanding triple nucleotide repeat. If the number of repeating CAG sequences goes above a certain critical number, then the person with that genotype will develop the symptoms of Huntington disease later in life. There is more about the inheritance pattern of this disease in topic 6.2.2.

Not all mutations are harmful

Many mutations are beneficial and have helped to drive evolution through natural selection. Different alleles of a particular gene are produced via mutation.

The mutation that gave rise to blue eyes arose in the human population 6000–8000 years ago. Such a mutation may be harmful in areas where the sunlight intensity is high, as the lack of iris pigmentation could lead to lens cataracts. However, in more temperate zones, it could enable people to see better in less bright light.

Early humans in Africa would have had black skin, the high concentrations of melanin protecting them from sunburn and skin cancer. When humans migrated to temperate regions, a paler skin would be an advantage, enabling vitamin D to be made with a lower intensity of sunlight. In such areas, people with fairer skin would have an advantage and be selected, as vitamin D not only protects us from rickets, it protects us from heart disease and cancer.

Some mutations appear to be neutral, being neither beneficial nor harmful, such as those that, in humans, cause:

- inability to smell certain flowers, including freesias and honeysuckle
- differently shaped ear lobes.

> **LEARNING TIP**
> Not all genes code for polypeptides. Some code for RNA that regulates the expression of other genes. Some apparently silent mutations in such genes could be harmful by incorrectly regulating the expression of another gene.

Questions

1. Explain how the degenerate nature of the genetic code reduces the effects of point mutations.

2. The following is a sequence of DNA bases on a piece of coding strand DNA:
   ATG TTT CCT GTT AAA TAC CAT CAG CAG CGC TAG CAC

   Below are six possible mutations to this piece of DNA:
   (i) ATG TTT CCT GTT AAA TAA CAT CAG CAG CGC TAG CAC
   (ii) ATG TTT CCT ATT AAA TAC CAT CAG CAG CGC TAG CAC
   (iii) ATT TTT CCT GTT AAA TAC CAT CAG CAG CGC TAG CAC
   (iv) ATG TTC CTG TTA AAT ACC ATC AGC AGC GCT AGC AC-
   (v) ATG TTT CCT GTT AAA TAC CAT CAG CAG CAG CGC TAG CAC
   (vi) ATG TTT CCT GTT AAA TAC CAT CAG CAG CGC TGG CAC

   In each case identify the type of mutation and explain the effect it will have on the translated protein. You will need to use information in Figure 1 and recall that most proteins begin with the amino acid methionine.

3. For the normal sequence of DNA base triplets shown in question 2, write down:
   (a) the sequence of bases on the corresponding DNA template strand
   (b) the sequence of RNA bases on the piece of mRNA transcribed from this gene.

4. Sometimes during translation, a tRNA molecule may combine with the wrong amino acid. How may this affect the protein being assembled at a ribosome?

② Regulation of gene expression

By the end of this topic, you should be able to demonstrate and apply your knowledge and understanding of:

* the regulatory mechanisms that control gene expression at the transcriptional level, post-transcriptional level and post-translational level

---

KEY DEFINITIONS

exon: the coding, or expressed, region of DNA.
intron: the non-coding region of DNA.
operon: a group of genes that function as a single transcription unit; first identified in prokaryote cells.
transcription factor: protein or short non-coding RNA that can combine with a specific site on a length of DNA and inhibit or activate transcription of the gene.

Regulation of gene expression at the transcriptional level

In prokaryotic cells

Enzymes that catalyse the metabolic reactions involved in basic cellular functions are synthesised at a fairly constant rate. Enzymes that may only be needed under specific conditions are synthesised at varying rates, according to the needs of the cell.

The *lac* operon

The bacterium *E. coli* normally metabolises glucose as respiratory substrate. However, if glucose is absent and the disaccharide lactose is present, lactose *induces* the production of two enzymes:

* lactose permease, which allows lactose to enter the bacterial cell

* β-galactosidase, which hydrolyses lactose to glucose and galactose.

LEARNING TIP

Although the enzymes to metabolise lactose are only induced if there is no glucose available, it is the lactose and not the absence of glucose that induces the enzymes.

The *lac* **operon** consists of a length of DNA, about 6000 base pairs long, containing an operator region *lacO* next to the structural genes *lacZ* and *lacY* that code for the enzymes β-galactosidase and lactose permease, respectively.

Next to the operator region, *lacO*, is the promoter region, P, to which the enzyme RNA polymerase binds to begin transcription of the structural genes *lacZ* and *lacY*.

The operator region and promoter region are the control sites.

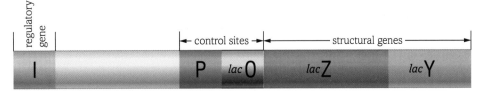

Figure 1 The *E. coli lac* operon and its regulatory gene.

A small distance away from the operon is the regulatory gene, I, that codes for a repressor protein (LacI). When this regulatory gene is expressed, the repressor protein produced binds to the operator, preventing RNA polymerase from binding to the promoter region. The repressor protein therefore prevents the genes *lacZ* and *lacY* from being transcribed (expressed), so the enzymes for lactose metabolism are not made. The genes are 'off'.

When lactose is added to the culture medium, once all the glucose has been used, molecules of lactose bind to the LacI repressor protein molecules; this alters the shape of the LacI repressor protein, preventing it from binding to the operator. The RNA polymerase enzyme can then bind to the promoter region and begin transcribing the structural genes into mRNA that will then be translated into the two enzymes. Thus lactose induces the enzymes needed to break it down.

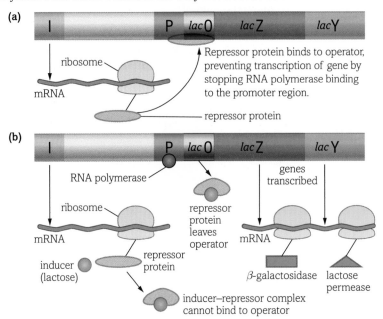

Figure 2 How the *lac* operon works (a) in the absence of lactose, and (b) in the presence of lactose.

INVESTIGATION

Observing the effect of the *lac* operon

If *E. coli* are grown in media without lactose and then lactose is added together with ONPG – an unnatural substrate for *β*-galactosidase – after a while, the lactose induces the enzyme, which converts ONPG to a yellow product that can be easily seen. If this is repeated using *E. coli* bacteria that have been grown in lactose, the yellow product of ONPG conversion appears much faster. If it is carried out using *E. coli* grown in the absence of lactose and not given any lactose, ONPG is not converted and no yellow colour is seen.

In eukaryotic cells

Every cell in a eukaryotic organism has the same genome but, because different cells use it differently, they function differently. This is the basis of cell differentiation (topic 2.6.4). In neurones, the genes being expressed differ to some extent from those being expressed in a liver or kidney cell, although all cells express the basic 'housekeeping' genes.

Transcription factors are proteins, or short non-coding pieces of RNA, that act within the cell's nucleus to control which genes in a cell are turned on or off. Transcription factors slide along a part of the DNA molecule, seeking and binding to their specific promoter regions. They may then aid or inhibit the attachment of RNA polymerase to the DNA, and activate or suppress transcription of the gene. They are essential for the regulation of gene expression in eukaryotes, making sure that different genes in different types of cells are activated or suppressed. Some transcription factors are involved in regulating the cell cycle (topic 2.6.1). Tumour suppressor genes and proto-oncogenes help regulate cell division via transcription factors. Mutations to these genes can lead to uncontrolled cell division or cancer.

About 8% of genes in the human genome encode transcription factors. Many genes have their promoter regions some distance away, along the unwound length of DNA but, because of how the DNA can bend, the promoter region may not be too far away spatially.

Post-transcriptional gene regulation

Introns and exons

Within a gene there are non-coding regions of DNA called **introns**, which are not expressed. They separate the coding or expressed regions, which are called **exons**.

All the DNA of a gene, both introns and exons, is transcribed. The resulting mRNA is called primary mRNA. Primary mRNA is then edited and the RNA introns – lengths corresponding to the DNA introns – are removed. The remaining mRNA exons, corresponding to the DNA exons, are joined together. Endonuclease enzyme may be involved in the editing and splicing processes.

Some introns may themselves encode proteins, and some may become short non-coding lengths of RNA involved in gene regulation. Some genes can be spliced in different ways. A length of DNA with its introns and exons can, according to how it is spliced, encode more than one protein.

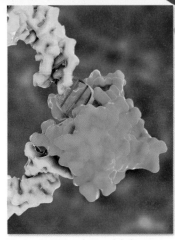

Figure 3 Molecular model showing a transcription factor (green) bound to part of the promoter region of a length of DNA.

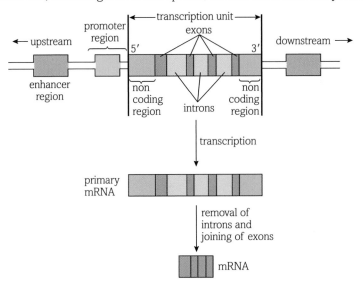

Figure 4 Removal of introns and joining of exons during splicing of primary mRNA to produce mRNA that will be translated into a protein.

Post-translational level of gene regulation

Post-translational regulation of gene expression involves the activation of proteins.

Many enzymes are activated by being **phosphorylated**.

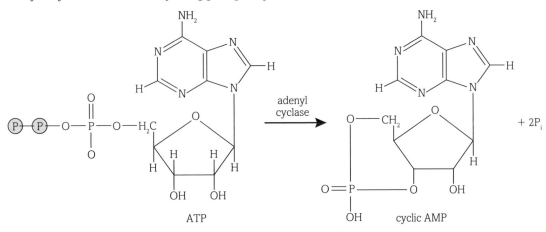

Figure 5 Formation of cAMP from ATP, catalysed by the enzyme adenyl cyclase.

Cyclic AMP, cAMP, is an important second messenger involved in this activation. The annotated diagram shown in Figure 6 explains how this happens.

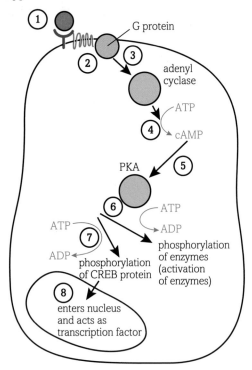

1. A signalling molecule, such as the protein hormone glucagon, binds to a receptor on the plasma membrane of the target cell.

2. This activates a transmembrane protein which then activates a G protein.

3. The activated G protein activates adenyl cyclase enzymes.

4. Activated adenyl cyclase enzymes catalyse the formation of many molecules of cAMP from ATP.

5. cAMP activates PKA (protein kinase A).

6. Activated PKA catalyses the phosphorylation of various proteins, hydrolysing ATP in the process. This phosphorylation activates many enzymes in the cytoplasm, for example those that convert glycogen to glucose.

7. PKA may phosphorylate another protein (CREB, cAMP response element binding).

8. This then enters the nucleus and acts as a transcription factor, to regulate transcription.

Figure 6 How cAMP activates enzymes and may also stimulate transcription.

Questions

1. Match the components of the *lac* operon system and regulator gene with the correct functions.

Component	Function
A structural gene	1 produces repressor protein
B regulator gene	2 binds to repressor
C promoter	3 codes for enzyme to metabolise lactose
D operator	4 binds to RNA polymerase

2. Explain the advantage to *E. coli* of having a *lac* operon system to induce the formation of enzymes to metabolise lactose

3. In the *lac* operon system, which molecule is the inducer?

4. Explain the functions of transcription factors in eukaryotic cells.

5. Describe the role of cAMP in activating enzymes.

6. Explain why a piece of mRNA is shorter than the gene from which it was transcribed.

7. What is the evidence from the investigation into the *lac* operon in *E. coli*, using ONPG, that although ONPG can act as a substrate for β-galactosidase, it does not induce the enzyme?

(3) Genetic control of body plan development

By the end of this topic, you should be able to demonstrate and apply your knowledge and understanding of:

* the genetic control of the development of body plans in different organisms

* the importance of mitosis and apoptosis as mechanisms controlling the development of body form

Homeobox gene sequences

The large and ancient family of genes, the homeotic genes, are involved in controlling the anatomical development, or morphogenesis, of an organism, so that all structures develop in the correct location, according to the body plan. Several of these genes contain **homeobox sequences**, and they are sometimes called homeobox genes.

Each homeobox sequence is a stretch of 180 DNA base pairs (excluding introns) encoding a 60-amino acid sequence, called a **homeodomain sequence**, within a protein. The homeodomain sequence can fold into a particular shape and bind to DNA, regulating the transcription of adjacent genes. These proteins are transcription factors and act within the cell nucleus. The shape that these homeodomain-containing proteins fold into is called H-T-H. It consists of two α helices (H) connected by one turn (T). Part of the homeodomain amino acid sequence recognises the TAAT sequence of the enhancer region (a region that initiates or enhances transcription) of a gene to be transcribed.

Homeobox gene sequences are very similar and are highly conserved

In 1984, scientists working in a team headed by Walter J. McGinnis demonstrated that the homeobox sequence, first identified in 1983 within the homeotic genes of the fruit fly, *Drosophila melanogaster*, also exists in the mouse. Moreover, the base sequences in these homeobox sequences were very similar in both species. This informed scientists that these gene sequences are crucial for the regulation of development and differentiation in organisms.

Also in 1984, Edward de Robertis discovered a subset of homeobox genes, ***Hox* genes**, in the African clawed frog, *Xenopus*. This discovery has led to a new branch of biology called

evolutionary development or 'evo-devo'. *Hox* genes are found only in animals. Homeobox genes are found in animals, plants and fungi.

Molecular evidence indicates that homeobox genes are present in Cnidaria, which means these genes arose before the Palaeozoic era, which began 541 million years ago, and before bilaterally symmetrical organisms evolved. This indicates that these genes first arose in an early ancestor, which gave rise to each of these types of organism, and have been **conserved**. The sequences are also similar in all organisms studied to date and that similarity extends across wide evolutionary distances.

How *Hox* genes control body plan development in animals

The *Hox* genes regulate the development of embryos along the anterior–posterior (head–tail) axis. They control which body parts grow where. If *Hox* genes are mutated, abnormalities can occur such as the antennae on the head of *Drosophila* developing as legs (see Figure 1), or mammalian eyes developing on limbs.

Figure 1 SEM of *Drosophila melanogaster* with antennapedia mutation (×115).

Hox genes are arranged in clusters and each cluster may contain up to 10 genes. In tetrapods (four-limbed vertebrates) including mammals and therefore humans, there are four clusters. At some stage during evolution, the *Hox* clusters have been duplicated (see Figure 2).

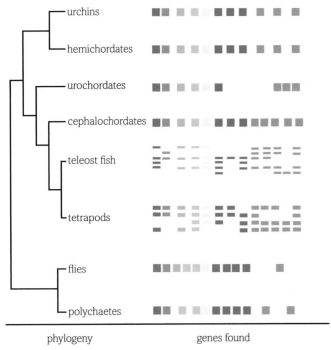

phylogeny   genes found

Figure 2 *Hox* gene clusters in several phyla.

In early embryonic development, *Hox* genes are active and are expressed in order along the anterior–posterior axis of the developing embryo. The sequential and temporal (in time) order of the gene expressions corresponds to the sequential and temporal development of the various body parts, a phenomenon known as colinearity (see Figure 3).

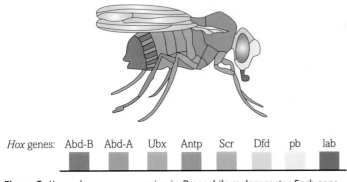

Hox genes: Abd-B   Abd-A   Ubx   Antp   Scr   Dfd   pb   lab

Figure 3 Homeobox gene expression in *Drosophila melanogaster*. Each gene in the cluster corresponds to a part of the insect's body, as shown by the corresponding colours.

Hox genes encode homeodomain proteins that act in the nucleus as transcription factors and can switch on cascades of activation of other genes that promote mitotic cell division, apoptosis, cell migration and also help to regulate the cell cycle (Book 1, topic 2.6.1).

Hox genes are similar across different classes of animals; a fly can function properly with a chicken *Hox* gene inserted in place of its own.

LEARNING TIP

All *Hox* genes are homeobox genes, but not all homeobox genes are *Hox* genes. Rather in the way that all mobile phones are examples of technology, but not every piece of technological equipment is a mobile phone.

Hox genes are a subset of homeobox genes and are found only in animals. Homeobox genes are found in plants, fungi and animals.

DID YOU KNOW?

Another family of homeobox-containing genes, called $NKX_{2.5}$, have evolved within the animal kingdom. $NKX_{2.5}$ genes are involved in controlling heart development. They are also known as 'tinman', after the character in The Wizard of Oz.

DID YOU KNOW?

In humans, retinoic acid helps regulate the activity of *Hox* genes. Vitamin A (retinol), found in liver, is a potent source of retinoic acid and too much eaten during early pregnancy can cause fetal abnormalities. In large quantities it is toxic to adults.

How are the regulators regulated?

Hox genes are regulated by other genes called gap genes and pair-rule genes. In turn, these genes are regulated by maternally supplied mRNA from the egg cytoplasm.

Mitosis

From zygote to embryo to fully formed adult, there are many mitotic cell divisions. Mitosis (Book 1, topic 2.6.2) is part of the cell cycle that is regulated with the help of homeobox and *Hox* genes. It ensures that each new daughter cell contains the full genome and is a clone of the parent cell.

During cell differentiation some of the genes in a particular type of cell are 'switched off' and not expressed.

In 1962, Leonard Hayflick showed that normal body cells divide a limited number of times (around 50 times – known as the Hayflick constant) before dying.

Apoptosis in the development of body form

In 1965, John Foxton Ross Kerr re-examined and researched the idea of programmed cell death, first described in 1842 by Carl Vogt. In 1972, the term '**apoptosis**' was used for programmed cell death. Apoptosis is different from cell death due to trauma, which involves hydrolytic enzymes.

The sequence of events during apoptosis

Figure 4 shows the sequence of events during apoptosis:

1. Enzymes break down the cell cytoskeleton

2. The cytoplasm becomes dense with tightly packed organelles

3. The cell surface membrane changes and small protrusions called blebs form

4. Chromatin condenses, the nuclear envelope breaks and DNA breaks into fragments

5. The cell breaks into vesicles that are ingested by phagocytic cells, so that cell debris does not damage any other cells or tissues. The whole process happens quickly.

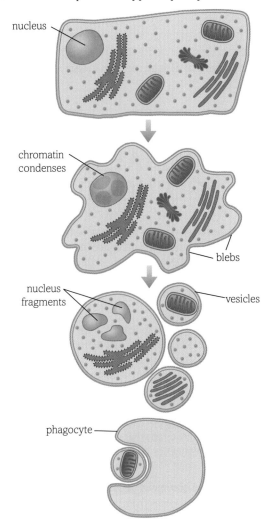

Figure 4 The sequence of events during apoptosis.

Control of apoptosis

Many cell signals help to control apoptosis. Some of these signalling molecules may be released by cells when genes that are involved in regulating the cell cycle and apoptosis respond to internal cell stimuli and external stimuli such as stress. These signalling molecules include cytokines from cells of the immune system (Book 1, topic 4.1.6), hormones, growth factors and nitric oxide. Nitric oxide can induce apoptosis by making the inner mitochondrial membrane more permeable to hydrogen ions and dissipating the proton gradient. Proteins are released into the cytoplasm where they bind to apoptosis inhibitor proteins, allowing apoptosis to occur.

Apoptosis and development

Apoptosis is an integral part of plant and animal tissue development. Extensive proliferation of cell types is prevented by pruning through apoptosis, without release of any hydrolytic enzymes that could damage surrounding tissues.

During limb development, apoptosis causes the digits to separate from each other (see Figure 5).

Apoptosis removes ineffective or harmful T-lymphocytes during the development of the immune system.

In children aged between 8 and 14 years, 20–30 billion cells per day apoptose; in adults, about 50–70 million cells per day apoptose.

The rate of cells dying should equal the rate of cells produced by mitosis:

* not enough apoptosis leads to the formation of tumours

* too much apoptosis leads to cell loss and degeneration.

Cell signalling plays a crucial role in maintaining the right balance.

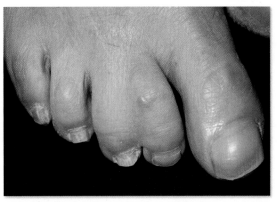

Figure 5 Incomplete separation of two toes (syndactyly) due to lack of apoptosis.

> **DID YOU KNOW?**
>
> One organ that hardly ever develops tumours is the mammalian heart. There is no cell division occurring in the heart after birth; it increases in size by increase in cell size. Sometimes a tumour forms in the fetal heart, but its growth is arrested after birth.

Questions

1. Distinguish clearly between the terms 'homeobox sequence' and 'homeodomain sequence'.

2. Describe the function of proteins that contain a homeodomain sequence.

3. 'Homeobox gene sequences are highly conserved.' Explain the significance of this phrase.

4. Explain the difference between homeobox genes and *Hox* genes.

5. What is meant by 'colinearity'?

6. Explain why the process of apoptosis does not damage nearby cells.

7. Why should the rate of apoptosis in an adult equal the rate of mitosis?

8. Suggest why pregnant mothers are advised not to eat liver.

THINKING BIGGER

EPIGENETICS

Our view of inheritance is changing. In life and development, the 'nature/nurture debate' is largely defunct, because genes and environmental effects are intertwined – therefore it is nature via nurture. Gene expression can be changed by environmental factors. Hence modern biology is changing our understanding of genetics, disease and inheritance.

EPIGENETICS – HOW GENE FUNCTION MAY BE CHANGED WITHOUT GENES UNDERGOING MUTATION

In a recent scientific study (Dias B.G., *Nature Neuroscience*, 2013), male mice were conditioned, using mild electric shocks, to fear a harmless odour. Offspring subsequently fathered by these males inherited a fear of this odour without being exposed to the electric shocks. This appears to contradict our understanding of how genes work and to indicate the inheritance of acquired characteristics.

During the Dutch Hunger Winter, many people in the Western Netherlands starved. Babies conceived when the mother had enough food, but who were deprived of nutrients during the last third of gestation were born small and stayed small throughout their lives, despite later access to unlimited food. Children born to mothers malnourished during the early part but not the latter part of pregnancy, had normal birth weights but higher than normal rates of obesity in later life; these children also showed a greater incidence of other health problems. Events that occur during the first three months of fetal development stay with people throughout their lives and can even affect the health of their offspring.

Scientists also found that abused children, even after being cared for by loving adoptive parents, may later develop depression and be at risk of self harm. Sometimes identical twins show differences in characteristics largely determined by genes, such as schizophrenia.

These are all examples of epigenetics. Events within a living organism can switch genes off, sometimes permanently; the switched-off genes are then inherited. DNA is not altered by mutation – a change to the nucleotide base sequence – but methyl (CH_3) groups may be added (methylation) to certain groups within a DNA molecule. A methyl group added to cytosine in a promoter region binds to a protein that then attracts other proteins that help switch off the gene, or causes the DNA to coil and deny access to binding sites by transcription factors. When DNA with a methylated C base is replicated, the new strands also have methylated cytosine at the same location.

Acetylation (addition of acetyl groups) of histone proteins around which DNA in chromosomes is wound can also switch genes off. Until recently, histone proteins were thought to be just for 'packaging' DNA and not to have any function related to gene expression.

The genome was once regarded as an inflexible code; scientists now know that the epigenome operating above (epi) the level of the genome can change how DNA is expressed throughout a person's life. DNA is like a script, and the epigenetic code can be read in different ways. Many switches for genes act more like volume dials than like on/off switches.

Mendel and Darwin defined the nineteenth century as the era of evolution and genetics. Watson and Crick defined the twentieth century as the era of DNA and understanding how genetic and evolution interact. In the 1950s, John Gurdon disproved the theory that when cells differentiate some genes are lost: he transferred a nucleus from a frog skin cell to an egg cell where the nucleus had been destroyed. The egg then developed into a frog, showing that the skin cell had all the genes but that some were switched off. In the twenty-first century, epigenetics will shed even more light on how genes work.

Epigenetics can have an enormous impact on human health. It is implicated in a range of diseases from schizophrenia to rheumatoid arthritis, cancer and chronic pain. A few types of drug can already treat certain cancers by interfering with epigenetic processes; this may be by smothering the methyl groups of epigenetically modified genes with another chemical.

Epigenetics has also given us the means of reprogramming differentiated cells to switch back on certain 'off' genes, making induced pluripotent stem cells.

When John Gurdon carried out his experiments with frogs, he did not know about epigenetic mechanisms such as methylation and acetylation. However, Darwin knew nothing about genes and Mendel knew nothing about DNA, but all three examined their data and saw things that no-one else saw. This insight led to people being able to view the world in new ways; this is called a paradigm shift.

Reference
● Nessa Carey. *The epigenetics revolution*, (2012) Icon Books Ltd.

Where else will I encounter these themes?

Book 1   5.1   5.2   5.3   5.4   5.5   5.6

Let's start by considering the nature of the writing in the article.

> 1. Explain, with reasons, the type of audience you think this book is aimed at.

Now let us look at the biological concepts linked to the information in this article.

> 2. Suggest why events happening to a fetus during the first three months of gestation in humans have a lasting effect on that individual.
>
> 3. In which type of cells/organisms is DNA associated with histone proteins, forming chromatin?
>
> 4. Explain how progress in the field of epigenetics has altered our perceptions about the function of histone proteins.
>
> 5. Before John Gurdon's research using frogs in the 1950s and 1960s, the accepted theory was that differentiated cells had lost certain genes or that certain genes were permanently turned off. A prediction or hypothesis to be tested, arising from this theory, would be that adult frogs could not develop from an egg cell with a nucleus taken from the cell of an adult frog. However, this is impossible to prove because if the experiment failed to produce such frogs it could be because the technique or equipment were not good enough. Instead Gurdon disproved the hypothesis. Explain how Gurdon's results disproved the hypothesis, and discuss how this is good scientific methodology.
>
> 6. Explain, with reference to examples, what is meant by 'a paradigm shift'.

Activity

Write a short article for a magazine aimed at the general public, explaining what epigenetics is.

DID YOU KNOW?

Audrey Hepburn, the iconic and very thin actress best known for her role in *Breakfast at Tiffany's*, suffered poor physical health, including being underweight, because she was severely undernourished as a child. She was a survivor of the Dutch Hunger Winter.

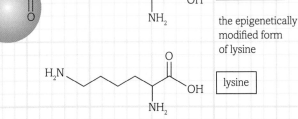

(a) cytosine → 5-methyl cytosine

(b)

(c) acetyl-lysine — the epigenetically modified form of lysine; lysine

Figure 1 (a) The methylation of cytosine. (b) The histone octamer (eight molecules of histone proteins) around which DNA is wrapped to form a nucleosome. Nucleosomes form chromatin. Acetylation of lysine amino acid molecules in the histone protein tails is an epigenetic modification. (c) Some histone modifications push gene expression up and some push it down.

Practice questions

1. Read the following statements.
   (i) Mutations that occur during DNA replication just before mitosis are not passed to offspring.
   (ii) Point mutations may cause a frameshift.
   (iii) Deletion of a triplet of nucleotide base pairs from a length of DNA causes a frameshift.

   Which statement(s) is/are true? [1]

   A. (i) only

   B. (i) and (ii) only

   C. (i) and (iii) only

   D. (iii) only

2. Which of the following statements is correct? [1]

   A. A homeobox sequence is a sequence of 60 base triplets within a gene.

   B. *Hox* genes are types of homeobox genes found only in animals.

   C. During apoptosis the cell dies and is destroyed by the action of hydrolytic enzymes.

   D. A lack of apoptosis leads to cell loss and degeneration.

3. Which are the control sites in the *lac* operon? [1]

   A. Genes Z and Y

   B. Regulator gene and repressor protein

   C. Operator region and promoter region

   D. RNA polymerase and beta galactosidase

4. Read the following statements.
   (i) Transcription factors may be proteins or short lengths of non-coding RNA (ncRNA).
   (ii) Transcription factors regulate gene expression in eukaryotic cells.
   (iii) Transcription factors act within the cell nucleus and can activate or inhibit transcription.

   Which statement(s) is/are true? [1]

   A. (ii) only

   B. (ii) and (iii) only

   C. (i), (ii) and (iii)

   D. (i) and (iii) only

5. Which of the following is an example of post-translational gene expression regulation? [1]

   A. Removal of introns from primary DNA

   B. Attachment of a transcription factor to a promoter region and preventing the binding of RNA polymerase

   C. Phosphorylation of proteins

   D. Binding of lactose to the repressor protein

   [Total: 5]

6. Fragile X syndrome is the most commonly inherited form of intellectual disability in humans.

   The gene involved, *FMR1*, is present near one end of the X chromosome and encodes an RNA-binding protein, FMRP, which carries mRNA molecules from the nucleus to the sites of protein assembly. FMRP also helps to regulate when the instructions in mRNA are used to assemble proteins involved in the production of synapses, particularly synapses in the brain.

   Within the base triplet sequence in the gene is a repeating base triplet, CGG. CGG codes for the amino acid arginine. Normally this triplet repeats between 5 and 40 times. If the repeat expands to between 55 and 200, this is called a premutation and that individual is a carrier. Carriers may have mild symptoms such as autism-like behaviour, anxiety and learning difficulties. In carriers, there may be overproduction of abnormal *FMR1* mRNA that attaches to some proteins and stops them performing their functions.

   During DNA replication, the repeating triplet may expand. The number of increases is always more in a female carrier. When the repeat has expanded to more than 230 CGG triplets (in some cases it expands to 1000), this is a full mutation.

   (a) Describe the effect of an expanding triple nucleotide repeat on the protein produced, when the gene *FMR1* is expressed. [3]

   (b) State precisely the destination in cells to which the protein FMRP carries mRNA molecules. [1]

   (c) Explain how attachment to a length of mRNA may inhibit a protein from performing its normal function. [3]

   (d) Explain why an increase in the number of CGG base triplets does not cause a frameshift. [2]

   (e) Name another genetic disorder caused by an expanding triple nucleotide repeat mutation. [1]

   [Total: 10]

7. 97% of the haemoglobin in human erythrocytes is haemoglobin A. Each molecule of haemoglobin A consists of two alpha and two beta polypeptide chains.

The sequence of codons on part of the mRNA transcribed from the gene coding for the beta haemoglobin polypeptide chain is shown below.

Position in chain	Normal mRNA codons	Amino acid
0	AUG	met (start)
1	GUG	valine
2	CAC	histidine
3	CUG	leucine
4	ACU	threonine
5	CCU	proline
6	GAG	glutamic acid
7	GAG	glutamic acid

- The DNA coding strand base triplet codes for glutamic acid are GAA and GAG.

- The DNA coding strand base triplet codes for valine are GTT GTC GTA GTG.

Sickle cell anaemia is caused by a mutation in the gene coding for the beta polypeptide chain of haemoglobin A.

If a person inherits two mutated alleles for this gene, they will suffer from sickle cell anaemia, where the haemoglobin becomes 'stringy' and erythrocytes may become abnormally shaped and block capillaries.

Mutated mRNA codons							
AUG	GUG	CAC	CUG	ACU	CCU	GUG	GAG

All polypeptide chains begin with methionine and often this amino acid then breaks away from the protein.

(a) Describe the normal shape of erythrocytes and explain how this shape helps them to carry out their function. [4]

(b) (i) Identify and describe the type of mutation that gives rise to sickle cell anaemia. [3]

   (ii) Describe the effect on the polypeptide chain of this mutation. [3]

(c) Haemoglobin protein has a quaternary structure. Explain what is meant by 'quaternary structure'. [1]

(d) Write out the sequence of base triplets on the length of DNA template strand from which this piece of normal mRNA was transcribed. [2]

[Total: 13]

8. O-nitrophenol galactoside, ONPG, is an unnatural substrate for the enzyme β-galactosidase. β-Galactosidase cleaves molecules of ONPG into galactose and O-nitrophenol, which is a yellow compound.

An investigation was set up using a normal strain of the bacterium *E. coli* and a mutant strain.

Tube	1	2	3	4
Mutant *E. coli* previously grown in medium containing glucose	✗	✗	✗	✓
E. coli previously grown in medium containing glucose	✓	✗	✓	✗
E. coli previously grown in medium containing lactose	✗	✓	✗	✗
Sugar present in nutrient medium of tube	lactose	lactose	none	lactose
ONPG	✓	✓	✓	✓
Observations	yellow colour after 20 minutes	yellow colour after 5 minutes	no yellow colour, even after 120 minutes	no yellow colour, even after 120 minutes

(a) Explain why the yellow colour appears faster in tube 2 than in tube 1. [4]

(b) Explain the absence of yellow colour:
   (i) in tube 3 [4]
   (ii) in tube 4. [4]

(c) What evidence is there from this investigation that although ONPG is a substrate for the enzyme β-galactosidase it does not induce the formation of the enzyme? [2]

(d) Suggest how ONPG can act as a substrate for the enzyme β-galactosidase, but not act as the inducer in the *lac* operon system of *E. coli*. [3]

(e) Suggest a possible location for the mutation present in the mutant strain of *E. coli* used in this investigation. [1]

[Total: 18]

PATTERNS OF INHERITANCE

Introduction

The evolutionary biologist Theodosius Dobzhansky, in a presidential address in 1964, said that nothing in biology makes any sense except in the light of evolution. In 1973, this was later included in one of his essays, published in *American Biology Teacher*.

Just over a century before that presidential address, Charles Darwin published his work *On the Origin of Species by Natural Selection*. Darwin had probably not read Mendel's work, or had missed its significance as he did not appreciate the role of maths in biology, and knew nothing about the role of genes and chromosomes in the inheritance of characteristics. However, he still postulated that offspring inherit characteristics from their parents and may inherit small and unpredictable alterations to those characteristics. He said that some of these alterations could promote survival and would spread within the population, whereas any changes having adverse effects would eventually disappear from the population. This is a simple but compelling theory. He added to this in 1871, in his book *The Descent of Man*, describing how mating preferences, or sexual selection, can also drive evolution. The central principle that all species on Earth today are interrelated and new species emerge from existing ones by random changes that are passed to offspring if they aid survival, has become the glue that binds biology together. It is the unifying theme of biology and the foundation of genetics.

In this chapter you will learn how genes, the environment and sexual reproduction contribute to phenotypic and genetic variation, and about the many patterns of inheritance and the genetic basis of continuous and discontinuous variation. You will be introduced to the ideas of population genetics and the factors that affect the evolution of a new species, and will consider the principles of artificial selection.

All the maths you need

To unlock the puzzles of this chapter you need the following maths:

- Using ratios, fractions and percentages
- Understanding simple probability
- Changing the subject of an equation
- Substituting numerical values into algebraic equations
- Solving algebraic equations

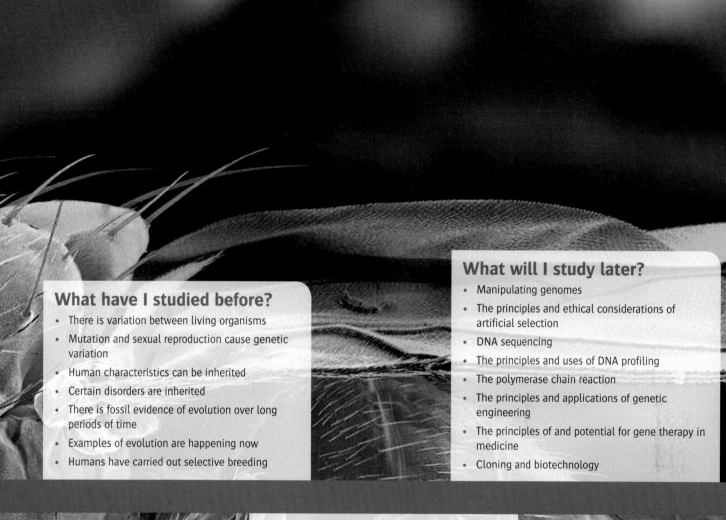

What have I studied before?

- There is variation between living organisms
- Mutation and sexual reproduction cause genetic variation
- Human characteristics can be inherited
- Certain disorders are inherited
- There is fossil evidence of evolution over long periods of time
- Examples of evolution are happening now
- Humans have carried out selective breeding

What will I study later?

- Manipulating genomes
- The principles and ethical considerations of artificial selection
- DNA sequencing
- The principles and uses of DNA profiling
- The polymerase chain reaction
- The principles and applications of genetic engineering
- The principles of and potential for gene therapy in medicine
- Cloning and biotechnology

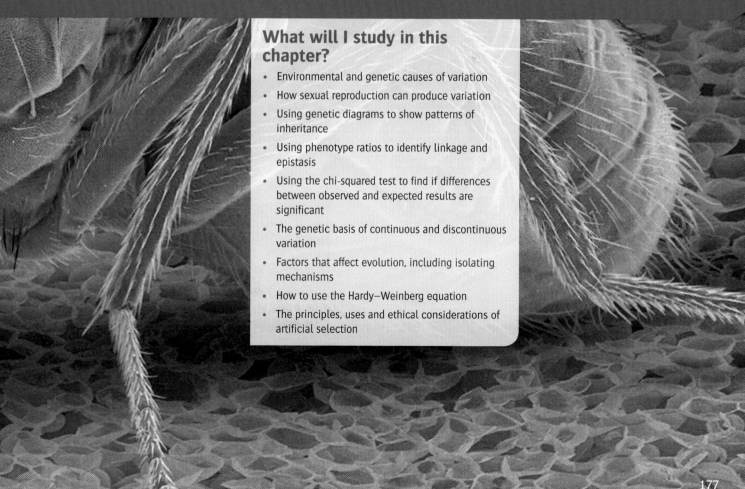

What will I study in this chapter?

- Environmental and genetic causes of variation
- How sexual reproduction can produce variation
- Using genetic diagrams to show patterns of inheritance
- Using phenotype ratios to identify linkage and epistasis
- Using the chi-squared test to find if differences between observed and expected results are significant
- The genetic basis of continuous and discontinuous variation
- Factors that affect evolution, including isolating mechanisms
- How to use the Hardy–Weinberg equation
- The principles, uses and ethical considerations of artificial selection

1 Genetic variation

By the end of this topic, you should be able to demonstrate and apply your knowledge and understanding of:

* the contribution of both environmental and genetic factors to phenotypic variation
* how sexual reproduction can lead to genetic variation within a species

---

KEY DEFINITIONS

genotype: genetic makeup of an organism.
phenotype: visible characteristic of an organism.

Causes of phenotypic variation

The appearance of a living organism – its **phenotype** – is influenced by both its **genotype** and its environment.

Genetic factors

Mutations have contributed to the process of evolution. A mutation is a change to the genetic material. This may involve changes to the structure of DNA, as described in topic 6.1.1, or changes to the number or gross structure of the chromosomes.

Sexual reproduction may also lead to genetic variation.

Gene mutations

Certain physical and chemical agents, described as mutagens, can increase the rate of mutation.

Type of mutagenic agent	Mutagen
Physical agents	• X-rays • gamma rays • UV light
Chemical agents	• benzopyrene (found in tobacco smoke) • mustard gas • nitrous acid • aromatic amines – in some synthetic dyes • reactive oxygen species – free radicals • colchicine
Biological agents	• some viruses • transposons – jumping genes, remnants of viral nucleic acid that have become incorporated into our genomes • food contaminants such as mycotoxins from fungi, e.g. aflatoxins in contaminated nuts, chemicals in charred meat, and alcohol

Table 1 Some examples of mutagens.

Mutations may be harmful, advantageous or neutral. Mutations that occur during gamete formation are also:

* persistent: they can be transmitted through many generations without change
* random: they are not directed by a need on the part of the organism in which they occur.

Chromosome mutations

Chromosome mutations may occur during meiosis. These include:

* deletion – part of a chromosome, containing genes and regulatory sequences, is lost
* inversion – a section of a chromosome may break off, turn through 180 degrees and then join again; although all the genes are still present, some may now be too far away from their regulatory nucleotide sequences to be properly expressed
* translocation – a piece of one chromosome breaks off and then becomes attached to another chromosome. This may also interfere with the regulation of the genes on the translocated chromosome
* duplication – a piece of a chromosome may be duplicated. Overexpression of genes can be harmful, because too many of certain proteins or gene-regulating nucleic acids may disrupt metabolism
* non-disjunction – one pair of chromosomes or chromatids fails to separate, leaving one gamete with an extra chromosome. When fertilised by a normal haploid gamete, the resulting zygote has one extra chromosome. Down syndrome, or trisomy 21, is caused by non-disjunction. Figure 1 shows another example of the consequences of non-disjunction.

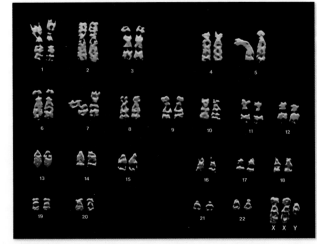

Figure 1 Karyotype showing the chromosomes taken from the cell nucleus of a male with Klinefelter syndrome, caused by an extra X chromosome.

- **aneuploidy** – the chromosome number is not an exact multiple of the haploid number for that organism. Sometimes chromosomes or chromatids fail to separate during meiosis (e.g. trisomy).
- **polyploidy** – if a diploid gamete is fertilised by a haploid gamete, the resulting zygote will be triploid (it has three sets of chromosomes). The fusion of two diploid gametes can make a tetraploid zygote. Many cultivated plants are polyploid (they have more than two sets of chromosomes).

Sexual reproduction

Genetic variation resulting from sexual reproduction has contributed to evolution.

Meiosis produces genetically different gametes. During meiosis (Book 1, topic 2.6.3), genetic variation may result from:

- allele shuffling (swapping of alleles between non-sister chromatids) during crossing over in prophase 1
- independent assortment of chromosomes during metaphase/anaphase 1
- independent assortment of chromatids during metaphase/anaphase 2.

Gametes produced by meiosis are individual and genetically dissimilar. They are also haploid, containing only one of each pair of homologous chromosomes and one allele for every gene.

The random fusion of gametes creates more genetic diversity. Any male gamete can potentially combine with any female gamete from an organism of the same species. The random fertilisation of gametes, that are already genetically unique, produces extensive genetic diversity among the resulting offspring.

> **DID YOU KNOW?**
> Within a species, the number of possible gametes with different chromosome combinations is 2^n, where n is the haploid chromosome number for that species. For humans, 2^{23} means more than 8 million different types of gamete. Fertilisation is random, therefore any one of these types from each parent could combine. This gives $(8 \times 10^6)^2$ or 64×10^{12} potential genetic combinations. This is far greater than the number of humans who have ever lived on Earth.

> **LEARNING TIP**
> Mutations may also occur during the replication of DNA before meiosis. However, remember that allele shuffling by crossing over, independent assortment and random fertilisation are *not* examples of mutations.

Environmental factors

Variation caused solely by the environment

Some phenotypic variation is caused by the environment and not passed on through genes. Examples include:

- speaking with a particular regional dialect. A person's offspring would not inherit the dialect through their genes, although they might learn to speak in this way by listening to other people
- losing a digit or limb, or having a scar following an injury.

Variation caused by the environment interacting with genes

If plants are kept in dim light after germination, or if the soil in which they are grown contains insufficient magnesium, then the leaves do not develop enough chlorophyll and are yellow or yellow-white. The plant is described as chlorotic, or suffering from chlorosis. The plant cannot photosynthesise. Chlorotic plants have the genotype for making chlorophyll, but environmental factors are preventing the expression of these genes.

Figure 2 Chlorosis in a japonica plant, *Fatsia japonica*. The yellow leaves are due to lack of chlorophyll, because of a mineral deficiency in the soil.

> **DID YOU KNOW?**
> Many genes in humans are involved in determining height, mass and intelligence. However, the full genetic potential of an individual cannot be reached if environmental factors such as nutrients and mental stimulation are missing or in short supply.

Questions

1. (a) What is a mutagen?
   (b) List two chemical, two physical and two biological types of mutagen.

2. When, during the cell cycle, is a mutation to the DNA most likely to occur?

3. Suggest when, during the cell cycle, a chromosome mutation is most likely to occur.

4. The diploid number of chromosomes in the crab-eating rat, *Ichthyomys pitteiri*, is 92. Calculate the number of potential genetic combinations that could be produced by sexual reproduction within this species. Express your answer in standard form.

5. Describe three types of chromosome mutation.

6. Explain how genetic variation and subsequent genetic diversity, leading to evolution, increased when organisms began to reproduce sexually.

Monogenic inheritance

By the end of this topic, you should be able to demonstrate and apply your knowledge and understanding of:

* genetic diagrams to show patterns of inheritance (continued in topics 6.2.3–6.2.7)

The foundation for genetics

In 1866, Gregor Mendel published the results of his investigations that would lay the foundation for the branch of biology known as genetics, which is now at the forefront of modern biology.

KEY DEFINITIONS

allele: a version of a gene.
heterozygous: not true-breeding; having different alleles at a particular gene locus on a pair of homologous chromosomes.
homozygous: true-breeding; having identical alleles at a particular gene locus on a pair of homologous chromosomes.
monogenic: determined by a single gene.

DID YOU KNOW?

When Mendel began his studies of inherited characteristics in pea plants, there was no knowledge of chromosomes or meiosis. Nevertheless, he showed that units of inheritance existed and he predicted their behaviour during the formation of gametes. His principles of inheritance form the basis for studying the transmission of characteristics through the generations.

Mendel studied an organism that was easy to grow and, although naturally self-fertilising, was easy to cross-fertilise artificially. He worked with seven characteristics of the pea plant, each characteristic having two distinctly contrasting traits: stem height, seed shape, seed colour, pod shape, pod colour, flower arrangement and flower colour.

Figure 1 Gregor Mendel.

Trait	Dominant trait	Recessive trait	F_1 results	F_2 results	F_2 ratio
Stem height	Tall	Dwarf	All tall	787 tall 277 dwarf	2.84:1
Flower colour	Purple	White	All purple	705 purple 224 white	3.15:1
Flower position	Axial	Terminal	All axial	651 axial 207 terminal	3.14:1
Pod shape	Full 	Constricted 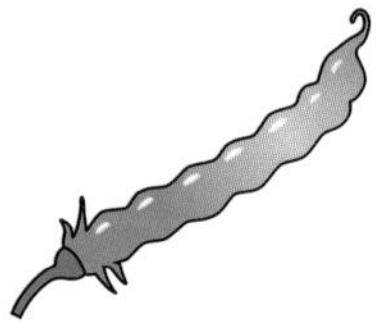	All full	882 full 299 constricted	2.95:1
Pod colour	Green 	Yellow 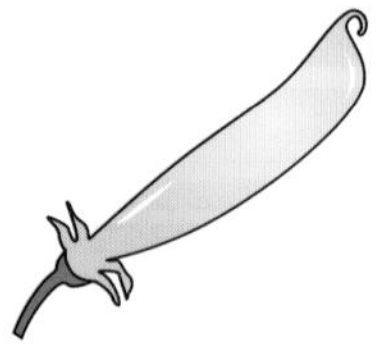	All green	428 green 152 yellow	2.82:1
Seed colour	Yellow 	Green 	All yellow	6022 yellow 2001 green	3.01:1
Seed shape	Round 	Wrinkled 	All round	5474 round 1850 wrinkled	2.96:1

Figure 2 The characteristics of pea plants, each having a contrasting pair of traits, studied by Gregor Mendel, together with his results.

Mendel obtained true-breeding strains, where the trait had appeared unchanged generation after generation, from local seed merchants. Mendel also kept accurate and quantitative records of data obtained, which he analysed.

The monohybrid cross

Mendel's simplest experiments involved only one characteristic with one pair of contrasting traits. He mated individuals from two parent strains, each of which showed a different phenotype. One parent was true-breeding for tall stems and the other was true-breeding for short stems. As in Figure 3, the parents were called the P_1 or parental generation. All of the offspring from this cross, the F_1 (first filial) generation, were phenotypically identical to one parent type. They were all tall.

When Mendel allowed members of the F_1 generation to self-fertilise, the resulting F_2 generation contained some short plants, but there were three times as many tall as short plants. Three-quarters were tall, and one quarter were short.

When Mendel crossed true-breeding plants showing the other six phenotypic variations, he obtained similar results.

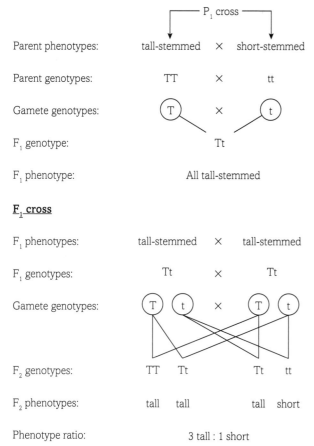

Phenotype ratio: 3 tall : 1 short

Figure 3 Genetic diagrams explaining the results of a monohybrid cross between tall and short pea plants.

In pea plants, the characteristic of height is **monogenic**; it is governed by one gene that has two distinct **alleles**, T/t.

One allele, t, when present in a **homozygous** individual giving the genotype tt, produces phenotypically short plants.

The other allele, T, when present in homozygous (TT) or **heterozygous** (Tt) individuals, produces phenotypically tall plants. The allele T is described as dominant (it codes for a dominant characteristic) and the allele t is recessive – coding for a recessive characteristic that will only be visible in the phenotype if there is no dominant allele.

Punnett squares

The genotypes and phenotypes resulting from the possible combinations of gametes during a monohybrid cross, showing the possible outcomes of monogenic inheritance, can be visualised in a Punnett square, named after its inventor, Reginald D. Punnett.

In a Punnett square (Figure 4), all possible gametes are assigned to a row, with those of the female parent in the vertical column and those of the male parent in the horizontal row. The genotypes of the next generation are predicted by combining the male and female gamete genotypes – a process that represents all possible random fertilisation events.

The phenotypes of the genotypes can also be predicted.

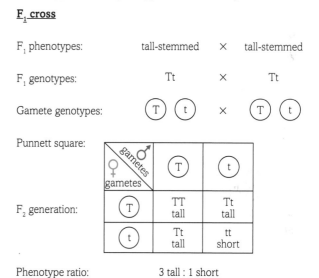

Phenotype ratio: 3 tall : 1 short

Figure 4 A Punnett square to explain the monohybrid cross between members of the F_1 generation, both heterozygous for the characteristic height. The phenotypic ratio is 3 tall : 1 short. The probability of obtaining tall-stemmed plants is 3/4 or 75%. The probability of obtaining homozygous tall-stemmed plants is 1/4 or 25%.

The test cross

How can we ascertain the genotypes of phenotypically similar individuals?

The genotypes of Mendel's pea plants are as follows:

- All short pea plants have the genotype tt, because shortness is a recessive characteristic, so individuals with that phenotype must have the genotype, tt. They are homozygous recessive.

- The tall pea plants in the F_1 generation all have the same genotype, Tt. They are all heterozygous. We can deduce this from the genotypes of their true-breeding parents (TT and tt) and the genotypes of the gametes (T and t) that must have combined to produce this generation.

- In the F_2 generation, some of the tall plants have the genotype TT and some have the genotype Tt. They both have the same phenotype, tall, and it is impossible to tell their genotype from their appearance.

Mendel devised a simple way to test the genotypes, called the test cross. This method is still used today.

The organism exhibiting the dominant phenotype (e.g. a tall-stemmed pea plant), but of unknown genotype (TT or Tt), is crossed with an individual showing the recessive phenotype, therefore being of homozygous recessive genotype (tt) (Figure 5). If any of the offspring have the recessive phenotype, then the dominant phenotype organism is heterozygous, Tt.

The expected ratio is 1 : 1 tall : short. The probability of obtaining each phenotype is 1 in 2, or 50%, or 0.5.

Phenotypes: tall-stemmed × short-stemmed

Genotypes: TT or Tt × tt

Gamete genotypes: all (T) or (T)(t) × all (t)

Punnett square:

♀ gametes / ♂ gametes	(T)
(t)	Tt 100% tall

or

♀ gametes / ♂ gametes	(T)	(t)
(t)	Tt 50% tall	tt 50% short

Figure 5 The use of a test cross during monogenic inheritance.

LEARNING TIP

If you are using genetic diagrams to solve a genetics problem or to explain an inheritance pattern, always put gamete genotypes in circles and link the phenotypes to their genotypes.

DID YOU KNOW?

Mendel worked with plants, but in animals his dominant/recessive patterns of monohybrid inheritance apply. In humans, some of the many single gene disorders are recessive, where both parents are symptomless carriers of the allele. Some, such as Huntington disease and achondroplasia, are dominant. Only one parent needs to be affected and carry the allele; the children have a 50% chance of inheriting the condition. A double dose of the dominant allele is usually a lethal combination, and embryos with the homozygous dominant genotype do not survive.

Questions

1. Figure 6 represents several gene loci on a chromosome. For each locus, describe the organism's genotype.

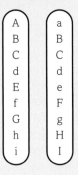

Figure 6

2. In a certain breed of domestic cat, brown coat is dominant and black coat is recessive. Show by means of genetic diagrams and/or Punnett squares, the possible outcomes of the following matings:
   (a) a heterozygous brown female and a black male
   (b) a homozygous brown female and a black male
   (c) two heterozygous brown cats
   (d) a homozygous brown female and a heterozygous brown male.
   In each case state the phenotype ratios and the phenotype probabilities.

3. Look at the Punnett square in Figure 4. What is the probability of obtaining, in the F_2 generation:
   (a) a homozygous tall stemmed plant?
   (b) a short stemmed plant?
   (c) a heterozygous tall stemmed plant?

4. About 1 in 500 people in the UK inherit a genetic condition called familial hypercholesterolaemia, FHC. This is caused by a dominant allele. Individuals with it have high blood plasma levels of cholesterol and are advised to take medication to reduce these levels. Affected individuals are heterozygous. Inheriting two dominant alleles is a lethal combination and the embryo usually does not survive. Use genetic diagrams to predict the possible offspring genotypes and phenotypes from the following partnerships:
   (a) two individuals both suffering from FHC
   (b) one individual with FHC and one not suffering from FHC.

5. In mice, grey coat is dominant; albino coat is recessive. A grey mouse mated several times in succession with an albino mouse. In the litters produced, 15 were albino and 18 were grey. Deduce the genotypes of the grey mouse and the albino mouse, and their grey and albino offspring. Show your working.

(3) Dihybrid inheritance

By the end of this topic, you should be able to demonstrate and apply your knowledge and understanding of:

* genetic diagrams to show patterns of inheritance (continued in topics 6.2.4–6.2.7)

KEY DEFINITION

dihybrid: involving two gene loci.

The simultaneous inheritance of two characteristics

Investigations that examine the simultaneous inheritance of two characteristics are **dihybrid** crosses.

In one dihybrid cross, Mendel examined the inheritance of seed shape and seed colour in pea plants. He crossed true-breeding pea plants with yellow and round seeds with true-breeding pea plants that had green and wrinkled seeds.

All the F_1 generation were hybrids, having the phenotype of yellow and round seeds. Each plant in the F_1 generation is heterozygous for both genes (seed colour and shape). Therefore, yellow and round are both dominant traits.

Figure 1 shows the independent assortment of two genes on two different homologous pairs of chromosomes.

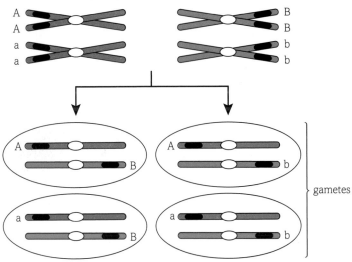

Figure 1 Formation of gametes by meiosis involving independent assortment of chromosomes.

Figure 2 shows the derivation of genotypes and phenotype of the F_1 generation and the resultant F_2 generation when crossing members of this F_1 generation.

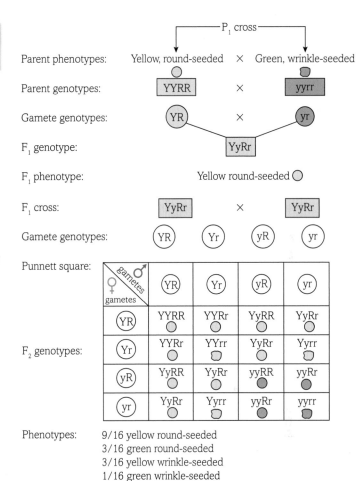

Figure 2 Genetic diagrams showing genotype and phenotype probabilities in the F_1 and F_2 generation produced during a dihybrid cross.

From the results of his dihybrid crosses, Mendel deduced that:

* the alleles of the two genes are inherited independently of each other, so each gamete has one allele for each **gene locus**
* during fertilisation, any one of an allele pair can combine with any one of another allele pair.

Predicting ratios of traits in the F_2 generation

If we consider the two crosses for seed colour and seed shape as two independent monohybrid crosses, with the two sets of traits being inherited independently, we can predict the outcome of allowing the members of the F_1 generation to self-fertilise. The chances of the traits for seed colour being inherited are not influenced by the chances of the traits for seed shape being inherited.

Assuming that seed colour and seed shape are two separate monogenic characteristics:

- for the characteristic of colour: in the F_2 generation we would predict $\frac{3}{4}$ would be yellow and $\frac{1}{4}$ green, a ratio of 3:1
- for the characteristic of shape: we would predict that $\frac{3}{4}$ would be round and $\frac{1}{4}$ wrinkled; a ratio of 3:1.

The Punnett square in Figure 2 shows that $\frac{12}{16}$ $\left(\frac{3}{4}\right)$ possible gamete combinations give genotypes that produce the phenotype yellow and $\frac{4}{16}$ $\left(\frac{1}{4}\right)$ would produce the phenotype green. This is a 3:1 ratio of yellow : green.

Similarly, $\frac{12}{16}$ plants produce round seeds and $\frac{4}{16}$ produced wrinkled seeds, also a 3:1 ratio.

When two independent events occur simultaneously:

product of individual probabilities
= combined probability of occurrence

We can use this equation to calculate the probabilities of all the other phenotypes in this F_2 generation.

Phenotypes	Probability of them occurring in the F_2 generation
yellow and round	$\left(\frac{3}{4}\right) \times \left(\frac{3}{4}\right) = \frac{9}{16}$
yellow and wrinkled	$\left(\frac{3}{4}\right) \times \left(\frac{1}{4}\right) = \frac{3}{16}$
green and round	$\left(\frac{1}{4}\right) \times \left(\frac{3}{4}\right) = \frac{3}{16}$
green and wrinkled	$\left(\frac{1}{4}\right) \times \left(\frac{1}{4}\right) = \frac{1}{16}$

Table 1 Probability of different phenotypes occurring in the F_2 generation.

When Mendel counted his pea plants in the F_2 generation of the dihybrid cross, he obtained this 9:3:3:1 ratio.

When investigating dihybrid inheritance, Mendel, without knowing about genes or the process of meiosis, chose two characteristics, the genes for which are on different chromosomes. You will see in topic 6.2.7 that if the two genes are on the same chromosome, the inheritance pattern is different.

LEARNING TIP

Always follow the rules for constructing genetics diagrams and Punnett squares. Use symbols to represent the alleles of the genes and make sure the letter for the dominant trait is large and upper case, and the letter representing the recessive trait is small and lower case. Remember that because gametes are produced by meiosis, each gamete has only one allele for any particular gene.

DID YOU KNOW?

Many seeds produced for farmers, gardeners, horticulturists and market gardeners are F_1 hybrids. This means the plants will exhibit specific desirable phenotypic traits, but will not breed true. Therefore the gardeners will not be able to save seed from their crop to obtain the same desirable traits next year. They will need to buy the F_1 hybrid seeds, produced at plant breeding establishments, again the following year.

Questions

1. When Mendel cross-fertilised tall purple-flowered pea plants with short white-flowered pea plants and grew the resulting seeds, all the plants of the F_1 generation were tall with purple flowers.
   - (a) Draw a genetic diagram to show the genotypes and phenotypes of the F_1 generation resulting from this cross fertilisation.
   - (b) Construct a Punnett square to predict the probabilities of the phenotypes produced if members of this F_1 generation were allowed to self-fertilise.

2. In fruit flies, *Drosophila melanogaster*, grey body colour is dominant and ebony body colour is recessive. Long wings are dominant and vestigial (short stubby) wings are recessive.
   - (a) Predict the genotypes and phenotypes in the F_1 generation when ebony-bodied long-winged females, homozygous at both gene loci, are allowed to mate with grey-bodied vestigial-winged males, also homozygous at both gene loci. Use the symbols G/g to represent the body colour alleles, and +/vg to represent the wing type alleles.
   - (b) Construct a Punnett square to predict the phenotypic ratios of the F_2 generation if members of the F_1 generation are allowed to interbreed.

3. In a certain breed of guinea pig, black fur is dominant and white fur recessive. Long hair is dominant and short hair recessive. A breeder has a black long-haired male, but does not know its genotype. He mates it with four female guinea pigs that have white short-haired fur. In total, 13 baby guinea pigs are born; six have black short-haired fur and seven have black long-haired fur.
   - (a) What are the genotypes of:
     - (i) the original black long-haired male
     - (ii) the four females
     - (iii) the offspring?
   - (b) Show by means of genetic diagrams how you have arrived at your answers. Use the symbols B/b and L/l to represent the alleles of the genes.

4. Look at Figure 2.
   - (a) (i) How many different genotypes produce yellow round seeds?
     - (ii) Write down these genotypes.
   - (b) Write down the genotypes that produce:
     - (i) yellow wrinkled seeds
     - (ii) green round seeds
     - (iii) green wrinkled seeds.
   - (c) Show that the genotypes of the F_1 generation produced by crossing yellow wrinkle-seeded pea plants with green round-seeded pea plants are the same as for the F_1 generation produced by crossing yellow round-seeded pea plants with green wrinkle-seeded pea plants.

(4) Multiple alleles

By the end of this topic, you should be able to demonstrate and apply your knowledge and understanding of:

* genetic diagrams to show patterns of inheritance (continued in topics 6.2.5–6.2.7)

Genes with multiple alleles

So far we have considered characteristics with distinct traits where there are two alleles for each **gene locus** in question. Over time a huge number of changes can occur anywhere within a gene; as a result, many genes have more than two alleles. When three or more alleles at a specific gene locus are known, then the gene is said to have **multiple alleles**. However, any individual can only possess two alleles, one on each gene locus, in a pair of homologous chromosomes.

Human blood groups

The inheritance of human ABO blood groups is a good example of multiple alleles. It also demonstrates both dominance and **codominance** (topic 6.2.6) of the alleles involved.

The four blood groups (phenotypes) – A, B, AB and O – are determined by three alleles of a single gene on chromosome 9. The gene encodes an isoagglutinogen, I, on the surface of erythrocytes. The alleles present in the human **gene pool** are I^A, I^B and I^o. I^A and I^B are both dominant to I^o, which is recessive. I^A and I^B are codominant. If both I^A and I^B are present in the genotype, they will both contribute to the phenotype. Any individual will have only two of the three alleles within their genotype.

A man of blood group A and a woman of blood group B, both heterozygous at this gene locus, can produce children of any of these four blood groups, as shown in the genetic diagram in Figure 1.

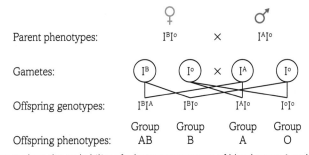

Figure 1 Genetic diagram to show the probability of a heterozygous man of blood group A and a heterozygous woman of blood group B producing children of each blood group.

Coat colour in rabbits

The coat colours in rabbits are:

* Wild type coat colour, called agouti: each hair has a grey base, a yellow band and a black tip.
* Albino: the condition where no pigment develops and the fur is white.
* Chinchilla: hairs are silvery grey, because they lack the yellow band.
* Himalayan: white, but with black feet, ears, nose and tail.

These coat colours are determined by one gene that has four alleles. There is a dominance hierarchy:

* Agouti, C, is dominant to all other alleles.
* Chinchilla, C^{ch}, is dominant to himalayan, C^h.
* Albino, c, is recessive to all other alleles.

Figures 2 and 3 show the results of crosses between rabbits of various coat colours.

Genotypes		Phenotypes
$I^A I^A$	$I^A I^o$	group A
$I^B I^B$	$I^B I^o$	group B
$I^A I^B$		group AB
$I^o I^o$		group O

Table 1 Blood group phenotypes produced by various genotypes.

i

Parent phenotypes:	agouti	×	albino
Parent genotypes:	CC	×	cc
Gametes:	Ⓒ	×	ⓒ
Offspring genotypes:		Cc	
Offspring phenotypes:		agouti	

ii

Parent phenotypes:	agouti	×	albino
Parents genotypes:	CC^{ch}	×	cc
Gametes:	Ⓒ Ⓒ^{ch}	×	ⓒ
Offspring genotypes:	Cc		C^{ch}c
Offspring phenotypes:	agouti		chinchilla

iii

Parent phenotypes:	agouti	×	albino
Parents genotypes:	CC^h	×	cc
Gametes:	Ⓒ Ⓒ^h	×	ⓒ
Offspring genotypes:	Cc		C^hc
Offspring phenotypes:	agouti		himalayan

Figure 2 The results of crossing albino and agouti rabbits.

i

Parent phenotypes:	albino	×	himalayan
Parents genotypes:	cc	×	C^hc
Gametes:	ⓒ	×	Ⓒ^h ⓒ
Offspring genotypes:	C^hc		cc
Offspring phenotypes:	himalayan		albino

ii

Parent phenotypes:	albino	×	himalayan
Parents genotypes:	cc	×	C^hC^h
Gametes:	ⓒ	×	Ⓒ^h
Offspring genotypes:		C^hc	
Offspring phenotypes:		all himalayan	

Figure 3 The results of crossing albino and himalayan rabbits.

(a)   (b)   (c)

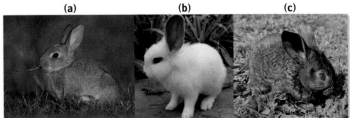

Figure 4 Rabbits' coat colour: (a) agouti; (b) albino; and (c) chinchilla.

Questions

1 In rabbits, what are the phenotypes corresponding to the following genotypes?

(a) CC^h   (b) CC   (c) C^{ch}C^h

(d) cc   (e) C^hC^h   (f) C^{ch}c

2 Complete the table for rabbit coat colours.

Phenotype	Possible genotypes
agouti	
chinchilla	
himalayan	
albino	

3 A rabbit breeder has a chinchilla male rabbit. He does not know its genotype. He has female rabbits of chinchilla, agouti, himalayan and albino phenotypes. Explain how he could find out the male rabbit's genotype.

4 A breeding pair of rabbits, one chinchilla and one himalayan, produce several litters of kittens. Of the 23 kittens, 16 are chinchilla, 4 are himalayan and 3 are albino. Deduce the genotypes of the two rabbit parents. Show how you have arrived at your answer.

5 A breeding pair of rabbits, both agouti, produces several litters of kittens. Of the 29 kittens, 21 are agouti and 8 are chinchilla. Deduce the genotypes of the two rabbit parents. Show how you have arrived at your answer.

6 Two sets of parents leave the maternity unit of a hospital with their babies. They think their babies have been mixed up. Mrs X is blood group A and her partner is blood group AB. Mrs Y is blood group A and her partner is blood group O. Baby 1 is blood group O and baby 2 is blood group B.

(a) Which baby belongs to which couple? Show how you have arrived at your answer.

(b) How many alleles controlling his ABO blood group will Mr Y have:

(i) in a goblet cell lining his trachea

(ii) in one of his spermatozoa?

7 A couple has four children, one of whom is adopted. The mother's blood group is group A, and the father's is group B. The children's blood groups are child 1: group A; child 2, group B; child 3, group O; and child 4, group A. Is it possible from this information to deduce which child is adopted? Explain your answer.

8 In a certain breed of cat, brown coat is dominant to black coat. Both black and brown are dominant to white. A brown female cat and a black male cat produce two litters of kittens. Their offspring show the following phenotypes, brown, black and white, in the ratio 2:1:1. Deduce the genotypes of the parent cats. Show how you have arrived at your conclusion.

⑤ Sex linkage

By the end of this topic, you should be able to demonstrate and apply your knowledge and understanding of:

* genetic diagrams to show patterns of inheritance (continued in topics 6.2.6–6.2.7)

sex-linked: gene present on (one of) the sex chromosomes.

Sex linkage in humans

In humans, sex is determined by one of the 23 pairs of chromosomes, called the **sex chromosomes**.

The other 22 pairs are called **autosomes**. Each of the autosomal pairs is fully **homologous** – they match for length and contain the same genes at the same loci.

The sex chromosomes are XY in males and XX in females. The X and Y chromosomes are not fully homologous. A small part of one matches a small part of the other, so that these chromosomes can pair up before meiosis.

Figure 1 Human sex chromosomes. Coloured SEM of the larger X chromosome and small Y chromosome. Males have one X and one Y; females have two X chromosomes.

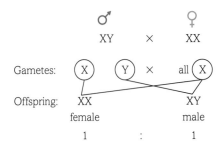

Figure 2 The inheritance of sex in humans. In this genetic diagram the symbols X and Y refer to whole chromosomes.

The human X chromosome contains over 1000 genes that are involved in determining many characteristics, or metabolic functions, not concerned with sex determination, and most of these have no partner alleles on the Y chromosome. If a female has one abnormal allele on one of her X chromosomes, she will probably have a functioning allele of the same gene on her other X chromosome.

If a male inherits, from his mother, an X chromosome with an abnormal allele for a particular gene, he will suffer from a genetic disease, as he will not have a functioning allele for that gene. Males are functionally haploid, or hemizygous, for X-linked genes. They cannot be **heterozygous** or **homozygous** for X-linked genes.

Sex-linked characteristics in humans include haemophilia A and colour blindness.

In humans, other mammals and some other animals, females are the homogametic sex – they have XX sex chromosomes and all their gametes contain one X chromosome. However, this is not the case for all organisms. In birds, butterflies and moths, males are the homogametic sex and females, having non-matching sex chromosomes, are the heterogametic sex.

Haemophilia A

A person with haemophilia A is unable to clot blood fast enough. Injuries may cause bleeding or an internal haemorrhage.

One of the genes on the non-homologous region of the X chromosome codes for a blood-clotting protein called factor 8. A mutated form of the allele codes for non-functioning factor 8.

A female with one abnormal allele and one functioning allele produces enough factor 8 to enable her blood to clot normally when required. However, this female is a carrier for the disease. If such a female passes the X chromosome containing the faulty allele to her son, he will have no functioning allele for factor 8 on his Y chromosome. As a result, he will suffer from haemophilia A.

Figure 3 shows the inheritance pattern for haemophilia A, where the father does not have haemophilia A and the mother is a symptomless carrier. Genotypes for sex-linked genes are represented by symbols that show they are situated on the X chromosome. The symbol H is used to represent the normal functioning allele for factor 8 and the symbol h represents the abnormal allele.

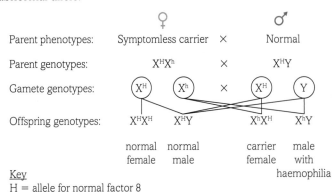

Key
H = allele for normal factor 8
h = allele for non-functioning factor 8

Figure 3 Inheritance of haemophilia A.

Colour blindness

One of the genes involved in coding for a protein involved in colour vision is on the X chromosome, but not on the Y chromosome. A mutated allele of this gene may result in colour blindness – an inability to distinguish between red and green. A female with one abnormal allele and one functioning allele will not suffer from colour blindness, but a male with an abnormal allele on his X chromosome will not have a functioning allele on his Y chromosome and will therefore suffer from red–green colour blindness. The inheritance pattern is the same as for haemophilia A – that of a recessive sex-linked disorder.

Figure 4 Ishihara test cards used to diagnose colour blindness. People with red–green colour blindness cannot see the numbers.

Sex linkage in cats

One of the genes, C, for coat colour in cats is sex-linked. It is on the non-homologous region of the X chromosome.

- The allele C^O produces orange (ginger) fur.
- The allele C^B produces black fur.

These alleles are codominant, as cats with the genotype $X^{C^O} X^{C^B}$ are tortoiseshell and have patches of black fur and patches of orange fur. Both the orange and black alleles contribute to the phenotype, but the orange allele is only expressed in cells where the X chromosome bearing the black coat colour allele is inactivated, and vice versa. Male cats may be either black or ginger but not tortoiseshell, as they only have one X chromosome.

Figure 5 Tortoiseshell cat.

Inactivation of X chromosomes in female mammals

It may appear that females have twice the number of X-linked genes being expressed as do males. However, a mechanism prevents this disparity. In every female cell nucleus, one X chromosome is inactivated. Determination of which member of the pair of X chromosomes becomes inactivated is random and happens during early embryonic development.

DID YOU KNOW?

Very rarely a male tortoiseshell cat is born. He is infertile and small. He has Klinefelter syndrome and has the genotype XXY instead of XY. The extra X chromosome is the result of non-disjunction during meiosis in one of his parents. One of his X chromosomes is inactivated in each cell nucleus.

Questions

1 (a) Explain the significance of a small part of the X and Y chromosomes (the sex chromosomes) being homologous to each other.

(b) Explain why human males cannot pass sex-linked genetic diseases to their sons.

(c) Can haemophiliac males pass the disease to their grandsons?

2 Draw genetic diagrams to show the possible outcomes of the following matings between male and female cats:

(a) a tortoiseshell female and a black male

(b) a ginger female and a black male

(c) a black female and a ginger male

(d) a tortoiseshell female and a ginger male.

3 Female haemophiliacs occur rarely. What would be the genotype of such a female? What would be the most likely genotypes of the parents of such a female?

4 A woman has a brother who suffers from Duchenne muscular dystrophy (DMD), an X-linked genetic disorder. This is a neuromuscular condition caused by lack of a functioning large protein, called dystrophin, in the skeletal muscles. The functional protein interacts with the cells' cytoskeleton. DMD leads to progressive muscle weakness. The gene for dystrophin is enormous, containing 79 exons and 2.3 million base pairs, which equates to 1% of the X chromosome. She has a son who has DMD.

(a) If she becomes pregnant again, what is the probability that her second child will suffer from DMD?

(b) About two-thirds of DMD patients have a deletion of one of the exons in the dystrophin gene. Explain how such a deletion may affect the structure of the dystrophin protein.

5 In a breed of chicken, called Plymouth Rock, there is a dominant X-linked allele, B, of a gene. The presence of B causes the normally black feathers to have a white bar. Newly hatched chicks with this mutated allele have a white spot on their heads. Females are heterogametic and males are homogametic. If females with the white bars on their feathers are mated with all black feathered males, the female chicks will have black heads and male chicks will have white spots on their heads. This enables breeders to quickly sex the chicks and keep the females to breed for egg laying. Draw genetic diagrams to explain the information here.

6 Codominance

By the end of this topic, you should be able to demonstrate and apply your knowledge and understanding of:

* genetic diagrams to show patterns of inheritance (continued in topic 6.2.7)

KEY DEFINITION

codominant: where both alleles present in the genotype of a heterozygous individual contribute to the individual's phenotype.

Codominant alleles

When both alleles of a gene in the genotype of a heterozygous individual contribute to that individual's phenotype, the alleles are described as **codominant**. The two alleles are responsible for two distinct and detectable gene products.

The phenotype of **heterozygotes** is different from the phenotype of the **homozygotes**.

Codominant inheritance in animals

Coat colour in shorthorn cattle is an example of codominant inheritance. The one gene for coat colour has two alleles: C^R (red) and C^W (white).

* Cattle that are homozygous for the red-coat allele, C^R, have a red (chestnut) coat.
* Cattle homozygous for the white-coat allele, C^W, have a white coat.
* Heterozygous cattle, genotype $C^R C^W$, have both red and white hairs – a roan coat.

(a)

(b)

Figure 1 Coat colour in shorthorn cattle: (a) roan bull and (b) red cow.

If red and white shorthorn cattle are interbred, all the offspring are roan. If roan cattle are mated, then the offspring will show all three phenotypes in the ratio 1 white : 2 roan : 1 red.

Figures 2 and 3 show genetic diagrams to explain these crosses.

Parent phenotypes:	cow (♀) red	×	bull (♂) white
Parent genotypes:	$C^R C^R$	×	$C^W C^W$
Gametes:	C^R	×	C^W
Offspring genotypes:		all $C^R C^W$	
Offspring phenotypes:		all roan	

Figure 2 Genetic diagram of a cross between a red shorthorn cow and a white shorthorn bull.

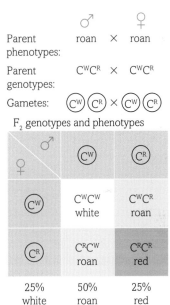

	♂		♀
Parent phenotypes:	roan	×	roan
Parent genotypes:	$C^W C^R$	×	$C^W C^R$
Gametes:	C^W C^R	×	C^W C^R

F_2 genotypes and phenotypes

♀ \ ♂	C^W	C^R
C^W	$C^W C^W$ white	$C^W C^R$ roan
C^R	$C^R C^W$ roan	$C^R C^R$ red

25% white	50% roan	25% red

Figure 3 Genetic diagram showing the probable outcome from interbreeding two roan shorthorn cattle.

Codominant inheritance in humans

MN blood groups

The MN blood group system is controlled by a single gene with two alleles, G^M and G^N. The gene codes for a particular protein on the surface of erythrocytes. The G^M allele codes for one version of the protein and the G^N allele codes for a slightly different version of the protein. These alleles are codominant.

The children of a couple, one being blood group M, and the other having blood group N, will all have blood group MN.

Figure 4 shows the probability of inheritance of the MN blood group system in children of a couple, where both are heterozygous, having blood group MN.

Parent phenotypes: group MN × group MN

Parent genotypes: MN × MN

Gamete genotypes: (M)(N) × (M)(N)

Punnett square to predict F$_2$ genotypes

♀ gametes \ ♂ gametes	(M)	(N)
(M)	MM	MN
(N)	MN	NN

F$_2$ phenotypes: 1 group M
2 group MN
1 group N

Figure 4 Genetic diagram showing the probability of inheritance of the MN blood group system in children of a couple, where both are heterozygous, having blood group MN.

ABO blood groups

Topic 6.2.4 describes the inheritance of the ABO blood groups. You will notice from that information that the alleles I^A and I^B are codominant to each other. An individual of genotype I^{A}I^B expresses both and has both types of isoagglutinogen protein on their erythrocytes. The inheritance of these blood groups also shows dominance, as both I^A and I^B are dominant to the allele I^o.

Sickle cell anaemia

Sickle cell anaemia is caused by a mutation in the gene that codes for the β-globin chain of haemoglobin (topic 6.1.1). The mutant allele is given the symbol Hb^S and the normal allele is given the symbol Hb^N. In heterozygous people, at least half the haemoglobin in their red blood cells is normal and half is abnormal. However, heterozygous people do not suffer from sickle cell anaemia.

If we consider the type of haemoglobin as the phenotype, then these alleles are considered as codominant. However, if we take sickle cell anaemia to be the phenotype, the Hb^S allele is considered to be recessive, as this disorder has a recessive inheritance pattern.

Codominant inheritance in plants

Some types of camellia have red flowers, and some have white flowers. If these two types are crossed, the offspring will have red and white spotted flowers.

Figure 5 Heterozygous camellia flower.

Both alleles of the gene for petal pigment, P^R and P^W, are expressed in the phenotype of the heterozygotes.

LEARNING TIP

For a person to suffer from sickle cell anaemia, they must be homozygous for the faulty allele. Heterozygotes are symptomless carriers.

Questions

1 In a particular breed of duck, some homozygous individuals have black feathers and other homozygous individuals have white feathers. A breeder has white ducks and a black drake. When the ducklings are hatched they are all grey in colour, usually described as blue. When the breeder allows a blue drake to mate with several of the blue ducks, 13 of the 25 ducklings are blue, six are white and six are black.

(a) Explain by means of genetic diagrams the outcomes of the breeder's observations.

(b) Predict the outcome of crossing a blue drake with (i) white ducks and (ii) black ducks.

2 A gardener buys a red and white spotted camellia. At the end of the summer she collects seeds from the plant and germinates these the following spring. She is disappointed that not all of the plants have red and white spotted flowers. Out of 63 plants, 31 have red and white spotted flowers, 16 have red flowers and 15 have white flowers.

(a) Explain, with the help of annotated genetic diagrams, why these plants did not breed true and why some of the offspring had red flowers and some had white flowers.

(b) Suggest how she could have obtained plants from her original camellia that all had red and white spotted flowers.

3 A farmer has a small herd of white shorthorn cows and buys a roan shorthorn bull. Predict and explain the probabilities of the calves being red, white or roan.

4 Predict and explain the probabilities of a couple, where the mother is blood group MN and the father is blood group N, having children of blood groups M, N and MN.

5 Patients with Tay Sachs disease are homozygous for a mutated allele of a gene that normally encodes an enzyme that regulates the metabolism of lipids in the plasma membranes of brain cells. As a result, lipids accumulate in their brains, causing progressive mental incapacity, blindness and loss of motor control. The condition is fatal. Heterozygotes do not suffer from the disease; they are symptomless. However, they have about half the levels of the enzyme than are present in people homozygous for the normal functioning allele of this gene.

(a) Suggest why heterozygotes have half the normal amount of the enzyme.

(b) Only homozygotes with two copies of the faulty allele will have symptoms of Tay Sachs disease, so this has a recessive inheritance pattern.

(i) On what level can these alleles, however, be considered codominant?

(ii) Imagine you are a genetic counsellor. How would you explain to a couple who have a baby with Tay Sachs disease that it is not the 'fault' of the parent who had a brother who died of Tay Sachs disease?

(7) Autosomal linkage

By the end of this topic, you should be able to demonstrate and apply your knowledge and understanding of:

* genetic diagrams to show patterns of inheritance

* the use of phenotypic ratios to identify linkage (autosomal and sex linkage) and epistasis

---

KEY DEFINITION

autosomal linkage: gene loci present on the same autosome (non-sex chromosome) that are often inherited together.

Linkage

When Mendel investigated the simultaneous inheritance (topic 6.2.3) of two characteristics, he chose seven characteristics for which the genes were on different chromosomes. Hence the genes assorted independently.

There are many more genes in a genome than there are chromosomes. When two or more gene loci are on the same chromosome, they are said to be **linked**.

The chromosome, not the gene, is the unit of transmission during sexual reproduction, therefore linked genes are *not* free to undergo independent assortment; they are usually inherited together as a single unit. Topic 6.2.5 describes sex linkage. This topic is concerned with **autosomal linkage**, where genes are linked by being on the same autosome.

DID YOU KNOW?

In eukaryotic organisms, the number of linkage groups is the same as the number of pairs of autosomes. For humans, there are 22 linkage groups.

Inheritance of autosomally linked genes with no crossing over

If linked genes are not affected by crossing over of non-sister chromatids during prophase 1 of meiosis, then they are always inherited as one unit. Figure 1 shows the formation of gametes under these circumstances.

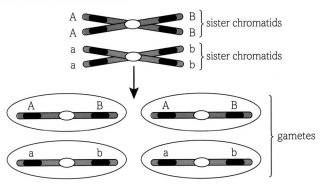

Figure 1 Gamete formation where genes are linked on a single pair of homologous chromosomes and no crossing over or subsequent exchange of alleles occurs.

Example

In sweet pea plants, the genes for flower colour and pollen grain shape are on the same chromosome. Each gene has two alleles:

* the gene for flower colour has two alleles: P for purple flowers, p for red flowers
* the gene for pollen grain shape has two alleles: L for long grains, l for short grains.

Figure 2 shows the phenotype ratios of the F_1 offspring produced when true-breeding purple-flowered, long-grained plants, homozygous at both gene loci, are crossed with true-breeding red-flowered, short-grained plants, homozygous at both gene loci. All the F_1 plants are heterozygous and phenotypically the same, exhibiting the dominant characteristics.

Figure 2 also shows the predicted phenotype ratios of the F_2 generation offspring resulting from interbreeding members of the F_1 generation. Because there is no independent assortment, we expect only two types of gamete.

The phenotype ratio in the F_2 generation is 3 purple-flowered long-grained : 1 red-flowered short-grained plants. The genotype ratio is 1 homozygous dominant : 2 heterozygous : 1 homozygous recessive. These ratios are the same as the Mendelian monohybrid inheritance ratios, with only two phenotypes in the F_2 generation. If the genes were not linked, we would expect four distinct phenotypes in the F_2 generation in the 9:3:3:1 ratio.

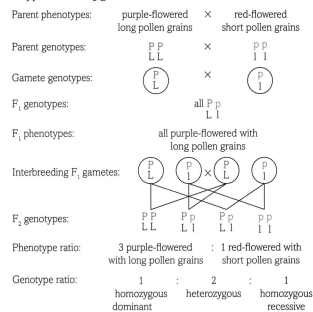

Figure 2 Genetic diagrams showing the inheritance of autosomally linked genes and the resulting phenotype ratios.

Inheritance of autosomally linked genes with crossing over

Figure 3 shows gamete formation where there is crossing over between two non-sister chromatids during prophase 1.

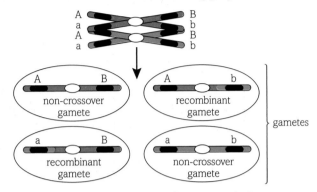

Figure 3 How recombinant gametes can form as a result of crossing over during meiosis 1.

Example

When members of the F₁ from the example described in Figure 2 are allowed to interbreed, the following results, shown in Table 1, are obtained.

Phenotype	Number
Purple-flowered with long pollen grains	920
Red-flowered with short pollen grains	274
Purple-flowered with short pollen grains	45
Red-flowered with long pollen grains	44

Table 1 Phenotypes in the F₂ generation.

The expected phenotypic ratio is 3 purple-flowered long-grained plants : 1 red-flowered short-grained plants.

From this sample of 1283 F₂ plants, we expect to see 962 purple-flowered long-grained plants and 321 red-flowered short-grained plants. We would not expect any of the recombinant phenotypes, those with purple flowers and short pollen grains, or with red flowers and short pollen grains.

These recombinant gametes have been produced by crossing over during meiosis 1, as explained in Figure 3. Figure 4 shows how recombinant gamete fusion may produce the unexpected

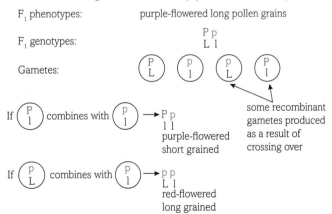

Figure 4 Diagram explaining how recombinant gametes may join at fertilisation to produce recombinant phenotypes.

phenotypes in the F₂ generation, described in Table 1. Here there are four phenotypes in the F₂ generation, but they are not in the 9:3:3:1 ratio that we would expect if the genes were not linked.

The further apart the two gene loci are on a chromosome, the greater is the chance of **recombinant gametes** forming.

Questions

1 In fruit flies, *Drosophila melanogaster*, there are two mutant alleles of closely linked genes on one of the autosomes. The genes are for eye colour and thickness of wing vein. The normal alleles code for red eyes, RE+, and thin wing veins, TV+. The mutant alleles code for brown eyes, re, and thick (heavy) wing veins, tv. Note: although there are genes for eye colour on the X chromosome, the gene referred to here is on an autosome.

When brown-eyed, thick wing-veined male fruit flies were allowed to mate with red-eyed, thin wing-veined female fruit flies, all the offspring had red eyes and thin wing veins. The original parent fruit flies were removed from the culture tubes before the offspring pupae hatched. When these F₁ generation fruit flies were allowed to interbreed, the following offspring phenotypes were seen:

- 588 flies with red eyes and thin-veined wings
- 186 flies with brown eyes and thick-veined wings
- 11 flies with red eyes and thick-veined wings
- 12 flies with brown eyes and thin-veined wings.

(a) Explain, with the aid of genetic diagrams, these observations.

(b) Predict the possible phenotypes that would be seen in the offspring if the red-eyed flies with thick wing veins were crossed with the brown-eyed flies that have thin wing veins.

(c) Predict the outcome of crossing each type of recombinant offspring with flies having brown eyes and thick-veined wings.

2 In rabbits, black fur (B) is dominant to brown fur (b). Another gene, C, codes for full colour and its mutated allele, c, codes for chinchilla. Both genes are on the same autosome. Rabbits that were heterozygous for both traits and had the phenotype black, full colour fur were crossed with brown chinchilla rabbits. The following phenotypes were observed in the resulting offspring:

- 31 brown chinchilla
- 34 black full colour
- 16 brown full colour
- 19 black chinchilla.

Explain, with the aid of genetic diagrams, the phenotype ratios seen in the progeny of this cross.

(8) Epistasis

By the end of this topic, you should be able to demonstrate and apply your knowledge and understanding of:

* the use of phenotypic ratios to identify linkage (autosomal and sex linkage) and epistasis

> **KEY DEFINITION**
>
> **epistasis:** interaction of non-linked gene loci where one masks the expression of the other. From the Greek, *ephistanai,* meaning 'stoppage'.

Genes can interact

In some cases different genes, at different loci on different chromosomes, interact to affect one phenotypic characteristic. When one gene masks or suppresses the expression of another gene, this is termed **epistasis**.

The genes in question may work together antagonistically (against each other) or in a complementary fashion.

Because the gene loci are not linked, they assort independently during gamete formation. Epistasis reduces the number of phenotypes produced in the F_2 generation of dihybrid crosses and therefore it reduces genetic variation.

Genes working antagonistically

Recessive epistasis

The homozygous presence of a recessive allele at the first locus prevents the expression of another allele at a second locus. The alleles at the first locus are epistatic to those at the second locus, which are **hypostatic** to those at the first locus.

An example of recessive epistasis is the inheritance of flower colour in *Salvia*. Two gene loci, A/a and B/b, on two different chromosomes, are involved.

Figure 1 Purple flower of *Salvia* plant being pollinated by a bee.

If a pure-breeding pink-flowered variety of *Salvia*, genotype AAbb, is crossed with a pure-breeding white-flowered variety, genotype aaBB, all the offspring of the F_1 generation have purple flowers. Their genotype is AaBb.

Interbreeding members of the F_1 generation results in plants that bear purple, pink and white flowers, in the ratio of 9:3:4. This is a modified version of the dihybrid 9:3:(3:1) ratio.

The homozygous aa is epistatic to both alleles of the B/b gene. Neither the allele B for purple nor the allele b for pink, when in the homozygous state, can be expressed if no dominant A allele is present.

Figure 2 shows genetic diagrams to explain these results.

Parent phenotypes:	pink-flowered *Salvia*	×	white-flowered *Salvia*
Parent genotypes:	AAbb	×	aaBB
Gametes:	(Ab)		(aB)
F_1 genotypes:		all AaBb	
F_1 phenotypes:		all purple-flowered	

F_1 gametes: (AB) (Ab) (aB) (ab) × (AB) (Ab) (aB) (ab)

Punnett square

♀ gametes \ ♂ gametes	AB	Ab	aB	ab
AB	AABB purple	AABb purple	AaBB purple	AaBb purple
Ab	AABb purple	AAbb pink	AaBb purple	Aabb pink
aB	AaBB purple	AaBb purple	aaBB white	aaBb white
ab	AaBb purple	Aabb pink	aaBb white	aabb white

F_2 phenotypes: 9 purple, 3 pink, 4 white

Figure 2 Recessive epistasis in *Salvia*.

Dominant epistasis

The inheritance of feather colour in chickens is an example of dominant epistasis. There is an interaction between two gene loci, I/i and C/c. The hypostatic gene, C/c, codes for coloured feathers. The I allele of the epistatic gene, I/i, prevents the formation of colour, even if one C allele is present.

Individuals carrying at least one dominant allele, I, have white feathers, even if they also have one dominant allele for coloured feathers.

Birds that are homozygous for the recessive allele, c, are also white, as this mutated allele does not cause pigment to be made.

Pure-breeding White Leghorn chickens have the genotype IICC. Pure-breeding White Wyandotte chickens have the genotype iicc (see Figure 3).

(a) 　　**(b)**

Figure 3 (a) White Leghorn chicken and (b) White Wyandotte chicken.

If White Leghorn chickens, genotype IICC, are crossed with White Wyandotte chickens, genotype iicc, the offspring are all white. They are heterozygous at both gene loci, having the genotype IiCc.

If the progeny interbreed, they produce, in the F_2 generation, white-feathered chickens and coloured-feathered chickens in the ratio of 13:3, respectively. Figure 4 explains these observations.

| Parent phenotypes: | White Leghorn | × | White Wyandotte |

| Parent genotypes: | IICC | × | iicc |

| Parent gametes: | (IC) | × | (ic) |

F_1 genotypes:　　all IiCc

F_1 phenotypes:　　all white

F_1 gametes:　(IC) (Ic) × (iC) (ic)

Punnett square to show F_2 genotypes and phenotypes

♀ gametes ＼ ♂ gametes	IC	Ic	iC	ic
IC	IICC white	IICc white	IiCC white	IiCc white
Ic	IICc white	IIcc white	IiCc white	Iicc white
iC	IiCC white	IiCc white	iiCC coloured	iiCc coloured
ic	IiCc white	Iicc white	iiCc coloured	iicc white

Phenotype ratio of F_2 generation:　13 white : 3 coloured
This is also a modified 　　(9 : 3 : 3 : 1) ratio

Figure 4 Genetic diagrams to explain the effect of dominant epistasis on the F_2 produced from interbreeding members of the F_1 generation obtained from crossing true-breeding White Leghorn chickens with true-breeding White Wyandotte chickens.

194

Genes working in a complementary fashion

As scientists are discovering more about how genes work and interact with each other, epistasis is more often explained in terms of the genes working to code for two enzymes that work in succession, catalysing sequential steps of a metabolic pathway.

Coat colour in mice

In mice, the gene locus C/c determines that the coat will have colour. The genotype CC or Cc produces coloured fur. However, in the recessive homozygous state, genotype cc, no pigment develops and the mice are albino.

A/a determines what that colour is by determining the distribution of pigment. The dominant allele, A, produces agouti colour – each hair has black pigment and a yellow band producing an overall greyish colour. The recessive allele, a, when homozygous, produces black fur as the yellow band from each hair is lacking. However, if there are two c alleles, no colour will develop, as there is no pigment to be distributed.

Genotypes	AACC	AACc	AAcc	AaCC	AaCc
Phenotype	agouti	agouti	albino	agouti	agouti

Genotypes	Aacc	aaCC	aaCc	aacc
Phenotype	albino	black	black	albino

Table 1 The phenotypes of various genotypes of mice.

Figure 5 shows the phenotype ratios from crossing two mice, heterozygous at both gene loci.

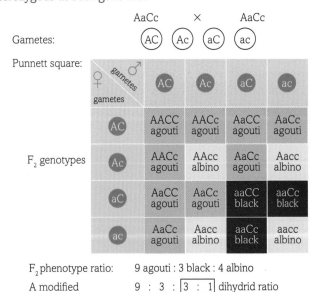

| | AaCc | × | AaCc |

Gametes:　(AC) (Ac) (aC) (ac)

Punnett square:

♀ gametes ＼ ♂ gametes	AC	Ac	aC	ac
AC	AACC agouti	AACc agouti	AaCC agouti	AaCc agouti
Ac	AACc agouti	AAcc albino	AaCc agouti	Aacc albino
aC	AaCC agouti	AaCc agouti	aaCC black	aaCc black
ac	AaCc agouti	Aacc albino	aaCc black	aacc albino

F_2 genotypes

F_2 phenotype ratio:　9 agouti : 3 black : 4 albino
A modified　9 : 3 : [3 : 1] dihydrid ratio

Figure 5 The phenotype ratio of offspring from crossing two mice, heterozygous at both gene loci, genotypes AaCc. The phenotype ratio of 9:3:4 is a modified 9:3:3:1 Mendelian dihybrid phenotype ratio.

This can be regarded as an example of recessive epistasis. However, it can be explained if we consider what each gene governs.

- In the presence of a C allele, the black pigment can be made from a colourless substance.

- In the presence of an A allele, this black pigment is deposited during the development of hair in a pattern, combined with a yellow band on each hair, that produces the agouti coloration.

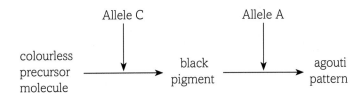

F_2 ratio	Genotype	Phenotype	Final phenotype ratio
$\frac{9}{16}$	A–B–	purple	$\frac{9}{16}$ purple
$\frac{3}{16}$	A–bb	white	
$\frac{3}{16}$	aaB–	white	$\frac{7}{16}$ white
$\frac{1}{16}$	aabb	white	

Key: the dash, –, denotes either a dominant or recessive allele of the gene in question.

Table 2 The results of a cross between white and purple sweet peas.

Flower colour in sweet peas

Two geneticists, William Bateson and Reginald Punnett, crossed two strains of true-breeding white-flowered sweet peas.

Figure 6 (a) Purple-flowered and (b) white-flowered sweet peas.

They were surprised when all the F_1 progeny produced purple flowers.

When they allowed these F_1 plants to interbreed by self fertilisation, the F_2 phenotypes contained white-flowered plants and purple-flowered plants in the ratio 9:7.

Two gene loci, A/a and B/b, may yield such results if one gene locus codes for an enzyme that catalyses the production of a colourless intermediate product from a colourless precursor substance, and the second gene locus codes for an enzyme that catalyses the production of a purple pigment from the intermediate product.

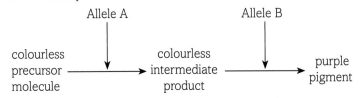

At least one dominant allele for both gene loci has to be present for the flowers to be purple.

The cross is shown as follows:

P_1	AAbb	×	aaBB
	(white)		(white)
F_1	all AaBb (purple)		

Combs of domestic chickens

Figure 7 shows four types of comb found in domestic chickens.

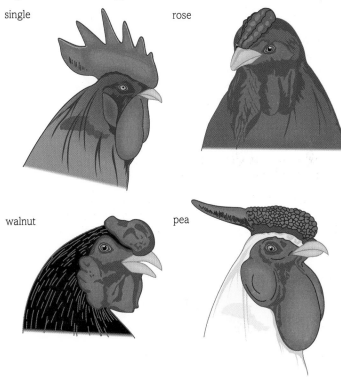

Figure 7 Comb shapes in domestic chickens.

Two gene loci, P/p and R/r, interact to affect comb shape. The effect of the P/p alleles depends upon which R/r alleles are present in the bird's genotype.

When true-breeding pea-combed chickens, genotype PPrr, are bred with true-breeding rose-combed chickens, genotype ppRR, the progeny, all PpRr, have walnut combs.

When the walnut-combed progeny are interbred, their progeny show four phenotypes: walnut comb, rose comb, pea comb and single comb, in the classic Mendelian dihybrid ratio of 9:3:3:1.

Genotype	Phenotype
P–R–	walnut comb
ppR–	rose comb
P–rr	pea comb
pprr	single comb

Table 3 Genotypes and corresponding phenotypes for comb shape in domestic chickens.

- At least one dominant allele for both gene loci has to be present in the bird's genotype for it to exhibit the phenotype walnut comb.
- At least one dominant R allele in the presence of two recessive p alleles produces a rose comb.
- At least one dominant P allele in the presence of two recessive r alleles produces a pea comb.
- Being recessive homozygous at both gene loci produces a single comb.

Figure 8 shows a Punnett square showing the distribution of genotypes and phenotypes within the F_2 generation produced by crossing two walnut-combed chickens, heterozygous at both gene loci.

gametes ♀ \ ♂	PR	Pr	pR	pr
PR	PPRR walnut	PPRr walnut	PpRR walnut	PpRr walnut
Pr	PPRr walnut	PPrr pea	PpRr walnut	Pprr pea
pR	PpRR walnut	PpRr walnut	ppRR rose	ppRr rose
pr	PpRr walnut	Pprr pea	ppRr rose	pprr single

Figure 8 Progeny from interbreeding two walnut-combed chickens.

LEARNING TIP

Epistatic ratios are modifications of the 9:3:3:1 Mendelian ratio from a dihybrid inheritance, where the two genes are unlinked. Epistasis also involves two unlinked gene loci.
A 9:3:4 ratio may suggest recessive epistasis.
A 12:3:1 and a 13:1 ratio both suggest dominant epistasis.
However, epistasis may occur by complementary gene action. A 9:7 ratio, or a 9:3:4 ratio may suggest epistasis by complementary gene action.
Complementary gene action may also produce a 9:3:3:1 ratio.

DID YOU KNOW?

At least two genes determine eye colour in humans. D/d and E/e, found at separate unlinked loci, interact. When the rear layer of the iris contains melanin and the front layer does not, the iris looks blue. As more and more melanin is produced in the front layer, the iris looks progressively darker blue, brown and black. Four dominant alleles, DDEE, produce dark brown/black eyes; three dominant alleles, DDEe or DdEE produce medium brown eyes; two dominant alleles, DDee, DdEe or ddEE produce light brown eyes. One dominant allele, Ddee or ddEe, produces dark blue eyes and recessive alleles produce pale blue eyes.
Some people have green eyes and some have grey or violet eyes, which indicates that other gene loci are also involved.

Questions

1. Another example of dominant epistasis may produce a phenotype ratio in the F_2 generation of 12:3:1.

   For example, in summer squash vegetables, the dominant allele, D, of the epistatic gene locus, D/d, masks the expression of the hypostatic gene locus, E/e, which codes for pigment. The dominant allele, E, leads to the production of yellow pigment. The recessive allele, e, when present in the homozygous condition, leads to production of green pigment. In homozygous dd individuals, the presence of at least one E allele leads to yellow fruits, whereas the presence of two e alleles leads to green fruits. The presence of at least one D allele, regardless of what alleles are present at the E/e gene locus, leads to white fruits.

   (a) Write down the phenotypes of squash fruits with the following genotypes:

   DDEE; DdEE; DdEe; DDee; Ddee; ddEE; ddEe; ddee

   (b) Show, by means of genetic diagrams, the genotypes and phenotype ratio of the F_2 generation when white-fruited plants, heterozygous at both gene loci, are crossed.

2. Construct a Punnett square to show the distribution of genotypes and the 9:7 phenotype ratios among the F_2 offspring of the sweet peas, described above.

3. Which two genotypes should a poultry breeder cross to produce chickens, so that 25% of the progeny have walnut combs, 25% have rose combs, 25% have pea combs and 25% have single combs?

4. A black cat and a white cat breed and produce several litters of kittens. Eight are white, four are black and four are brown. A pure-breeding brown cat of the same breed and a pure-breeding black cat together produce several litters of brown kittens. There are two gene loci involved in determining their coat colour. Suggest an explanation for these observations.

5. (a) Write down all the genotypes that produce:
   (i) purple *Salvia* flowers
   (ii) pink *Salvia* flowers
   (iii) white *Salvia* flowers.

   (b) Predict the phenotype ratios from crossing purple *Salvia* plants, heterozygous at both gene loci, with:
   (i) pink-flowered *Salvia* plants homozygous at both gene loci
   (ii) pink-flowered *Salvia* plants heterozygous at the A/a locus
   (iii) true-breeding white-flowered plants
   (iv) white-flowered plants heterozygous at the B/b locus.

6. A poultry breeder has a white cockerel, whose mother was a White Wyandotte and father was a White Leghorn. The white cockerel mates with a brown hen whose grandparents were White Wyandottes and White Leghorns. The brown hen lays and hatches 8 eggs, 3 of which hatch into brown chicks and 5 hatch into white chicks. What are the genotypes of:

   (a) the white cockerel
   (b) the brown hen?

7. Predict the phenotype ratio you would expect if you cross pollinated purple-flowered sweet peas, heterozygous at both gene loci, with white-flowered sweet pea plants not homozygous at both gene loci.

8. Predict the phenotype ratio among offspring of six couples all with light brown eyes, genotype DdEe.

⑨ Using the chi-squared test

By the end of this topic, you should be able to demonstrate and apply your knowledge and understanding of:

* using the chi-squared (χ^2) test to determine the significance of the difference between observed and expected results

---

KEY DEFINITION

chi-squared test: statistical test designed to find out if the difference between observed and expected data is significant or due to chance.

Is the difference between observed and expected results significant?

If we obtain results that are not quite as expected, we need to know whether the difference is just due to chance or whether the difference between what we observe and what we expect is significant. If it is significant, it may be that the inheritance pattern is different to what we thought and we need to rethink how to explain our observations.

Consider the predicted outcome from a dihybrid cross (see topic 6.2.3) where the two genes are unlinked, in terms of the expected phenotype ratio in the F$_2$ generation. We expect the 9:3:3:1 ratio.

Figure 1 gives an overview of Mendel's results with a genetic explanation.

If the sample collected is large, then we should obtain results that are close to those that we expect to see. However, if the sample is not so large, we may not get an exact 9:3:3:1 ratio. Also, some ovules may not become fertilised and some seeds may not develop in the pod. There is also the influence of the 'genetic lottery' as to which gametes fuse at fertilisation.

We can use the **chi-squared** (χ^2) test when:
- the data are in categories (e.g. different phenotypes or discrete variables) and are not continuous
- we have a strong biological theory to use to predict expected values
- the sample size is large
- the data are only raw counts (percentages or ratios cannot be used)
- there are no zero scores in the raw count data.

The null hypothesis

Statistical tests cannot be used to directly test a hypothesis; instead they test a null hypothesis. If the null hypothesis is not supported, then we accept the original hypothesis.

The null hypothesis states:

> *'There is no statistically significant difference between the observed and expected data. Any difference is due to chance.'*

Applying the chi-squared test

In an investigation like the one shown in Figure 1, 288 F$_2$ generation plants were grown. The types of seeds they produced were examined. The results are shown in Table 1.

Phenotype	Number of plants showing this trait
yellow and round	169
green and round	54
yellow and wrinkled	51
green and wrinkled	14

Table 1 Observed numbers in different phenotype categories.

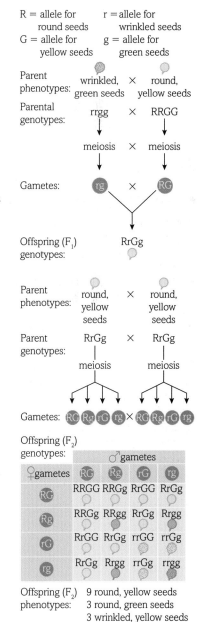

R = allele for round seeds
r = allele for wrinkled seeds
G = allele for yellow seeds
g = allele for green seeds

Parent phenotypes: wrinkled, green seeds × round, yellow seeds

Parental genotypes: rrgg × RRGG

meiosis × meiosis

Gametes: rg × RG

Offspring (F$_1$) genotypes: RrGg

Parent phenotypes: round, yellow seeds × round, yellow seeds

Parent genotypes: RrGg × RrGg

meiosis × meiosis

Gametes: RG Rg rG rg × RG Rg rG rg

Offspring (F$_2$) genotypes:

♀gametes	♂ gametes			
	RG	Rg	rG	rg
RG	RRGG	RRGg	RrGG	RrGg
Rg	RRGg	RRgg	RrGg	Rrgg
rG	RrGG	RrGg	rrGG	rrGg
rg	RrGg	Rrgg	rrGg	rrgg

Offspring (F$_2$) phenotypes:
9 round, yellow seeds
3 round, green seeds
3 wrinkled, yellow seeds
1 wrinkled, green seed

Figure 1 A genetic explanation for one of Mendel's dihybrid crosses.

Out of 288 individuals, assuming a 9:3:3:1 ratio, the expected numbers in each phenotype category would be:

- $\frac{9}{16}$ of 288 = 162 would be yellow and round
- $\frac{3}{16}$ of 288 = 54 would be green and round
- $\frac{3}{16}$ of 288 = 54 would be yellow and wrinkled
- $\frac{1}{16}$ of 288 = 18 would be green and wrinkled.

The formula for calculating the value of chi-squared, χ^2, is:

$$\chi^2 = \text{the sum of } \frac{[\text{each observed number (O)} - \text{each expected number (E)}]^2}{\text{each expected number (E)}}$$

This is usually written as:

$$\chi^2 = \sum \frac{(O - E)^2}{E}$$

In this equation:

- The differences may be positive or negative, so they are squared. This prevents any negative values cancelling out any positive values.
- Dividing by E takes into account the size of the numbers.
- The 'sum of' sign (Σ) takes into account the number of comparisons being made.

The procedure is:

1. Calculate the value of χ^2.

2. Determine the number of degrees of freedom (= number of categories − 1).

3. Determine the value of p from a distribution table. The probability value of 0.05 identifies the level that could occur by chance just 5 times in 100 (5%), or 1 time in 20. We need to know that the probability of our deviation being the result of chance is greater than 5%.

4. Decide whether the difference is significant at the $p = 0.05$ level of probability.

Example

1. To calculate the value of χ^2, it is best to use a table, such as Table 2.

Category	Observed (O)	Expected (E)	O − E	(O − E)²	$\frac{(O - E)^2}{E}$
yellow, round	169	162	7	49	$\frac{49}{162} = 0.30$
green, round	54	54	0	0	$\frac{0}{54} = 0.00$
yellow, wrinkled	51	54	−3	9	$\frac{9}{54} = 0.17$
green, wrinkled	14	18	−4	16	$\frac{16}{18} = 0.88$
				$\chi^2 =$	1.35

Table 2

2. The number of degrees of freedom
   = (number of categories − 1) = (4 − 1) = 3.

3. If we look up the value of χ^2 in a distribution table (see Figure 2), we can see that p values range from 0.01 (1%) to 0.99 (99%).

Number of classes	Degrees of freedom	χ^2							
2	1	0.00	0.10	0.45	1.32	2.71	3.84	5.41	6.64
3	2	0.02	0.58	1.39	2.77	4.61	5.99	7.82	9.21
4	3	0.12	1.21	2.37	4.11	6.25	7.82	9.84	11.34
5	4	0.30	1.92	3.36	5.39	7.78	9.49	11.67	13.28
6	5	0.55	2.67	4.35	6.63	9.24	11.07	13.39	15.09
Probability that deviation is due to chance alone		0.99 (99%)	0.75 (75%)	0.50 (50%)	0.25 (25%)	0.10 (10%)	0.05 (5%)	0.02 (2%)	0.01 (1%)

Accept null hypothesis (any difference is due to chance and not significant) ← → Reject null hypothesis; accept experimental hypothesis (difference is significant, not due to chance)

Critical value of χ^2 0.05 p level; this is the level at which we are 95% certain the result is not due to chance, agreed on by statisticians as a cut-off point

Figure 2 Part of a χ^2 (chi-squared) distribution table.

4. Figure 2 shows that the critical value of χ^2 for three degrees of freedom is 7.82. Our value of χ^2 is smaller than this, therefore the difference between our observed and expected data is due to chance and is not significant. We accept the null hypothesis. If the value of χ^2 were greater than the critical value, then we would reject the null hypothesis.

LEARNING TIP

Get into the habit of writing out fully a sentence such as: 'The calculated value of chi-squared is smaller than the critical value of chi-squared at $p = 0.05$, therefore the difference is not significant and we accept the null hypothesis.'

Questions

In mice, one of the genes for coat colour has two alleles, Y/y. Allele Y produces yellow fur and is dominant to allele y that produces grey fur when in the homozygous state. When two heterozygous yellow-furred mice, genotype Yy, are crossed repeatedly, they produce several litters of young. The offspring phenotypes are 140 yellow and 68 grey.

1 What is the genotype of the grey mice?

2 What ratio of phenotypes would you expect if this was a normal Mendelian monohybrid cross?

3 Carry out a χ^2 test to find out if the difference between the observed and expected data is significant.

4 Suggest an explanation for the difference. (Hint: see topic 6.2.1.)

5 If a yellow mouse was crossed with a grey mouse, supposing the yellow mouse to be heterozygous, what phenotype ratio would you expect in the progeny of this test cross?

(10) Discontinuous and continuous variation

By the end of this topic, you should be able to demonstrate and apply your knowledge and understanding of:

* the genetic basis of continuous and discontinuous variation

The genetic basis of continuous and discontinuous variation

Within any population of organisms of the same species, there is genetic variation, as described in topic 6.2.1.

Discontinuous variation

Where phenotype classes are distinct and discrete, each clearly discernible from the others in a *qualitative* way, this is **discontinuous variation**. There are no or very few intermediates between the different phenotypes. For example, you are either male or female and have only one of the four possible ABO blood groups. Ear lobes are another example: they may be attached or free-hanging (Figure 1).

Characteristics that exhibit discontinuous variation are usually determined by the alleles of a single gene locus. They are **monogenic**. Sometimes the alleles of two genes interact to govern a single characteristic (see topic 6.2.8). In either case:

* different alleles at a single gene locus have large effects on the phenotype
* different gene loci have quite different effects on the characteristic.

In this chapter, you have already encountered many examples of discontinuous variation, such as pea plant stem height, flower colour and seed shape.

In tomato plants, many genes determine features of the plant's leaves. One gene locus codes for leaf shape and another determines whether the leaves have hairs. A third gene locus determines the presence or absence of chlorophyll.

Genes at different loci may interact to influence one characteristic and produce discontinuous variation, as in epistasis (see topic 6.2.8).

Continuous variation

Where the genetic variation between individuals, even if they are related, within a population shows a range with a smooth gradation between the many intermediates, it is described as **continuous variation**.

Examples include:

* birth mass, foot size, finger length, height, skin colour, hair colour, eye colour, mass and heart rate in humans
* cob length in maize plants
* leaf length in many plants
* tail length in mice
* red kernel colour in wheat.

Many genes are involved in determining such characteristics. Therefore such characteristics are described as **polygenic**.

The alleles of each gene may contribute a small amount to the phenotype, therefore the alleles have an *additive* effect on the phenotype. As a result, the phenotypic categories vary in a *quantitative* way. The greater the number of gene loci contributing to the determination of the characteristic, the more continuous the variation (the greater the range).

The study of the genetics of such inherited characteristics is called quantitative genetics. Many characteristic of crop plants are polygenic, so plant breeders need to apply knowledge of quantitative genetics.

KEY DEFINITIONS

continuous variation: variation that produces phenotypic variation where the quantitative traits vary by very small amounts between one group and the next.
discontinuous variation: genetic variation producing discrete phenotypes – two or more non-overlapping categories.

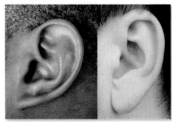

Figure 1 Attached and free-hanging ear lobes. This variation is discontinuous.

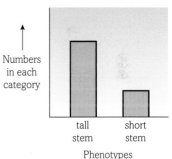

Figure 2 Graph showing the distribution of the two phenotypes obtained in the F$_2$ generation resulting from a monohybrid cross in pea plants.

Genetic analysis of the inheritance of such traits becomes more complicated as the number of gene loci increases above two. The example in the fact box shows the complexity of a trihybrid cross.

Figure 3 Wheat seeds.

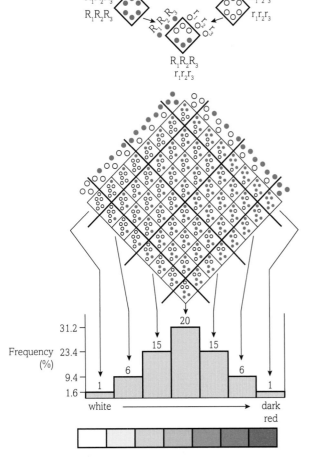

Figure 4 Results from an investigation carried out by H. Nilsson-Ehle into the inheritance of colour in wheat seeds, an example of polygenic variation. The more dominant alleles are inherited, the darker red is the seed. There is a smooth gradation between the phenotype classes.

Interaction between genes and environment

The environment has a greater effect on the expression of polygenes/polygenic characteristics than it does on monogenic characteristics.

For example, each person has a genetic potential for height and intelligence, but without proper nutrition and also, for intelligence, mental stimulation, these potentials will not be reached.

Questions

1 Warfarin was first used as a rat poison in the UK during the 1950s. It slows blood clotting by interfering with vitamin K metabolism and rats die from internal bleeding. By the end of the 1970s, rats resistant to warfarin were found throughout the UK. Resistance is controlled by a single gene locus. The allele R^{WR} confirms resistance and the allele R^{WS} makes rats susceptible. There are three possible genotypes and two phenotypes. Rats with genotypes $R^{WR} R^{WR}$ and $R^{WR} R^{WS}$ are both resistant and rats with genotypes $R^{WS} R^{WS}$ are susceptible.

Use a genetic diagram to indicate the genotypes and phenotypes of rats produced when two individuals, both heterozygous at this gene locus, interbreed.

2 Imagine that two gene loci, A/a and B/b, contribute to the cob length in a variety of corn. The dominant alleles of each gene contribute 4 cm to the length of the cob and each recessive allele contributes 2 cm to the cob length. Construct a Punnett square to predict the outcome of a cross between two F_1 individuals, both heterozygous at each gene locus. For each genotype, calculate the cob length (phenotype) and then plot a histogram showing the distribution of phenotypes in the F_2 generation.

3 Look at your answer to question 8 in topic 6.2.8.

(a) Plot a histogram to show the eye colour distribution among the F_2 offspring.

(b) Does the result suggest that the inheritance of eye colour in humans should be considered as an example of discontinuous or continuous variation?

(11) Factors affecting the evolution of a species

By the end of this topic, you should be able to demonstrate and apply your knowledge and understanding of:

* the factors that can affect the evolution of a species

KEY DEFINITIONS

directional selection: a type of natural selection that occurs when an environmental change favours a new phenotype and so results in a change in the population mean.

founder effect: when a small sample of an original population establishes in a new area; its gene pool is not as diverse as that of the parent population.

genetic bottleneck: a sharp reduction in size of a population due to environmental catastrophes such as earthquakes, floods, disease or human activities such as habitat destruction, overhunting or genocide, which reduces genetic diversity. As the population expands it is less genetically diverse than before.

stabilising selection: natural selection leading to constancy within a population. Intermediate phenotypes are favoured and extreme phenotypes selected against. Alleles for extreme phenotypes may be removed from the population. Stabilising selection reduces genetic variation within the population.

Natural selection

Mutations and migration introduce new alleles into populations. Some individuals within a population will be better adapted than others to the environment, due to differences in their genotypes and phenotypes. These individuals are more likely to survive and reproduce, passing on the advantageous alleles. Over time, **allele frequencies** within the population will change. This is natural selection. Natural selection may also maintain constancy of a species, as well as leading to new species.

There are three main types of selection: stabilising, directional and disruptive selection.

Stabilising selection

Stabilising selection normally occurs when the organisms' environment remains unchanged. It favours intermediate phenotypes.

In humans, babies of birth mass close to 3.5 kg are more likely to survive. Their offspring inherit alleles from them, also leading to this mean birth mass.

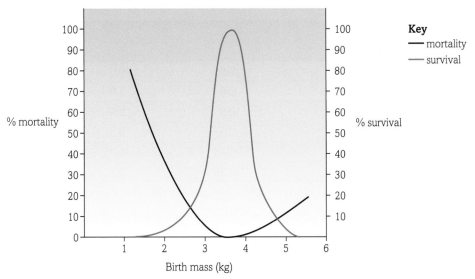

Figure 1 Relationship between birth mass and mortality in humans. The optimal birth weight for human babies was found, by Mary Karn and Sheldon Penrose, to be around 3.5 kg.

Directional selection

If the environment changes, for example by becoming colder, there may now be an advantage to being larger, so a new larger mass becomes the ideal and will be selected for. If more larger individuals survive and reproduce, they will be more likely to pass genes and alleles for larger size to their offspring. Over several generations, there is a gradual shift in the optimum value for the trait. Figure 2 shows the impact of **directional selection** on the mean value for a trait within a population.

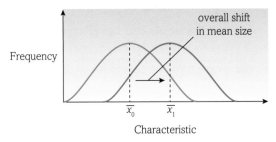

Figure 2 The impact of directional selection on the mean value of a trait within a population.

Directional selection is used by plant and animal breeders to produce desirable traits – see topic 6.2.14 on artificial selection.

In nature, within a population, periods of directional selection may alternate with periods of stabilising selection.

Genetic drift

If a population descends from a small number of parents, the gene pool will lack genetic variation. Some alleles resulting from mutation confer neither an advantage nor disadvantage on the individual, so there will be no selection pressure acting upon them. However, chance events may drastically alter the allele frequency.

Imagine a small population descended from one set of heterozygous parents. There are only two alleles, A and a, in the population (see Table 1).

If a catastrophic event occurs, such as an earthquake, flood, outbreak of a disease or a severe shortage of food, which leads to the death of many of the already small population, one of the alleles may disappear from this population.

When the population recovers and increases in size, it will have less genetic diversity than before and may lack particular alleles. The allele(s) in question did not disappear due to selection pressures, but due to genetic drift.

Possible genotypes of two offspring	Allele frequency	
	A	a
AA and AA	1.0	0.0
aa and aa	0.0	1.0
AA and aa	0.5	0.5
Aa and Aa	0.5	0.5
AA and Aa	0.75	0.25
aa and Aa	0.25	0.75

Table 1 The allele frequencies of all possible pairs of offspring produced by two heterozygous parents (Aa × Aa).

Genetic drift can arise after a **genetic bottleneck** or as the result of the **founder effect**.

Genetic bottleneck

When a population size shrinks and then increases again, it is said to have gone through a genetic bottleneck.

After this event, the genetic diversity within that population will be reduced. There may be loss of some advantageous alleles or a disproportionate frequency of deleterious (harmful) alleles, putting that population's chances of long-term survival at risk. Sometimes, after a genetic bottleneck, a population shrinks to such a small size that its fertility is affected, leading to the species becoming endangered and then extinct.

However, if the ones that survive are those that have a particular advantage, for example resistance to a particular pathogen, then a bottleneck could improve the gene pool whilst also shrinking genetic diversity.

Species that have been selectively bred for certain traits have also been through a genetic bottleneck – see topic 6.2.14.

Founder effect

If a new population is established by a very small number of individuals who originate from a larger, parent population, the new population is likely to exhibit loss of genetic variation.

Some groups of migrating humans, not fully genetically representative of the parent population, have set up populations in new areas. If they have remained isolated from other human populations, for example because of religious and cultural differences or due to geographic isolation, then the new population will have a small gene pool. This has happened in Iceland, the Faroe Islands, Pitcairn Island, Easter Island and among the Amish people of North America. Founder effect is a special case of genetic drift.

LEARNING TIP

Genetic bottlenecks and the founder effect do not cause mutations or the emergence of harmful alleles, but they contribute to the increase of the frequency of mutations and harmful alleles within the resulting populations.

Questions

1 Male pot-bellied seahorses (*Hippocampus abdominalis*) (a type of fish) select larger females to mate with. The males incubate the eggs, but larger females produce more eggs to be fertilised. The females of this species do not seem to discriminate between size of mating partners. Smaller seahorses can hide more easily from predators.

    (a) If you randomly sampled a population of these seahorses, what proportions of large, small and intermediate sized individuals would you expect to find?

    (b) Is this an example of stabilising, directional or disruptive selection?

2 Suggest how shrinkage of a population size may affect the reproductive rate of that population.

3 About 12 000 years ago on Earth, a mass extinction event occurred that eliminated about 75% of the large mammal species. A small number of cheetahs survived.

    (a) Suggest why many of the world's cheetahs today have poor sperm quality, kinked tails and are all susceptible to the same infectious diseases.

    (b) How many of these factors impact on the survival of this species?

4 Maple syrup urine disease is a rare genetic disorder affecting about 1 in 180 000 births in the general human population. However, the incidence of this disorder is higher among the Amish people. Suggest an explanation for this observation.

5 Suggest why the population of Iceland has a higher than normal rate of breast cancer.

(12) The Hardy–Weinberg principle

By the end of this topic, you should be able to demonstrate and apply your knowledge and understanding of:

∗ the use of the Hardy–Weinberg principle to calculate allele frequencies in populations

---

> **KEY DEFINITION**
>
> **population:** members of a species, living in the same place and at the same time, that can interbreed.

Population genetics

Many topics in this chapter describe the inheritance of specific traits in individuals that may be determined by one, a few or many alleles. Population genetics attempts to study the changes in allele frequencies within a **population**, over time.

If a species is to succeed and not become extinct, it needs genetic variation between the individuals in its populations. Individuals inherit their genomes from their parents and pass on some of their genetic material to their offspring.

Population genetics studies the variation in the alleles and genotypes within the **gene pool** and how their frequencies vary over time. Factors affecting allele frequencies within populations, and hence the genetic diversity within a gene pool, include:

- population size
- mutation rate
- migration
- natural selection (whether stabilising, disruptive or directional)
- changes to the environment – e.g. adverse environments can lead to a genetic bottleneck
- isolation of a population from other populations of the same species (founder effect)
- non-random mating
- genetic drift
- gene flow.

If two populations of the same species become so genetically different that they can no longer interbreed and produce fertile offspring, they have undergone speciation and formed two new species.

The Hardy–Weinberg principle

The Hardy–Weinberg principle, developed by two mathematicians, is a fundamental concept in population genetics. It describes and predicts a balanced equilibrium in the frequencies of alleles and genotypes within a breeding population. It can also be used to determine the frequencies of those carrying a recessive allele (heterozygotes) for a genetic disorder with a recessive inheritance pattern, if we know the incidence of affected babies born each year in that population.

The principle assumes that:

- the population is large enough to make sampling error negligible
- mating within the population occurs at random
- there is no selective advantage for any genotype and hence no selection
- there is no mutation, migration or genetic drift.

> **WORKED EXAMPLE 1**
>
> Imagine we are studying the frequency of alleles and genotypes for a gene that has two loci, A and a.
> - The frequency of the dominant allele, A, is represented by the symbol, p.
> - The frequency of the recessive allele, a, is represented by the symbol q.
>
> Within the population, as there are only two alleles in the gene pool, the sum of $p + q = 100\%$ of the alleles.
> Therefore $p + q = 1$.
> The probabilities of the various genotypes arising from random matings of heterozygous individuals, Aa × Aa, are shown in Table 1
>
> | Gametes | A | a |
> |---------|-----|-----|
> | A | AA | Aa |
> | a | Aa | aa |
>
> **Table 1** Possible genotypes of offspring.
> - The probability that a sperm and egg both contain the dominant, p, allele is $p \times p = p^2$
> - The probability that a sperm and egg both contain the recessive allele, q, is $q \times q = q^2$
> - The probability that the gametes will each carry different alleles of a gene is $(p \times q)(p \times q) = 2pq$
>
> There can only be three genotypes in the population, so their frequencies must also add to 1 or 100%.
> Therefore we have another equation:
>
> $$p^2 + 2pq + q^2 = 1$$
>
> These two equations show that allele frequencies determine genotype frequencies.

WORKED EXAMPLE 2

Using the Hardy–Weinberg principle

Within a population some people suffer from cystic fibrosis, a recessive genetic disorder.

- Affected individuals are homozygous, genotype q^2.
- Normal people with no mutated allele are homozygous, genotype p^2.
- Symptomless carriers are heterozygotes, and the frequency of their genotype in the population is $2pq$.

In a population, 1 in 1600 people suffer from cystic fibrosis. These have the genotype q^2.

Therefore the frequency of q^2 is $1/1600 = 0.000625$

Therefore q = the square root of $0.000625 = 0.025$

If $p + q = 1$, then $p = 1 - 0.025 = 0.975$

Therefore $p^2 = 0.951$

The frequency of symptomless carriers, as given by $2pq = 2 (0.975 \times 0.025) = 0.049$

This is equivalent to about 5%, which means that approximately 1 in 20 of the population are symptomless carriers of the allele for cystic fibrosis.

Figure 1 Patient with cystic fibrosis receiving physiotherapy. Cystic fibrosis involves a mutation for chloride ion channel proteins in the plasma membranes of epithelial cells (see Book 1, Chapter 2.5). Patients need treatment to remove excess thick and sticky mucus from their lungs.

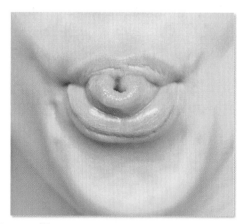

Figure 2 Tongue rolling.

Questions

1. What factors can alter the amount of genetic variation within a population?

2. In a population, 64% of people can roll their tongue and 36% cannot. Tongue rolling is a dominant trait. Calculate the percentage of people in the population that are heterozygous for the characteristic of tongue rolling.

3. Within a population of an African equatorial country, the incidence of sickle cell disease is about 1 in 500 births. Those who suffer from the disease inherit two recessive alleles for abnormal beta polypeptide chains of haemoglobin. Most sufferers of sickle cell disease die before reaching reproductive age, unless they receive regular treatment and blood transfusions. Heterozygotes (symptomless carriers) are protected from the most severe type of malaria.

   (a) Calculate the percentage of people in this population who are symptomless carriers of the disease.

   (b) Suggest why the recessive allele for this potentially fatal genetic disorder has quite a high frequency within this population.

4. The M/N blood group system is determined by a pair of codominant alleles, M and N. In a sample of people within a population, 26% of people are blood group M, and 25% are blood group N. Suggest why the Hardy–Weinberg principle does not need to be used to calculate:

   (a) the frequency of the genotype MN

   (b) the frequency of the M and N alleles.

(13) Isolating mechanisms

By the end of this topic, you should be able to demonstrate and apply your knowledge and understanding of:

* the role of isolating mechanisms in the evolution of new species

Speciation

Over time, one species may evolve into another, or it may evolve into two new species.

For a species to evolve into two species, it must be split into two isolated populations. If this happens, then any mutations that occur in one population are not transmitted by interbreeding to the other population. In each location, there will be different selection pressures and each population will accumulate different allele frequencies. Hence each population can evolve along its own lines.

At times during the evolutionary process, the two populations will be different, but still able to interbreed. They are then called sub-species.

When there have been sufficient genetic, behavioural and physiological changes in the two populations so that they can no longer interbreed, they are then separate species.

The process by which new species are formed is called **speciation**.

Isolating mechanisms

There are two main types of isolating mechanism: geographical and reproductive.

Geographical isolation

If populations are separated and isolated from each other by geographical features such as lakes, rivers, oceans and mountains, these also act as barriers to gene flow between the populations.

The isolated populations, being subject to different selection pressures in the two different environments, then undergo independent changes to the allele frequencies and/or chromosome arrangements within their gene pools. These genetic changes may be the result of mutation, selection and genetic drift.

As a result of natural selection, each population becomes adapted to its environment.

This type of speciation is called **allopatric speciation** ('allopatric' means 'in different countries').

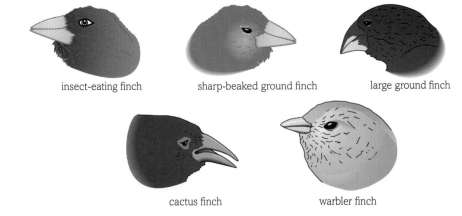

insect-eating finch   sharp-beaked ground finch   large ground finch

cactus finch   warbler finch

Figure 1 The heads of some species of Galapagos Island finches.

Reproductive isolation

Biological and behavioural changes within a species may lead to reproductive isolation of one population from another.

If a mutation leads to some organisms in a population changing their foraging behaviour and becoming active at dawn, dusk or at night rather than during the day, enabling them to exploit a new niche, the members of the diurnal population will be unlikely to mate with members of either the crepuscular (active at dawn and dusk) or the nocturnal populations.

Genetic changes can also lead to reproductive isolation. A change in chromosome number may:

- prevent gamete fusion
- make the zygotes less viable, so that they fail to develop
- lead to infertile hybrid offspring with an odd number of chromosomes, so that chromosome pairing during meiosis cannot occur.

Mating between members of the reproductively isolated populations may also be prevented by mutations leading to changes in:

- courtship behaviour, e.g. time of year for mating or courtship rituals that precede mating
- animal genitalia or plant flower structure.

Speciation resulting from reproductive isolation is called **sympatric** (meaning 'in the same country') **speciation**.

Figure 2 When horses and donkeys interbreed, sterile hybrid offspring result. If the mother is a horse and the father a donkey, the offspring is a mule.

Figure 3 Different species of duck rarely interbreed, due to differences in their courtship behaviour.

Figure 4 Interbreeding between different species of plants that exist in the same environment may be prevented, because the pollen of one species cannot germinate on the stigmas of other species.

DID YOU KNOW?

Scientists think that new species arise within a few thousand years (rapid in evolutionary and geological terms) and then remain unchanged for millions of years.

LEARNING TIP

Evolution is still occurring. Examples of evolution in action include tolerance to heavy metals in plants, antibiotic resistance in bacteria, insecticide resistance in insects and warfarin resistance in rats. Some animals also show cultural evolution – for example, captured dolphins learn to tail walk, and when they are re-released into the wild, other dolphins copy this behaviour; chimpanzees taught to peel oranges then teach other chimpanzees in their troupe how to peel oranges.

Questions

1   The chromosome number of horses, *Equus ferus caballus*, is 64. The chromosome number of donkeys, *Equus africanus asinus*, is 62. Suggest why mules are infertile.

2   In some species of British songbirds, scientists have observed that members of the species that live in towns have developed differences in their song, to enable them to be heard above the background noise of traffic by potential mates. If the changes in song develop to the extent that their rural counterparts no longer recognise and respond to the song, two new species of bird will have been formed. Explain which sort of speciation this is.

3   Suggest how certain human activities may speed the process of evolution among other species.

4   Marine iguanas are found only in the Galapagos Islands. Unlike other iguanas, they can forage for marine algae in the sea. They have webbed feet and flattened tails that they use for swimming, and claws that help them grip rock surfaces. There are also land iguanas on the Galapagos Islands. Marine and land Galapagos iguanas are classified as subspecies.

   (a) Suggest why the marine and land Galapagos Island iguanas are not classified as separate species.

   (b) These Galapagos iguanas have evolved from iguanas found on the mainland. What type of speciation was involved in their evolution?

Artificial selection

By the end of this topic, you should be able to demonstrate and apply your knowledge and understanding of:

* the principles of artificial selection and its uses
* the ethical considerations surrounding the use of artificial selection

The principles and uses of artificial selection

Humans have been practising animal and plant breeding via **artificial selection** for about 10 000 years, since the beginning of settled agriculture.

Whereas when natural selection is operating the environment is the agent of selection, during artificial selection, humans are the agents of selection. Breeders select individuals with the desired traits and allow them to interbreed, whilst at the same time preventing those without the desired characteristics from breeding.

Considering the number of species of organisms on the planet, humans have domesticated very few. Included are cereals, potatoes, vegetables and fruits, cattle, pigs, sheep and goats, horses, oxen, dogs, cats, pigeons and poultry.

Desirable characteristics in plants include increased yield, and pest and disease resistance. In livestock, desirable characteristics include docility, placidity and the ability to be trained. Animals that normally live in social groups (herds) with a dominance hierarchy may be able to be trained to accept a human as the pack leader, and to tolerate being penned with other animals.

Artificial selection produces new breeds of organisms.

Organism	Desirable traits/some examples of their use by humans
cereal (e.g. wheat, barley, maize, millet) rice fruits and vegetables	increased yield; shorter maturation time; resistance to pests, infection, frost, drought, flooding and wind; improved flavour
cattle	milk, meat and leather
sheep and goats	wool and meat
horses	haulage and transport; racing and military use
pigs	meat
pigeons	flight capacity, plumage
poultry	eggs, meat and feathers
dogs	hunting, guarding, racing, retrieving and companionship
cats	pest control and companionship

Table 1 Examples of organisms that humans have bred using artificial selection.

New breeds may be produced by selective breeding programmes. Breeders may grow many plants of a particular type under the conditions they wish these plants to withstand – such as low temperature or high salinity. They will then select those individuals that grow best under these conditions and cross pollinate them; collect and sow the seeds and repeat this process over many generations. A selective breeding programme takes about 20 years.

Inbreeding depression and hybrid vigour

At each stage of selective breeding, the individuals with the desirable characteristics and no or few undesirable characteristics are selected. Inevitably, the genetic diversity in the gene pool of the selected breed is reduced. If related individuals are crossed, **inbreeding depression** can result. The chances of an individual inheriting two copies of a recessive harmful allele are increased.

Breeders sometimes outcross individuals belonging to two different varieties, to obtain individuals that are heterozygous at many gene loci. This property is termed **hybrid vigour**.

Selective breeding, whilst developing bigger and better varieties of crop plants and animal breeds, has reduced the organisms' genetic diversity. The number of commercially grown varieties of crop has greatly reduced over the last 50–100 years. All commercial varieties are genetically similar; if a pathogen was introduced, most plants would succumb to the infection. Breeders may need to outcross the cultivated varieties with varieties more like their wild ancestors to increase hybrid vigour. Samples of such wild ancestral types need to be conserved, often in gene banks.

Gene banks

Much of the wheat grown in the UK has a dwarfing allele introduced from a Japanese variety of wheat. If given extra fertiliser, the wheat does not grow taller and fall over in the wind, but uses the extra nutrients to increase seed size and yield. However, if the environmental temperature rises above 30 °C, the effect of this allele is changed and yield is decreased. If climate change is likely to produce higher temperatures during the British summer, a new breed of wheat will have to be developed. Wheat breeders are looking, in a gene bank, for different dwarfing alleles. Gene banks store genomes, but in their organisms. Examples of gene banks include:

- rare breed farms
- wild populations of organisms
- crops in cultivation
- botanic gardens and zoos
- seed banks
- sperm banks
- cells in tissue culture
- frozen embryos.

DID YOU KNOW?

If the environmental temperature increases to 30 °C or above when wheat crops are flowering, then their metabolism is disrupted and seeds do not form.

Ethical considerations of artificial selection

Ethical considerations include the following:

- Domesticated animals retain many juvenile characteristics, making them friendly, docile and playful, but less able to defend themselves. The loss of their nervous disposition can also make them easy prey.
- Livestock animals, such as pigs, selected to have more lean meat and less fat, might succumb to low environmental temperatures during winter if they were not housed.

Dogs have been domesticated for many thousands of years and used by humans for hunting, companionship, protection, herding, transport, as guide dogs and to help deaf or disabled people, as well as for their aesthetic qualities.

- The traits in dogs, considered desirable by humans, might put the dogs at a selective disadvantage if they had to survive in the wild.
- Some breeds, through inbreeding from a limited number of pedigree dogs, have susceptibility to disease (see Table 2).
- Some coat colours, selected because humans like the look of them, would also fail to camouflage the animals.

Breed of dog	Conditions to which the breed is susceptible
Boxer	cancer and heart disease
Labrador retriever	chronic skin itchiness; abnormality of hip joints and elbow joints; shoulder pain and lameness
German shepherd	heart disease; cancer; elbow and hip dysplasia; skin infections; lack of digestive enzymes
Cocker spaniel	inflammation of ear; glaucoma (increased pressure inside the eyes)
Bulldog and Pekingese	breathing problems; hip and elbow problems; difficulty whelping
Dalmatian	congenital deafness; heart disease; skin itchiness and infections
Doberman	heart failure; spinal cord deformity and paralysis
Great Dane	heart disease; bone cancer; twisting of the stomach; shoulder pain and lameness; spinal cord deformity and paralysis
West Highland Terrier	dry eye; skin irritation and infections

Table 2 Different breeds of dog are susceptible to various conditions, due to artificial selection.

Australian Cattle Dog   Australian Shepherd   Basset Hound   Border Collie

Bulldog   Chow Chow   Dalmatian   Labrador Retriever

Figure 1 Breeds of dogs, produced by artificial selection, all descend from wolves, but are now very different from the wolf and from each other.

Questions

1 Outline how a plant breeder would carry out a selective breeding programme to develop a new variety of frost-resistant potatoes.

2 What is 'inbreeding depression'?

3 Explain the value of gene banks in helping breeders introduce hybrid vigour into a variety of crop plant.

4 Outline the ethical considerations surrounding the use of artificial selection.

5 Dogs, such as German shepherds, Belgian Malinois and Labrador retrievers, are trained to discover IEDs (improvised explosive devices) and are used in war zones. They learn to recognise the smell of certain chemicals used in making the devices. Their handlers are taught how to look after them properly and a strong bond builds between human handler and dog. However, dogs have died in explosions. Discuss the ethical considerations around such use of dogs.

THE GUT MICROBIOME

Scientists have known for quite a long time that the bacteria in our intestines help us digest food, but recent research has shown that the microbes that live in and on us are vital for our wellbeing.

WE RELY ON MORE GENES THAN THOSE OF OUR OWN GENOME

We have many species of bacteria and Archaea inhabiting our gastric tract. They are regarded as mutualistic symbionts and they help us digest food and produce some nutrients such as vitamin K. These gut-dwelling bacteria are termed the gut microbiome.

Scientists have recently found that we have about 100 species of gut bacteria, and different people may harbour different proportions of these species. The actual number of bacteria is about 10 times the total number of cells in our bodies, but weighing in total between 200 g and 1500 g. Bacteria also live on our skin and in our noses and vaginas. We are a veritable ecosystem. Different types of bacteria can inhabit only specific areas of our bodies.

The bacteria we house have genes that are expressed. Human gut bacteria produce many chemicals and some of them act within us to help regulate our appetite. It is possible that people who have too often been exposed to antibiotics or who do not eat enough dietary fibre have an imbalance of gut bacteria, resulting in a deficiency of appetite-regulating chemicals that contributes to obesity.

A study published in *Nature* in August 2013 showed that people with more diverse gut microbiomes showed fewer signs of metabolic syndrome including obesity, high blood pressure and insulin resistance. Lack of certain types of gut bacteria may contribute to some autoimmune diseases such as rheumatoid arthritis, Crohn's disease and multiple sclerosis. Some gut bacteria can modify the production of neurotransmitters, hence some imbalances of gut bacteria diversity can contribute to depression.

Researchers at Imperial College, London, and at Johns Hopkins University in the USA, have shown that the gut biome in mice helps to regulate their blood pressure. Mice treated with antibiotics to knock out their gut bacteria showed a significant rise in blood pressure. This indicates that the gut microbiome is also involved in regulating aspects of the host's physiology by making chemicals that stimulate the production of specific enzymes involved in a metabolic pathway that leads to a reduction in blood pressure.

Some people suffer from unrestrained growth of a particular gut-dwelling bacterium, *Clostridium difficile*. This anaerobic bacterium is found in everyone and normally kept in check by other intestinal bacteria. When certain antibiotics are taken and some of the other gut bacteria are killed, *C. difficile* can multiply unchecked and produce toxins that cause severe diarrhoea and inflammation of the bowel. This may be fatal. These spore-forming bacteria can be spread on the hands of hospital staff, patients and their visitors, as well as on surfaces such as toilets, bedpans, door handles, curtains, clothes and floors.

Scientists have found that transplanting a small amount of faeces from a healthy person to a person suffering from a *C. difficile* infection can abolish the infection where antibiotics have failed to do so. Pharmaceutical companies are now producing capsules to make this transplant of bacteria easier to undertake, as the capsules can be swallowed. However, if someone receives a transplant of gut bacteria, they also need to adopt a fibre-rich diet that promotes the growth of those bacteria. This involves increasing their dietary fibre, mostly in the form of more fruit and vegetables.

Humans have about 20 000 genes in their genome. The collective genome of the gut microbiome may contain as many as 100 times more genes than our own genome. Many of these microbial genes code for proteins that we need, so it appears that our microbiomes influence our phenotypes. Our resident bacteria also help our immune system to develop and function properly.

Because we rely on more than just our own genes, we may perhaps regard ourselves as superorganisms.

Source
- 'Me, myself, us. The human microbiome – looking at humans as ecosystems that contain many collaborating species could change the practice of medicine', *The Economist*, 18 August 2012

Where else will I encounter these themes?

DID YOU KNOW?

Babies are born head first and facing downwards. As they come through the birth canal and out of the mother's body, they collect bacteria from her anus. These bacteria start the colonisation of the infant's gut. Other bacteria from the environment also enter the infant's gut and this colonisation is essential for the health of the baby. Babies born by Caesarean section do not collect bacteria during birth and there is a delay in their gut being colonised.

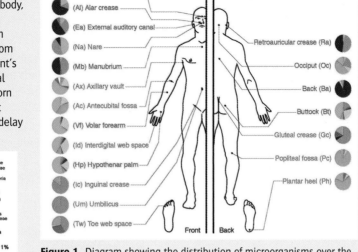

Figure 1 Diagram showing the distribution of microorganisms over the skin of the human body.

This text is adapted from an article in *The Economist*. Let's start by considering the nature of the writing in the article.

1. From the content and style of writing, suggest – with reasons – the type of audience at which this article is aimed.

Now let us look at the biological concepts underlying the information in this article.

2. Explain why the bacteria living in our intestines and on our skin are regarded as symbionts rather than pathogens.

3. Under what circumstances may some of these gut-dwelling bacteria become pathogenic?

4. Explain why, although we have 10 times as many bacterial cells as our own cells, the total mass of bacterial cells inhabiting our bodies is only around 1 kg.

5. Explain why overexposure to antibiotics may play a part in the increase in incidence of obesity.

6. Describe the experimental evidence that products of some genes from the gut bacteria may be involved in helping to lower and regulate the host's blood pressure.

7. Suggest how a faecal transplant from a healthy person can treat an infection by *C. difficile*.

8. A superorganism is an organism consisting of many organisms. Examples include ant, termite and bee colonies. Suggest why humans may be considered to be superorganisms.

Activity

Write a short article (500–800 words) for a health magazine, explaining how our gut microbiome can affect our health and also explaining what the Human Microbiome Project is.

Remember that you are addressing an audience, many of whom may not possess much biological knowledge.

You will need to research, using the Internet, to find out about the Human Microbiome Project.

Practice questions

1. Which of the following is a physical mutagenic agent? [1]
   A. Benzopyrene
   B. Colchicine
   C. Gamma rays
   D. Viruses

2. Which of the following is not a chromosome mutation? [1]
   A. Aneuploidy
   B. Non-disjunction
   C. Translocation
   D. Frameshift

3. Which scientist is known as the father of genetics? [1]
   A. Charles Darwin
   B. Edward Jenner
   C. Gregor Mendel
   D. Reginald Punnett

4. Read the following statements.
   (i) The expected F_2 phenotype ratio for a dihybrid cross is 9:3:3:1.
   (ii) Chromosomes that determine whether we are male or female are called autosomes.
   (iii) Where alleles are codominant, the phenotype of heterozygotes is different from that of homozygotes.

   Which statement(s) is/are true? [1]
   A. (i), (ii) and (iii)
   B. (i) and (iii) only
   C. (i) and (ii) only
   D. (i) only

5. Read the following statements.
   (i) The ABO blood group in humans is determined by three genes.
   (ii) Any individual can only have two alleles of a particular gene locus in their genome.
   (iii) The human sex chromosomes, X and Y, are not fully homologous.

   Which statement(s) is/are true? [1]
   A. (i), (ii) and (iii)
   B. (ii) and (iii) only
   C. (i) and (iii) only
   D. (iii) only

   [Total: 5]

6. Manx cats do not have a tail. This abnormality is due to a dominant allele M that leads to a spinal deformity. This allele, in the homozygous condition, is lethal. All Manx cats are heterozygous, genotype Mm, at this gene locus.

   A cat breeder has a male Manx cat and female Manx cat, He allows them to breed and after several litters this pair of cats has produced 22 Manx cats and 12 cats with tails.

   (a) What is meant by the terms:
      (i) allele [1]
      (ii) homozygous? [1]
   (b) Construct genetic diagrams to explain the observations described above. [6]
   (c) Predict the expected phenotype ratios if a Manx cat mated with one of the tailed cats. [2]

   Cats use their tails to help them balance and also for communication to other cats. Side swishing indicates anger, a female's tail held to one side indicates to a male cat that she is sexually receptive, a fluffed up tail indicates fear or aggression, a raised curved tail indicates a friendly greeting and a lowered tail indicates submission.

   (d) In some circumstances, such as during a war, many cats are abandoned by their owners and may become feral. How would you expect the frequencies of the M and m alleles, in a feral population originating from some abandoned pet Manx cats, to change over several generations of cats? [4]

   [Total: 14]

7. In 1984, Siamese pet rats made their first appearance at an American Fancy Rat and Mouse Association show. These rats have cream/beige-coloured coats with dark noses, ears, feet and tails. The darkest area is the tip of the tail.

   This coloration is due to acromelanism – a type of thermo-sensitivity. In the cooler areas of the rats' bodies, the fur develops a darker colour.

   Siamese rats are darker in winter than during summer.

   The Siamese rat kittens are born with an even creamy colour without the dark points.

   The phenotype cream coat colour is recessive to black coat colour.

   Rat breeders are advised to sometimes outcross their Siamese rats to black rats.

   The breeders' handbook states that if Siamese rats are bred to black rats, all the offspring will be black. If two of these black rats are bred together, one-quarter of their offspring will be Siamese.

   (a) Suggest a possible mechanism for acromelanism. [2]
   (b) Suggest why Siamese rat kittens are born without the dark-coloured noses, ears, feet and tails. [1]

(c) Explain the terms:
  (i) hybrid vigour [2]
  (ii) inbreeding depression. [2]

(d) Construct a genetic diagram to explain the observations of:
  (i) crossing Siamese rats with black rats [3]
  (ii) crossing the F_1 black rats together. [3]

(e) Explain why rat breeders are advised to sometimes outcross their Siamese rats with black rats. [4]

[Total: 17]

8. (a) Explain the terms:
  (i) sex linkage
  (ii) autosomal linkage. [3]

(b) A gene on the X chromosome codes for eye colour in fruit flies, *Drosophila melanogaster*. When true-breeding red-eyed females mate with white-eyed males, all the progeny have red eyes.

When flies of this F_1 generation are allowed to interbreed, they produce 650 red-eyed females, 320 red-eyed males and 340 white-eyed males.

Construct annotated genetic diagrams to explain these observations. [7]

(c) A couple, both normal for blood clotting time, have a son who suffers from haemophilia A.
  (i) What is the most likely explanation for this child suffering from haemophilia? [2]
  (ii) Predict and explain the probability of their next child suffering from haemophilia? [3]

(d) Male pattern baldness in humans is determined by many genes on autosomes and also needs the trigger of dihydrotestosterone, a potent derivative of testosterone. Both parents contribute alleles that may cause male pattern baldness in their sons.
  (i) Explain why male pattern baldness is not a sex-linked characteristic. [2]
  (ii) Suggest why this type of baldness is extremely rare in females, even if they inherit alleles predisposing them to baldness. [2]

[Total: 19]

9. (a) Explain how sexual reproduction makes individuals genetically unique. [8]

(b) A rare disease called alkaptonuria causes patients' urine to turn black on exposure to air, but otherwise seems to do no harm. It was at first thought to be caused by infection, but in the early 1900s an English doctor, Archibald Garod, noticed that this disease was more common following marriages between cousins and that in susceptible families the ratio of unaffected children to affected children was almost exactly 3:1.

Garod proposed that this rare disorder, affecting about 1 in 200 000 people, was genetic and followed a Mendelian inheritance pattern. The substance that makes the urine black is homogenistic acid. In normal people this chemical is broken down and does not accumulate in urine. Garod postulated that patients with alkaptonuria lacked an enzyme as a result of a genetic defect.
  (i)   Suggest a function for the enzyme that is absent in patients with alkaptonuria. [1]
  (ii) Is the inheritance pattern for this condition dominant or recessive? Use genetic diagrams to explain your answer. [3]
  (iii) Explain why the incidence of this disease is more common among children of marriages between cousins. [2]

(c) The 'one gene makes one enzyme' hypothesis was proposed in 1941 by two scientists, Beadle and Tatum. Suggest three ways in which this hypothesis has had to be modified since then. [3]

[Total: 17]

10. (a) In many regions of the world, most adult humans are lactose intolerant. They lack the enzyme lactase and cannot digest lactose sugar in milk and dairy products, such as cheese and chocolate. These foods give them serious intestinal discomfort, flatulence and vomiting.

In all baby mammals, the gene coding for lactase is switched on, but once the babies are weaned, this gene is switched off.

In human societies where dairy produce has been an important part of the diet for many thousands of years, the frequency of lactose-intolerant individuals is low.
  (i) Suggest how natural selection has led to a higher frequency of lactose-tolerant individuals in European countries. [4]
  (ii) Suggest why lactose-intolerant individuals can eat yoghurt without suffering any adverse effects. [2]

(b) Adult cats are not able to tolerate milk. However, there is special milk, available to purchase, that has undergone a treatment involving heating, and is suitable for adult cats.
  (i) Suggest why adult cats should not be given milk to drink. [2]
  (ii) Suggest how the treatment the special milk for adult cats undergoes makes it suitable for adult cats to drink. [2]

[Total: 10]

CHAPTER
6.3

MANIPULATING GENOMES

Introduction

Modern humans have long wondered about our origins, how our bodies work, why we behave as we do and why, although very similar to each other, we are also all individual. Many academic disciplines have attempted to answer these questions. The Darwin–Wallace theory of evolution by natural selection caused a paradigm shift – a radical change in the way that humans viewed their place in nature. Knowledge of our genetic code, beginning with the discovery of the structure of DNA by Watson, Crick and Franklin in 1953, and leading to the complete sequencing of the human genome 50 years later, has changed our understanding of life on Earth.

Unlocking the secrets of chromosomes and genes has helped us understand how life came to be the way it is, and how our individuality is a product of both our genes and the effect of our environment on the expression of those genes – nature via nurture.

Genome sequencing gives information about the location of genes. Our ability to understand the coded language of DNA has provided more evidence and has allowed us to better understand the evolutionary relationships between the extinct and extant species on Earth; to solve crimes and convict criminals; to correct miscarriages of justice and exonerate, by using DNA-based evidence, some of those who have been wrongly convicted; to manipulate genomes and produce genetically-modified organisms; and to develop gene therapy with a potential to treat certain genetic disorders. It has also indicated that what was previously described as 'junk DNA' plays a part in regulating gene expression.

This branch of biology has led to a new branch of science called bioinformatics, which involves computer science, maths, statistics and engineering to process, store and understand biological data.

The capacity to manipulate genomes has many potential benefits, but the implications of such techniques are subject to much public debate and ethical issues to be considered. For a public debate to be meaningful, the public needs to be well informed of this topic.

All the maths you need

To unlock the puzzles of this chapter you need the following maths:

* Translate graphical and numerical information

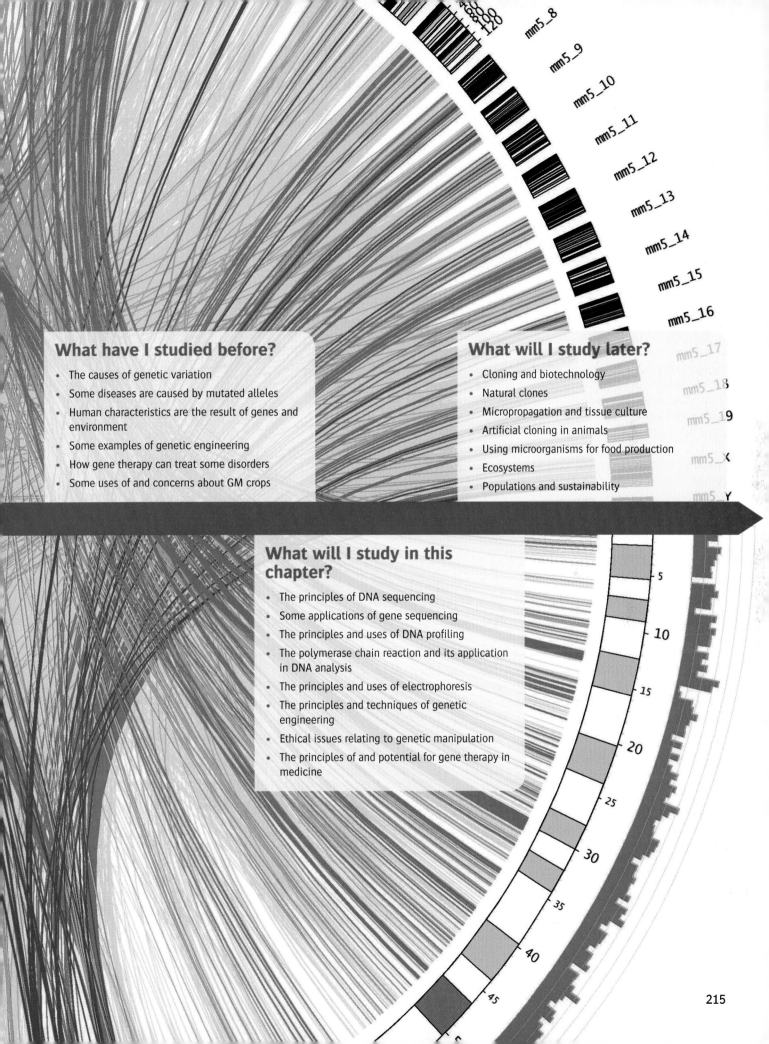

What have I studied before?

- The causes of genetic variation
- Some diseases are caused by mutated alleles
- Human characteristics are the result of genes and environment
- Some examples of genetic engineering
- How gene therapy can treat some disorders
- Some uses of and concerns about GM crops

What will I study later?

- Cloning and biotechnology
- Natural clones
- Micropropagation and tissue culture
- Artificial cloning in animals
- Using microorganisms for food production
- Ecosystems
- Populations and sustainability

What will I study in this chapter?

- The principles of DNA sequencing
- Some applications of gene sequencing
- The principles and uses of DNA profiling
- The polymerase chain reaction and its application in DNA analysis
- The principles and uses of electrophoresis
- The principles and techniques of genetic engineering
- Ethical issues relating to genetic manipulation
- The principles of and potential for gene therapy in medicine

① DNA sequencing

By the end of this topic, you should be able to demonstrate and apply your knowledge and understanding of:

* the principles of DNA sequencing and the development of new DNA sequencing techniques

Early DNA research

By the early 1970s, the structure of DNA was known, as were the sequences of base triplets that coded for the various amino acids. However, at this time it was difficult to work out the sequence of the nucleotide base triplets in genes. In 1969, a gene was isolated from a bacterial chromosome. In 1972, a Belgian molecular biologist sequenced a gene that codes for the protein coat of a virus, MS2. Both scientists worked from the mRNA transcribed from the gene, and not the raw DNA. RNA is unstable and this whole process was extremely slow and only suitable for very short genes.

In 1975, the British biochemist Fred Sanger developed a method that ultimately allowed scientists to sequence whole genomes.

Fred Sanger's DNA sequencing approach

Sanger's approach was to use a single strand of DNA as a template for four experiments in separate dishes. Each dish contained a solution with the four bases – A, T, C and G – plus an enzyme, DNA polymerase.

To each dish, a modified version of one of the DNA bases was added. The base was modified in such a way that, once incorporated into the synthesised complementary strand of DNA, no more bases could be added. Each modified base was also labelled with a radioactive isotope.

As the reaction progressed, thousands of DNA fragments of varying lengths were generated. The DNA fragments were passed through a gel by electrophoresis. Smaller fragments travelled further, so the fragments became sorted by length.

The nucleotide base at the end of each fragment was read according to its radioactive label (see Figure 1).

- If the first one-base fragments had thymine at the end, then the first base in the sequence is T.

- If the two-base fragments have cytosine at the end, then the sequence is TC.

- If the three-base fragment ends with guanine, then the base sequence is TCG.

single-stranded DNA sequence

Single strands of DNA are broken into chunks of every possible length, and the final base is tagged with a radioactive label.

The radioactive tag at the end of each length of DNA is read and the pieces are lined up in order of length to generate the sequence.

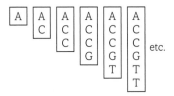

Figure 1 How chain termination sequencing works.

This method was efficient and safe. Sanger used it to sequence the genome of a phage virus (a virus that infects bacteria) called Phi-X174, the first DNA-based organism to have its genome sequenced. He had to count off the bases, one by one, from the bands in a piece of gel – a very time-consuming and therefore costly process.

In 1981, Sanger published his sequencing of the human mitochondrial genome, consisting of 37 genes and 16 569 base pairs. In 1984, scientists sequenced the 170 kilobase pair-long genome of the Epstein–Barr virus. In 1995, the genome of the bacterium *Haemophilus influenzae* was sequenced using this approach.

Cloning DNA

The gene to be sequenced was isolated, using restriction enzymes, from a bacterium.

The DNA was then inserted into a bacterial plasmid (the vector) and then into an *Escherichia coli* bacterium host that, when cultured, divided many times, enabling the plasmid with the DNA insert to be copied many times.

Each new bacterium contained a copy of the candidate gene. These lengths of DNA were isolated using plasmid preparation techniques and were then sequenced.

The first DNA sequencing machine

In 1986, the first automated **DNA sequencing** machine was developed at the California Institute of Technology, based on Fred Sanger's method. Fluorescent dyes instead of radioactivity were used to label the terminal bases. These dyes glowed when scanned with a laser beam, and the light signature was identified by computer. This method dispensed with the need for technicians to read **autoradiograms**.

High throughput sequencing

In the first decade of the twenty-first century, a variety of approaches was used to develop fast, cheap methods to sequence genomes. One of them is pyrosequencing, outlined in Figure 2.

Pyrosequencing

This method was developed in 1996 and uses sequencing by synthesis, not by chain termination as in the Sanger method.

It involves synthesising a single strand of DNA, complementary to the strand to be sequenced, one base at a time, whilst detecting, by light emission, which base was added at each step.

1. A long length of DNA to be sequenced is mechanically cut into fragments of 300–800 base pairs, using a nebuliser.

2. These lengths are then degraded into single-stranded DNA (ssDNA). These are the template DNAs and they are immobilised.

3. A sequencing primer is added and the DNA is then incubated with the enzymes DNA polymerase, ATP sulfurylase, luciferase, apyrase and the substrates adenosine 5' phosphosulfate (APS) and luciferin. Only one of the four possible activated nucleotides, ATP, TTP, CTP and GTP is added at any one time and any light generated is detected.

4. (a) One activated nucleotide (a nucleotide with two extra phosphoryl groups), such as TTP (thymine triphosphate, is incorporated into a complementary strand of DNA using the strand to be sequenced as a template. (b) As this happens, the two extra phosphoryls are released as pyrophosphate (PP$_i$). (c) In the presence of APS, the enzyme ATP sulfurylase converts the pyrophosphate to ATP. (d) In the presence of this ATP, the enzyme luciferase converts luciferin to oxyluciferin. (e) This conversion generates visible light which can be detected

by a camera. The amount of light generated is proportional to the amount of ATP available and, therefore, indicates how many of the same type of activated nucleotide were incorporated adjacently into the complementary DNA strand.

Unincorporated activated nucleotides are degraded by apyrase and the reaction starts again with another nucleotide.

One million reads occur simultaneously, so a 10-hour run generates 400 million bases of sequencing information. Software packages assemble these sequences into longer sequences.

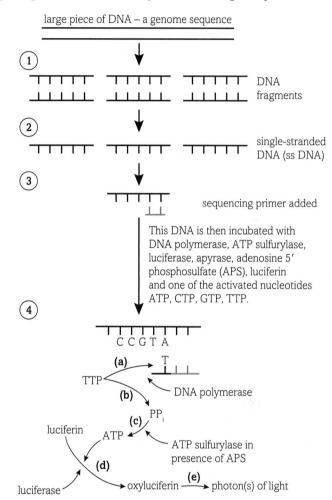

Figure 2 Diagram outlining the process of pyrosequencing.

Bioinformatics

A branch of biology called **bioinformatics** has grown out of this research, to store the huge amounts of data generated. It would have been impossible to store and analyse these data prior to computers and microchips. Software packages are specially designed for this purpose.

Questions

1　Figure 3(a) shows the base sequence found using chain termination sequencing and gel electrophoresis. What is the base sequence indicated by diagram (b)?

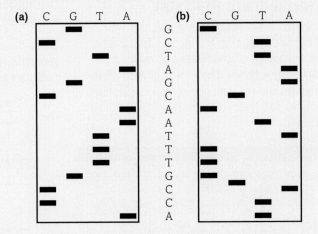

Figure 3 (a) and (b).

2　Figure 4 shows the brightness generated from each activated nucleotide added during high throughput sequencing. What is the sequence of nucleotide bases on the length of DNA being sequenced?

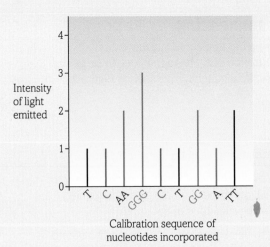

Figure 4 Brightness from each activated nucleotide in a single well during high throughput sequencing.

By the end of this topic, you should be able to demonstrate and apply your knowledge and understanding of:

∗ how gene sequencing has allowed for genome-wide comparisons between individuals and between species

∗ how gene sequencing has allowed for the sequences of amino acids in polypeptides to be predicted

∗ how gene sequencing has allowed for the development of synthetic biology

The Human Genome Project

Scientists predicted that the human genome would contain about 100 000 genes. In 1990, the Human Genome Project was launched, and the genome was sequenced by 2003. Scientists were surprised to learn that the human genome contained only about 24 000 genes – not many more than in the mouse genome.

Genome-wide comparisons between individuals and species

Whole genome sequencing determines the complete DNA sequence of an organism's genome – in the case of eukaryotic cells, that is the genetic material of the chromosomes, mitochondria and, if plants or algae, also of chloroplasts. Sequenced genomes are stored in gene banks.

Comparisons between species

When the human genome was compared with those of other species, it became clear that few human genes are unique to us. Most of our genes have counterparts in other organisms. We share over 99% of our genes with chimpanzees. This verifies that genes that work well tend to be conserved by evolution. For example, pigs and humans have similar genes for insulin, which is why, prior to genetically-modifying bacteria to make insulin, pig insulin was used to treat patients with diabetes.

Sometimes, as evolution progresses, some genes are co-opted to perform new tasks. Tiny changes to a gene in humans called *FOXP2*, which is found in other mammals including mice and chimpanzees, mean that in humans this gene allows speech.

Many of the differences between organisms are not because the organisms have totally different genes, but because some of their shared genes have been altered and now work in subtly different ways. Some changes to the regulatory regions of DNA that do not code directly for proteins have also altered the expression of the genomes – regulatory and coding genes interact in such ways that, without increasing the number of genes, the numbers of proteins made may be increased.

Evolutionary relationships

Comparing genomes of organisms thought to be closely related species has helped confirm their evolutionary relationships or has led to new knowledge about the relationships and, in some cases, to certain organisms being reclassified.

The DNA from bones and teeth of some extinct animals can be amplified and sequenced, so that the animals' evolutionary history can be verified.

Recently, samples of extinct cave bear (*Ursus spelaeus*) genomes were sequenced using high throughput techniques, and the sequence data obtained was compared to those of dogs. Dogs and bears diverged about 50 million years ago and share 92% of their genome.

> **DID YOU KNOW?**
> Bacterial genomes can be extracted from ancient human bones and the evolution of certain human pathogens can therefore be studied.

Variation between individuals

All humans are genetically similar. Except for rare cases, where a gene has been lost by deletion of part of a chromosome, we all have the same genes, but we have different alleles. About 0.1% of our DNA is not shared with others. This sounds very small, but given that our genome contains three billion DNA base pairs, this means that there are three million places on the DNA lengths where our DNA sequences can differ, due to random mutations such as substitution.

The places on the DNA where these substitutions occur are called single nucleotide polymorphisms, or SNPs (pronounced 'snips'). Some have no effect on the protein, some can alter a protein or alter the way a piece of RNA regulates the expression of another gene.

Methylation of certain chemical groups in DNA plays a major role in regulating gene expression in eukaryotic cells. Methods to map this methylation of whole human genomes can help researchers to understand the development of certain diseases, for example certain types of cancer and why they may or may not develop in genetically similar individuals. The study of this aspect of genetics is called epigenetics.

Predicting the amino acid sequences of proteins

Determining the sequence of amino acids within a protein is laborious and time consuming. However, if researchers have the organism's genome sequenced and know which gene codes for a specific protein, by using knowledge of which base triplets code for which amino acids, they can determine the primary structure of proteins. The researchers need to know which part of the gene codes for exons and which codes for introns (topic 6.1.2).

Synthetic biology

Synthetic biology is an interdisciplinary science concerned with designing and building useful biological devices and systems. It encompasses biotechnology, evolutionary biology, molecular biology, systems biology and biophysics (see Figure 1). Its ultimate goals may be to build engineered biological systems that store and process information, provide food, maintain human health and enhance the environment.

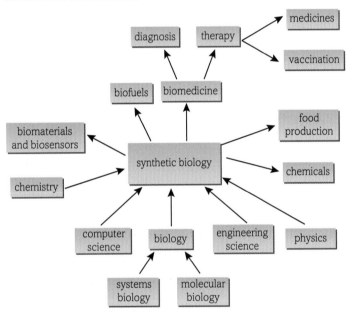

Figure 1 Aspects of synthetic biology.

The sequences of DNA found by analysing genomes provide potential building blocks for synthetic biologists to build devices.

LEARNING TIP
Do not confuse gene sequencing with DNA profiling.

Table 1 gives some examples of synthetic biology applications.

Example	Description of application
Information storage	Scientists can encode vast amounts of digital information onto a single strand of synthetic DNA. One project has encoded the complete works of William Shakespeare onto a strand of synthetic DNA.
Production of medicines	*Escherichia coli* and yeast have both been genetically engineered to produce the precursor of a good antimalarial drug, artemisinin, previously only available by extracting it from certain parts of the *Artemisia* plants at particular times in the plant's life cycle.
Novel proteins	Designed proteins have been produced, for example one that is similar to haemoglobin and binds to oxygen, but not to carbon monoxide.
Biosensors	Modified bioluminescent bacteria, placed on a coating of a microchip, glow if air is polluted with petroleum pollutants.
Nanotechnology	Material can be produced for nanotechnology – e.g. amyloid fibres for making biofilms – for functions such as adhesion.

Table 1 Some applications of synthetic biology.

Bioethics

Synthetic biology raises issues of ethics and biosecurity. Extensive regulations are already in place, due to 30–40 years of using genetically-modified organisms. There are many advisory panels and many scientific papers have been written on how to manage the risks. Synthetic biology is not about making synthetic life forms from scratch, but is about a potential for new systems with rewards and associated risks to be managed.

Questions

1. What is meant by the phrase 'genes that work well will be conserved'?

2. Outline the useful applications of interspecific comparisons of genome sequences.

3. Outline the useful applications of comparing genomes of individuals.

(3) DNA profiling

By the end of this topic, you should be able to demonstrate and apply your knowledge and understanding of:

* the principles of DNA profiling and its uses

The development of DNA profiling

In 1978, Alec Jeffreys was locating tandem repeat sequences of DNA. Tandem repeats are repetitive segments of DNA that do not code for proteins. They may be between 10 and 100 base pairs long and they all feature the same core sequence, GGGCAGGAXG, where X can be any one of the four nucleotide bases. Tandem repeats occur at more than 1000 locations in the genome and, in each of these places, they may be repeated a random number of times. Some types are highly variable and called variable number tandem repeats (VNTRs).

Jeffreys obtained some DNA from his lab technician and her parents and analysed it. He was surprised to find that the number of tandem repeats showed a family resemblance, but the DNA profile for each family member was unique. He realised that a person's DNA profile could confirm or refute paternity and maternity.

Figure 1 Professor Sir Alec Jeffreys.

DNA profiling

Procedure

There are slightly different procedures, but the principles underlying them all are:

1. DNA is obtained from the individual – either by a mouth swab, from saliva on a toothbrush, from blood or hair or, in the case of ancient remains, from bone.

2. The DNA is then digested with restriction enzymes. These enzymes cut the DNA at specific recognition sites. They will cut it into fragments, which will vary in size from individual to individual.

3. The fragments are separated by gel electrophoresis (topic 6.3.5) and stained. Larger fragments travel the shortest distance in the gel.

4. A banding pattern can be seen.

5. The DNA to which the individual's is being compared is treated with the same restriction enzymes and also subjected to electrophoresis.

6. The banding patterns of the DNA samples can then be compared.

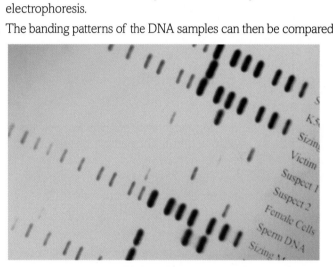

Figure 2 DNA fingerprint results from a rape investigation. The suspect's DNA is a match to the sperm DNA taken from the victim.

> **DID YOU KNOW?**
>
> In 1986, DNA profiling was first used to solve some murders in Leicestershire. Professor Alec Jeffreys compared the suspect's DNA with samples from the scene of the crime and established that the suspect was not guilty. The police then took DNA samples from more than 5000 local men. None matched the crime scene DNA. However, a man was overheard boasting that a friend had given DNA on his behalf. DNA from this man, Colin Pitchfork, was found to match that at the crime scene.

Types of DNA analysed

The first method involved restriction fragment length **polymorphism** analysis. This method is laborious and is no longer used.

Today, short tandem repeat (STR) sequences of DNA are used. These are highly variable short repeating lengths of DNA. The exact number of STRs varies from person to person.

STR sequences are separated by electrophoresis (topic 6.3.5). Each STR is polymorphic, but the number of alleles in the gene pool for each one is small. Thirteen STRs are analysed simultaneously, so although each STR is present in between 5% and 20% of individuals, the chances of two people sharing STR sequences at all the loci is 1×10^{18}. This is a greater number than the number of people on Earth. However, there are an estimated 12 million identical twins on Earth.

The technique is very sensitive, and even a trace of DNA left when someone touches an object can produce a result. Samples must be treated carefully to avoid contamination.

DNA can be stored for many years if a crime case is unsolved. It can then later be used to assess new evidence.

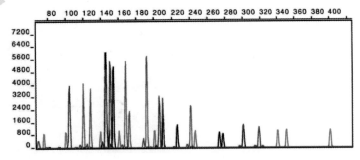

Figure 3 A DNA profile from STR analysis.

Applications of DNA profiling

Forensic science

DNA profiling has transformed forensic science. Not only has it brought about convictions, and established the innocence of many suspects and of people previously wrongly convicted, it has been used to:

- identify Nazi war criminals hiding in South America
- identify the remains of the Romanov family and to refute a person's claim to be the survivor, Anastasia
- identify remains found in Leicester as those of Richard III
- identify victims' body parts after air crashes, terrorist attacks or other disasters
- match profiles from descendants of those lost during World War I with the unidentified remains of the soldiers who fell on battlefields in Northern France.

Maternity and paternity disputes

Half of every child's genetic information comes from the mother and half from the father; hence half the short tandem repeat (STR) fragments come from the mother and half from the father. Comparing the DNA profiles of mother, father and child can therefore establish maternity and/or paternity.

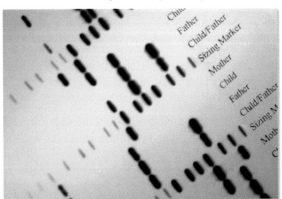

Figure 4 DNA profile in paternity dispute. Half the bands of the child match some of the mother's bands, and the other half match some of the father's bands.

Analysis of disease

Protein electrophoresis can detect the type of haemoglobin present and aid diagnosis of sickle cell anaemia. A varying number of repeat sequences for a condition such as Huntington disease can be detected by electrophoresis, as shown in Figure 5.

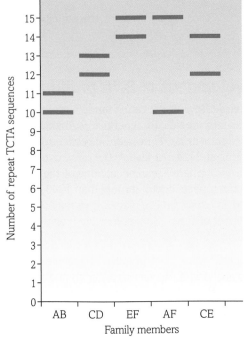

Figure 5 Each of the alleles for a polymorphic gene, within a family, carries a different number of copies of a repeat sequence, TCTA. The resulting different sizes of the alleles can be seen after electrophoresis. Note, this is for a condition, not Huntington disease, also caused by an expanding triple nucleotide repeat.

LEARNING TIP

DNA profiling does not look at genes that code for proteins, as we are all far too similar to each other in this respect.

Questions

1 Sometimes DNA is obtained from a blood sample. Even barely visible blood spots at a crime scene contain enough DNA to generate a profile and suspects can have blood taken. Suggest, with reasons, which part of the blood tissue the DNA is extracted from.

2 Suggest why scrapings from under the fingernails of a victim who had been violently attacked would be taken for forensic analysis.

3 The UK has the world's first national DNA database. In 2010, this contained over 5 million DNA profiles, mostly taken from people who had been suspects in investigations or convicted of crimes. A decision by the European Court of Human Rights has led to a law limiting the length of time that profiles can be stored on this database. Each profile consists of a series of 20 numbers plus an indicator that shows the sex of the person. Some people who were suspects but found innocent want their DNA removed from the database. Others think everyone should have their DNA profile taken and stored in the database. Discuss the pros and cons of a DNA database.

4 The polymerase chain reaction

By the end of this topic, you should be able to demonstrate and apply your knowledge and understanding of:

* the principles of the polymerase chain reaction (PCR) and its application in DNA analysis

Principles of the PCR

Modern DNA profiling can obtain results from as few as five cells. In theory, if a crime suspect had touched a surface and left a few cells behind, his or her DNA profile could be obtained. However, as we all leave skin cells behind when we touch a surface, a frequently touched surface could transfer innocent people's DNA onto the hands of a criminal and from there onto a crime scene. In 1983, Kary Mullis developed a technique, the **polymerase chain reaction (PCR)**, to amplify (increase the amount of) DNA, enabling it to be analysed. The PCR soon became incorporated into forensic DNA analysis and into the protocols for analysis of DNA for genetic diseases.

The PCR is artificial replication of DNA (Book 1, topic 2.3.2). It relies on the facts that:

* DNA is made of two antiparallel backbone strands

* each strand of DNA has a 5' end and a 3' end (Book 1, topic 2.3.1)

* DNA grows only from the 3' end

* base pairs pair up according to complementary base pairing rules, A with T and G with C.

The PCR differs from DNA replication in that:

* only short sequences, of up to 10 000 base pairs, of DNA can be replicated, not entire chromosomes

* it requires the addition of **primer** molecules to make the process start

* a cycle of heating and cooling is needed to separate the DNA strands, bind primers to the strands and for the DNA strands to be replicated.

The PCR process

The PCR is a cyclic reaction (see Figure 2).

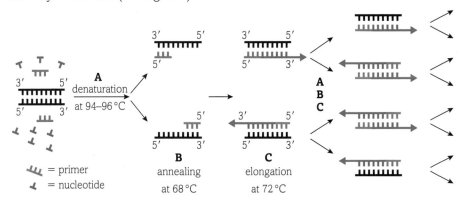

= primer
= nucleotide

A denaturation at 94–96 °C
B annealing at 68 °C
C elongation at 72 °C

Figure 2 Stages of the polymerase chain reaction.

At first the process was quite time consuming as the DNA was heated to denature it and then cooled to around 35 °C to anneal the primers and allow the DNA polymerase to work. Later, DNA polymerase was obtained from the thermophilic bacterium *Thermophilus aquaticus*. This enzyme is called Taq polymerase and is stable at high temperatures.

Figure 1 Kary Mullis.

The steps are:

1. The sample of DNA is mixed with DNA nucleotides, primers, magnesium ions and the enzyme Taq DNA polymerase.

2. The mixture is heated to around 94–96 °C to break the hydrogen bonds between complementary nucleotide base pairs and thus denature the double-stranded DNA into two single strands of DNA (A in Figure 2).

3. The mixture is cooled to around 68 °C, so that the primers can anneal (bind by hydrogen bonding) to one end of each single strand of DNA (B in Figure 2). This gives a small section of double-stranded DNA at the end of each single-stranded molecule.

4. The Taq DNA polymerase enzyme molecules can now bind to the end where there is double-stranded DNA. Taq polymerase is obtained from a bacterium that lives at high temperatures; 72 °C is the optimum temperature for this enzyme.

5. The temperature is raised to 72 °C, which keeps the DNA as single strands (C in Figure 2).

6. The Taq DNA polymerase catalyses the addition of DNA nucleotides to the single-stranded DNA molecules, starting at the end with the primer and proceeding in the 5' to 3' direction.

7. When the Taq DNA polymerase reaches the other end of the DNA molecule, then a new double strand of DNA has been generated.

8. The whole process begins again and is repeated for many cycles.

The amount of DNA increases exponentially: $1 \rightarrow 2 \rightarrow 4 \rightarrow 8 \rightarrow 16 \rightarrow 32 \rightarrow 64 \rightarrow 128$ and so on.

Figure 3 An on-site DNA amplifier kit that uses the PCR to replicate DNA.

Applications of the PCR

Since its inception, the PCR has been improved and elaborated on in many ways. It is used to amplify DNA samples for sequencing. It is commonly used for a wide variety of applications including:

- **Tissue typing**: donor and recipient tissues can be typed prior to transplantation to reduce the risk of rejection of the transplant.

- **Detection of oncogenes**: if the type of mutation involved in

a specific patient's cancer is found, then the medication may be better tailored to that patient.

- **Detecting mutations**: a sample of DNA is analysed for the presence of a mutation that leads to a genetic disease. Parents can be tested to see if they carry a recessive allele for a particular gene; fetal cells may be obtained from the mother's blood stream for prenatal genetic screening; during IVF treatment, one cell from an eight-cell embryo can be used to analyse the fetal DNA before implantation.

- **Identifying viral infections**: sensitive PCR tests can detect small quantities of viral genome amongst the host cells' DNA. This can be used to verify, for example, HIV or hepatitis C infections.

- **Monitoring the spread of infectious disease**: the spread of pathogens through a population of wild or domestic animals, or from animals to human populations, can be monitored, and the emergence of new more virulent sub-types can be detected.

- **Forensic science**: small quantities of DNA can be amplified for DNA profiling (topic 6.3.3), to identify criminals or to ascertain parentage.

- **Research**: amplifying DNA from extinct ancient sources such as Neanderthal or woolly mammoth bones, for analysis and sequencing. In extant organisms, tissues or cells can be analysed to find out which genes are switched on or off.

LEARNING TIP

Remember that DNA polymerase cannot bind to single-stranded DNA, which is why primers are needed for the PCR, but not for natural DNA replication.

Questions

1. Suggest a role for the magnesium ions added during the PCR.

2. Explain why the DNA polymerase enzymes used in the PCR are obtained from *T. aquaticus*.

3. Compare the PCR with natural DNA replication (describe similarities as well as differences).

4. Explain why primers are needed for the PCR, but not for natural DNA replication.

5. After three cycles of PCR, one length of DNA is amplified to eight lengths. How many cycles are needed to make:
   (a) one million copies of the original length of DNA?
   (b) four million copies of the original length of DNA?

(5) Electrophoresis

By the end of this topic, you should be able to demonstrate and apply your knowledge and understanding of:

* the principles and uses of electrophoresis for separating nucleic acid fragments or proteins

Principles of electrophoresis

Separating DNA

Electrophoresis is used to separate different sized fragments of DNA. It can separate fragments that differ by only one base pair, and is widely used in gene technology to separate DNA fragments for identification and analysis.

The technique uses an **agarose** gel plate covered by a buffer solution. Electrodes are placed in each end of the tank so that, when it is connected to a power supply, an electric current can pass through the gel. DNA has an overall negative charge, due to its many phosphate groups, and the fragments migrate towards the anode (positive electrode). Fragments of DNA all have a similar surface charge, regardless of their size.

You can carry out this process in your school or college.

> **KEY DEFINITION**
>
> **electrophoresis:** process used to separate proteins or DNA fragments of different sizes.

INVESTIGATION

Gel electrophoresis

An electrophoresis kit is available for use in schools and colleges, so you can try this.

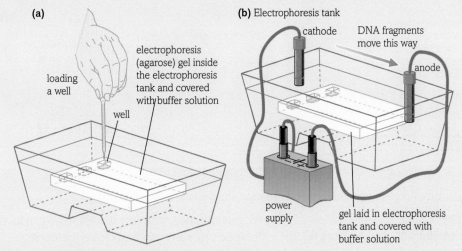

(a) loading a well · electrophoresis (agarose) gel inside the electrophoresis tank and covered with buffer solution · well

(b) Electrophoresis tank · cathode · DNA fragments move this way · anode · power supply · gel laid in electrophoresis tank and covered with buffer solution

(c) Electrophoresis gel showing separated DNA fragments, revealed by flooding with a DNA-binding dye.

1 2 3

Figure 1 Carrying out electrophoresis.

1. The DNA samples are first digested with restriction enzymes (topic 6.3.6) to cut them, at specific recognition sites, into fragments. This is carried out at 35–40 °C and may take up to an hour.

2. While the restriction enzymes are cutting the DNA, the tank is set up. The agarose gel is made up and poured into the central region of the tank, whilst combs are in place at one end. Once the gel is set, buffer solution is added so that the gel is covered and the end sections of the tank contain buffer solution. Now the comb can be carefully removed, leaving wells at one end of the gel.

3. A loading dye is added to the tubes containing the digested DNA.

4. The digested DNA plus loading dye is added to wells in the electrophoresis gel. To do this, a pipette is used and this is held, in the buffer solution, just above one of the wells. The loading dye is dense and carries the DNA down into the well. The pipette should not be placed right into the well, otherwise you might pierce the bottom of the well.

5. Once all the wells have been loaded with the different DNA samples, the electrodes are put into place and connected to an 18 V battery. This is then left to run for up to 6–8 hours. Alternatively, a higher voltage power pack can be used and the gel run for a much shorter time (less than 2 hours); do not use a higher voltage unless the current is limited to 5 mA or less, otherwise there is a risk of severe electric shock from the electrodes or gel.

6. The DNA fragments move through the gel at different speeds. Smaller fragments travel faster, so in a fixed period they travel further.

7. At the end of the period, the buffer solution is poured away and a dye is added to the gel. This dye adheres to the DNA and stains the fragments.

Separating proteins

The principle for separating proteins is the same as for separating DNA fragments, but is often carried out in the presence of a charged detergent such as sodium dodecyl sulfate (SDS), which equalises the surface charge on the molecules and allows the proteins to separate as they move through the gel, according to their molecular mass. In some cases the proteins can be separated according to mass and then, without SDS, according to their surface charge.

This technique can be used to analyse the types of haemoglobin proteins for diagnosis of conditions such as:

- sickle cell anaemia, where the patient has haemoglobin S and not the normal haemoglobin A

- aplastic anaemia, thalassaemia and leukaemia, where the patients have higher than normal amounts of fetal haemoglobin (haemoglobin F), and lower then normal amounts of haemoglobin A.

Using DNA probes

A DNA probe is a short (50–80 nucleotides) single-stranded length of DNA that is complementary to a section of the DNA being investigated. The probe may be labelled using:

- a radioactive marker, usually with ^{32}P in one of the phosphate groups in the probe strand. Once the probe has annealed (bound), by complementary base pairing, to the piece of DNA, it can be revealed by exposure to photographic film.

- a fluorescent marker that emits a colour on exposure to UV light. Fluorescent markers may also be used in automated DNA sequencing (topic 6.3.1).

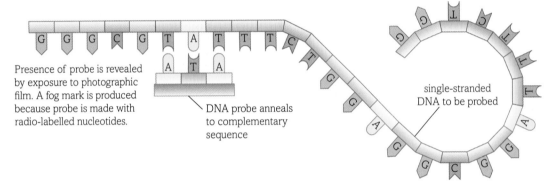

Presence of probe is revealed by exposure to photographic film. A fog mark is produced because probe is made with radio-labelled nucleotides.

DNA probe anneals to complementary sequence

single-stranded DNA to be probed

Figure 2 The action of a DNA probe.

Probes are useful in locating specific DNA sequences, for example:

- to locate a specific gene needed for use in genetic engineering (topic 6.3.6)

- to identify the same gene in a variety of different genomes from different species when conducting genome comparison studies (topic 6.3.2)

- to identify the presence or absence of a specific allele for a particular genetic disease or that gives susceptibility to a particular condition.

Microarrays

Scientists can place a number of different probes on a fixed surface, known as a DNA microarray. Applying the DNA under investigation to the surface can reveal the presence of mutated alleles that match the fixed probes, because the sample DNA will anneal to any complementary fixed probes.

The sample DNA must first be broken into smaller fragments, and it may also be amplified using the polymerase chain reaction (PCR). A DNA microarray can be made with fixed probes, specific for certain sequences found in mutated alleles that cause genetic diseases, in the well.

Reference and test DNA samples are labelled with fluorescent markers. Where a test subject and a reference marker both bind to a particular probe, the scan reveals fluorescence of both colours, indicating the presence of the particular sequence in the test DNA (see Figure 3).

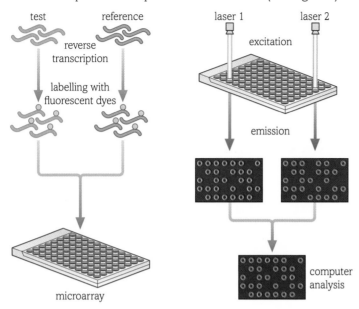

Figure 3 How a DNA microarray may be used.

LEARNING TIP

Electrophoresis uses agarose gel. It is similar to agar, on which we grow bacteria, but not exactly the same.

Questions

1. What types of bonds are formed during annealing?

2. Explain how fragments of DNA are separated from one another during electrophoresis.

3. Explain how electrophoresis is similar to thin layer chromatography.

4. Suggest why, prior to electrophoresis, the digestion of DNA is carried out at 35–40 °C.

5. Describe how a probe can be used to identify specific DNA sequences.

6 Genetic engineering

By the end of this topic, you should be able to demonstrate and apply your knowledge and understanding of:

* the principles of genetic engineering

* the techniques used in genetic engineering

KEY DEFINITIONS

DNA ligase: enzyme that catalyses the joining of sugar and phosphate groups within DNA.
electroporation: method for introducing a vector with a novel gene into a cell; a pulse of electricity makes the recipient cell membrane more porous.
plasmids: small loops of DNA in prokaryotic cells.
recombinant DNA: a composite DNA molecule created *in vitro* by joining foreign DNA with a vector molecule such as a plasmid.
restriction enzymes: endonuclease enzymes that cleave DNA molecules at specific recognition sites.
vector: in gene technology, anything that can carry/insert DNA into a host organism; examples of such vectors include plasmids, viruses and certain bacteria.

The principles of genetic engineering

Genetic engineering is also known as **recombinant DNA** technology, because it involves combining DNA from different organisms. It is also called genetic modification. Genes are isolated from one organism and inserted into another organism, using suitable **vectors**.

The following stages are necessary:

1. The required gene is obtained.

2. A copy of the gene is placed inside a vector.

3. The vector carries the gene into a recipient cell.

4. The recipient expresses the novel gene.

Techniques in genetic engineering

A variety of approaches may be used for each stage outlined above:

1. Obtaining the required gene

* mRNA can be obtained from cells where the gene is being expressed. An enzyme, reverse transcriptase, can then catalyse the formation of a single strand of complementary DNA (cDNA) using the mRNA as a template. The addition of primers and DNA polymerase can make this cDNA into a double-stranded length of DNA, whose base sequence codes for the original protein.

* If scientists know the nucleotide sequence of the gene, then the gene can be synthesised using an automated polynucleotide synthesiser.

* If scientists know the sequence of the gene, they can design polymerase chain reaction (PCR) primers to amplify the gene from the genomic DNA.

* A DNA probe can be used to locate a gene within the genome and the gene can then be cut out using **restriction enzymes**.

2. Placing the gene into a vector

* **Plasmids** can be obtained from organisms such as bacteria and mixed with restriction enzymes that will cut the plasmid at specific recognition sites.

- The cut plasmid has exposed unpaired nucleotide bases, called **sticky ends**.

- If free nucleotide bases, complementary to the sticky ends of the plasmid, are added to the ends of the gene to be inserted, then the gene and cut plasmid should anneal (bind). **DNA ligase** enzyme catalyses the annealing.

- A gene may be sealed into an attenuated (weakened) virus that could carry it into a host cell.

3. Getting the vector into the recipient cell

DNA does not easily cross the recipient cell's plasma membrane. Various methods can be used to aid the process:

- Heat shock treatment – if bacteria are subjected to alternating periods of cold (0 °C) and heat (42 °C) in the presence of calcium chloride, their walls and membranes will become more porous and allow in the recombinant vector. This is because the positive calcium ions surround the negatively charged parts of both the DNA molecules and phospholipids in the cell membrane, thus reducing repulsion between the foreign DNA and the host cell membranes.

- **Electroporation** – a high voltage pulse is applied to the cell to disrupt the membrane.

- **Electrofusion** – electrical fields help to introduce DNA into cells.

- Transfection – DNA can be packaged into a bacteriophage, which can then transfect the host cell.

- T_1 (recombinant) plasmids are inserted into the bacterium *Agrobacterium tumefaciens*, which infects some plants and naturally inserts its genome into the host cell genomes.

4. Direct method of introducing gene into recipient

If plants are not susceptible to *A. tumefaciens*, then direct methods can be used. Small pieces of gold or tungsten are coated with the DNA and shot into the plant cells. This is called a 'gene gun'.

Reverse transcriptase

Retroviruses, such as HIV, which contain RNA that they inject into the host genome, have reverse transcriptase enzyme that catalyses the production of cDNA (complementary DNA) using their RNA as a template. This is the reverse of transcription. These enzymes are useful for genetic engineering, as outlined above.

Restriction enzymes

Bacteria and Archaea have restriction enzymes, called restriction endonucleases, to protect them from attack by phage viruses. These enzymes cut up the foreign viral DNA, by a process called restriction, preventing the viruses from making copies of themselves. The prokaryotic DNA is protected from the action of these endonucleases by being **methylated** at the recognition sites.

The restriction endonucleases are named according to the bacterium from which they have been obtained. The first one used was EcoR1 – it was obtained from *E. coli* and was restriction endonuclease number 1.

Restriction enzymes are useful to molecular biology and biotechnology as molecular scissors, as they recognise specific sequences within a length of DNA and cleave the molecule there.

Some make a staggered cut leaving sticky ends. Others make a cut that produces **blunt ends** (see Figure 1).

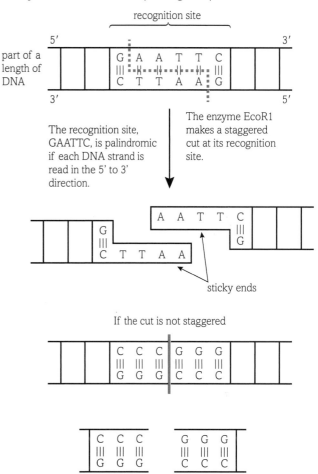

Figure 1 How restriction endonucleases produce sticky ends or blunt ends.

Figure 1 shows that these enzymes always recognise a palindromic sequence; reading the two strands of DNA in the same orientation (e.g. 5' to 3'), the sequence of bases is the same. Some restriction endonucleases need magnesium ions as cofactors.

Ligase enzymes

DNA ligase enzyme is used in molecular biology to join DNA fragments. It catalyses condensation reactions that join the sugar groups and phosphate groups of the DNA backbone.

These enzymes catalyse such reactions during DNA replication in cells and are also used in the PCR.

Insulin from GM bacteria

Scientists can obtain mRNA from beta cells of islets of Langerhans in the human pancreas, where insulin is made.

1. Adding reverse transcriptase enzyme makes a single strand of cDNA and treatment with DNA polymerase makes a double strand – the gene.

2. Addition of free unpaired nucleotides at the ends of the DNA produces sticky ends.

3. Now, with the help of ligase enzyme, the insulin gene can be inserted into plasmids extracted from *E. coli* bacteria. These are now called recombinant plasmids, as they contain inserted DNA.

4. *E. coli* bacteria are mixed with recombinant plasmids and subjected to heat shock in the presence of calcium chloride ions, so that they will take up the plasmids.

Genetically modified bacteria are cultured in large numbers to produce insulin. Figure 2 outlines this process.

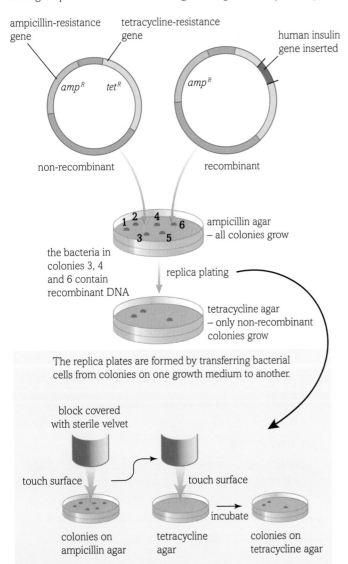

Figure 2 How transformed bacteria are identified by replica plating.

Safety

Because transformed (transgenic) bacteria have resistance to some antibiotics, we do not want them to 'escape' from laboratories into the wild. They also have a gene knocked out, which means they cannot synthesise a particular nutrient. They survive in the laboratory where they are given that nutrient in their growth medium, but will not survive outside of the laboratory.

> **LEARNING TIP**
>
> In a colony of bacteria on an agar plate, all the bacteria have come from one single bacterium that divided by binary fission, so each colony is a clone – all the bacteria in it are genetically identical.

Questions

1. mRNA from the beta cells of the pancreas can be obtained to make the insulin gene. Explain why the gene made in this way will not contain any introns.

2. What types of bond form between the complementary bases of sticky ends of a gene and the sticky ends of a plasmid destined to be the vector of the gene?

3. State the functions of each of the following:
   (a) reverse transcriptase
   (b) ligase
   (c) restriction endonuclease
   (d) sticky ends.

4. Describe three ways that a vector of a novel gene can be introduced into a recipient cell.

(7) Issues relating to genetic manipulation

By the end of this topic, you should be able to demonstrate and apply your knowledge and understanding of:

* the ethical issues (both positive and negative) relating to the genetic manipulation of animals (including humans), plants and microorganisms

Ethical issues of genetic manipulation

Humans have been genetically modifying plants for thousands of years. All the crops grown around the world have genomes vastly different from those of their wild relatives, as a result of human intervention and selective breeding (topic 6.2.14). Agriculture has changed the face of the landscape and produced domesticated breeds of animals and plants. However, selective breeding relies on a rather 'hit or miss' technique and may produce unexpected results.

In the 1970s, the techniques of recombinant DNA technology allowed scientists to splice new genes into plant genomes. This gave a more exact way of transforming crop plants, compared with inducing mutations or crossing different strains. Now specific genes conferring desirable traits can be excised from one organism and inserted into another, using a vector or a 'gene gun' (topic 6.3.6).

Some people are concerned about potential **hazards** and **risks** associated with genetic modification (GM). However, the potential benefits have to be recognised and weighed against the potential hazards. Anti-GM campaigners want the technology abolished, which would also abolish any choice and the opportunity to exploit potential benefits.

> **DID YOU KNOW?**
>
> In the 1920s, Herman Muller realised that bombarding crop plants with ionising radiation could induce mutations, some of which may produce desirable qualities in agricultural plants.

Modified organism	Potential benefit(s)	Potential hazard(s)
Microorganisms, e.g. *E. coli*	GM microorganisms can make human insulin to treat all diabetics (this was not possible using pig insulin), and human growth hormone to treat children with pituitary dwarfism.	Microorganisms could escape into the wild and transfer marker genes for antibiotic resistance to other bacteria. However, the GM bacteria are also modified so they cannot synthesise an essential nutrient and therefore cannot live outside the lab.
Plants Bt tobacco and Bt maize	In 1985, tobacco plants were genetically modified to produce the toxin normally produced by a bacterium, *Bacillus thuringiensis*. This toxin, Bt, had been used by organic farmers as a pesticide, as it is toxic to insects. Because the bacterial gene, *Bt*, was inserted into some crop plants, the GM plants produced the toxin, eliminating the need to spray it around the environment possibly contaminating other organisms.	Bt is toxic to monarch butterflies. However, these butterflies do not take nectar from tobacco plants or maize plants in the wild; they feed on milkweed. Despite many thousands of hectares of land in the USA being planted with Bt crops, the monarch butterfly has continued to thrive.
Soya beans This is an example of a first-generation GM plant, where the main advantage appeared to be to the production company making the herbicide.	GM soya beans, resistant to a herbicide (Round-Up Ready™), were produced, so that weeds competing with the soya plants could be killed with the herbicide.	Possible risks include the potential for the gene for herbicide resistance to pass into weeds, producing 'superweeds'. This does not appear to have happened to date.

Table 1 Some potential benefits and hazards of GM technology (*continued overleaf*).

Modified organism	Potential benefit(s)	Potential hazard(s)
Golden Rice™ A second-generation GM crop – bred to be nutritionally enhanced	About 500 000 children each year in India go blind, and some of them die, through lack of beta carotene, the precursor to vitamin A. Golden Rice™ (see Figure 2) was genetically modified to contain a gene from daffodils, so that beta carotene would be present in the rice grains. As rice is the staple food in this region, this seemed a great solution to a problem.	Some people were concerned that farmers would have to buy the seed every year, but the company that developed this rice has offered free licences to farmers so they can keep and replant rice seeds.
Plantains – a type of banana and a staple food in Kenya	A local biotechnology company in Kenya, Africa Harvest, is producing plantains that are nutritionally enhanced to contain more zinc. In areas where people eat very little meat, they may be deficient in zinc, an important enzyme cofactor and essential for regulating insulin secretion.	Some people fear eating food that contains foreign DNA and worry that the inserted genes will somehow be expressed in us. However, all the food we eat contains genes and we digest the DNA with specific enzymes, nucleases and nucleotidases.
Crop plants resistant to pests	The biotechnology company Africa Harvest is producing crops that are resistant to pests, so that when farmers sow these seeds, they do not need to use pesticides. Not only is this better for the environment, it is good for farmers. Every year in Africa, about 2000 people die through exposure to pesticides while applying them.	There were concerns that local farmers might not want the GM seed and would not have the choice to buy non-GM seed. However, Monsanto also sells non-GM seeds, but many farmers see the benefits of the GM seeds and do not want to be exposed to pesticides.
Pathogens	Viruses, genetically modified to have no virulence, can be used to make vaccines, as they still have the antigens on their surfaces. This reduces the chance of a vaccine making the recipient ill. Modified viruses can also be used as vectors in gene therapy (topic 6.3.8).	There have been some problems with the use of viruses in gene therapy, as the allele may be inserted into the genome in a way that increases the risk of cancer or interferes with gene regulation.
Mice	Since 1974, millions of GM mice have been bred for medical research and used to develop therapies for breast and prostate cancer. Other types of mice have had certain genes knocked out, so that researchers can find out the function of those genes.	Some people object to the use of animals for medical and pharmaceutical testing. However, in the UK, strict regulations govern the welfare of animals used in this way and many of us have benefitted from medical protocols developed using animals.
Pharmaceutical proteins	Genes for human pharmaceutical proteins, such as alpha antitrypsin to treat hereditary emphysema, can be inserted into goats or sheep, and the human protein they express into their milk is harvested. Transgenic mammals were used because this protein is too large for a bacterial cell to synthesise.	There are concerns for the welfare of the GM sheep and goats. However, these animals are valuable and likely to be well looked after.
Silk	Silk is one of the strongest materials known. Spiders are impossible to farm but genes for spider silk have been inserted into goats. These GM goats produce spider silk protein in their milk. Silk can be used for cables, sutures (stitches following surgery or wounds), artificial ligaments and bullet proof vests.	Concerns were raised about the welfare of the GM goats. However, these animals are valuable and likely to be well looked after and not be eaten.

Table 1 *(continued)* Some potential benefits and hazards of GM technology.

Figure 1 GM maize, *Zea mays*, growing in a field trial in Lincolnshire. It has been treated with glyphosate weedkiller, to which it is resistant. Glyphosate is neutralised on contact with the soil and does not cause groundwater pollution. Field trials are in guarded locations to prevent attacks by anti-GM protesters; no GM crops are grown commercially in the UK at this time.

Figure 2 Golden Rice™ and non-GM rice.

Questions

1. We all take risks every day, for example driving or using a mobile phone. Because we want to do the 'risky activity', we play down those risks. Because we may have no immediate need of GM food, as we have food security, should we reject it and deny it to those in areas of the world that could benefit from it? With reference to specific examples, discuss the potential benefits and risks of growing GM crop plants.

2. Many people argue that if something is not natural, it is bad. This is known as the naturalistic fallacy. List six natural phenomena that are harmful to us and six unnatural interventions that are beneficial to us. Share your ideas with others in your class.

(8) Gene therapy

By the end of this topic, you should be able to demonstrate and apply your knowledge and understanding of:

* the principles of, and potential for, gene therapy in medicine

---

> **KEY DEFINITIONS**
>
> **germ line gene therapy:** gene therapy by inserting functional alleles into gametes or zygotes.
> **somatic cell gene therapy:** gene therapy by inserting functional alleles into body cells.

> **DID YOU KNOW?**
>
> Soon after the discovery of the structure of DNA, in 1953, scientists started to consider the potential to treat genetic disorders by inserting functioning alleles into patients. In 1990, a four-year-old girl, Ashanta de Silva, became the first person to receive gene therapy. She suffered from a rare recessive immune disorder, adenine deaminase (ADA) deficiency, also called severe combined immunodeficiency (SCID). The inability to produce an enzyme leads to an accumulation of toxins that destroy T cells, leaving those affected vulnerable to infections and with a shortened life expectancy. A functioning allele was inserted into some of Ashanta de Silva's white blood cells, which were then put back into her body; this is an example of *ex vivo* therapy. Unfortunately, these white cells did not produce other modified white cells, so Ashanta receives gene therapy regularly. Ashanta has not been cured, but her condition is being managed.

The principle of gene therapy

The basic principle of gene therapy is to insert a functional allele of a particular gene into cells that contain only mutated and non-functioning alleles of that gene. If the inserted allele is expressed, then the individual will produce a functioning protein and no longer have the symptoms associated with the genetic disorder.

Knowledge gained from the Human Genome Project has led to further possibilities, such as using interference RNA to silence genes by blocking **translation**. Interference RNA has been used to treat cytomegalovirus infections in AIDS patients by blocking replication of the cytomegalovirus.

Somatic cell gene therapy

Some metabolic disorders such as cystic fibrosis occur when an individual inherits two faulty, recessive alleles for a particular gene. As a result, the differentiated cells where this gene should normally be expressed lack the protein product of that gene. If functioning alleles for this gene can be put into specific cells so that these cells then make the protein, these cells will function normally. Various methods are available for delivering the functioning alleles to the patient's body cells.

Somatic cell gene therapy affects only certain cell types. The alterations made to the patient's genome in those cells are not passed to the patient's offspring.

Liposomes

Patients with cystic fibrosis lack a functioning *CFTR* gene.

The alleles, which are lengths of DNA, can be packaged within small spheres of lipid bilayer to make **liposomes**. If these are placed into an aerosol inhaler and sprayed into the noses of patients, some will pass through the plasma membrane of cells lining the respiratory tract. If they also pass through the nuclear envelope and insert into the host genome, the host cell will express the CFTR protein – a transmembrane chloride ion channel (Book 1, topic 2.5.2).

Epithelial cells lining the respiratory tract are replaced every 10–14 days, so this treatment has to be repeated at regular short intervals.

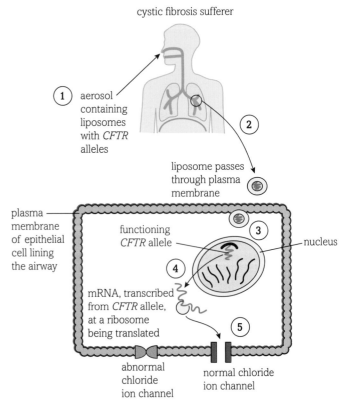

Figure 1 How gene therapy to treat cystic fibrosis can be carried out using liposomes as vectors.

Viruses

Viruses have been used as vectors. If a virus that usually infects humans is genetically modified so that it encases the functioning allele to be inserted into the patient, whilst at the same time being made unable to cause a disease, it can enter the recipient cells, taking the allele with it. In 1999, a patient taking part in a trial for such a technique died, and in 2002 some trials were interrupted after several patients developed leukaemia.

There are potential problems with using viruses as gene delivery agents:

- Viruses, even though not virulent, may still provoke an immune or inflammatory response in the patient.

- The patient may become immune to the virus (Book 1, Chapter 4.1), making subsequent deliveries difficult or impossible.

- The virus may insert the allele into the patient genome in a location that disrupts a gene involved in regulating cell division (Book 1, topic 2.6.1), increasing the risk of cancer.

- The virus may insert the allele into the patient's genome in a location that disrupts the regulation of the expression of other genes (topic 6.1.2).

Artificial chromosomes

Research is being carried out into the possibility of inserting genes into an artificial chromosome that would co-exist with the other 46 chromosomes in the target cells.

Germ line gene therapy

Germ line gene therapy involves altering the genome of gametes or zygotes. Not only will all the cells of that individual be altered, their offspring may also inherit the foreign allele(s). Thus, this type of therapy has the potential to change the genetic makeup of many people, the descendants of the original patient, none of whom would have given consent. There are also concerns about how the genes may be inserted – they may find their way into a location that could disrupt the expression or regulation of other genes or increase the risk of cancer.

Strict guidelines drawn up by regulatory bodies and ethics committees consider germ line gene therapy for humans to be ethically impermissible.

LEARNING TIP

Do not confuse pre-implantation genetic screening with genetic engineering/germ line gene therapy. If eggs are fertilised by IVF, the resulting embryos screened and only those without a genetic disorder implanted into the mother's uterus, there has *not* been any genetic manipulation.

DID YOU KNOW?

Some genetic diseases with severe consequences result from mutation to mitochondrial genes. We all inherit our mitochondria from our mothers, in the cytoplasm of the ovum. In some cases where a mother has a mitochondrial gene mutation, the nucleus from one of her ova can be transplanted into an enucleated ovum of a donor. The resulting ovum is then fertilised *in vitro* and implanted into the uterus of the mother. The resulting baby has the nuclear genetic material of mother and father, but the mitochondrial DNA of the donor.

Questions

1. Explain why, at present, gene therapy can only be used to treat recessive genetic disorders and not disorders, such as Huntington disease, caused by a dominant allele.

2. Look back at Book 1, Chapter 2.3 and re-read the Thinking Bigger spread. Suggest how CRISPR technology may lead to gene therapy for genetic diseases caused by dominant alleles.

3. Explain why the effects of gene transfer may be unpredictable.

4. Outline the problems of using viruses as vectors for somatic cell gene therapy.

5. Discuss the reasons why germ line gene therapy is not permitted for humans.

THE HUMAN GENOME PROJECT

Some surprising findings have come from the Human Genome Project.

LESSONS OF THE GENOME

When the acrimonious race to sequence the human genome reached its final furlong, the competing parties could agree on one thing – the 'Book of Man' was going to contain a lot of genes.

Simple animals such as the fruit fly and the nematode worm had 13 500 and 19 000 genes respectively. Scientists were expecting humans to have around 100 000 genes. The first published draft of the Human Genome sequence was a surprise – humans appeared to have only around 21 500–24 000 genes; just a few thousand more than a mouse.

In the 1940s, experiments carried out by George Beadle and Edward Tatum established that genes code for proteins and since that time the mantra was 'one gene codes for one protein'. However, there are hundreds of thousands of human proteins and only just over 20 000 genes. The mantra was wrong. Both genes and proteins are more versatile than had been assumed.

Single genes, in fact, contain recipes for many different proteins. Only the sections of genes known as exons actually carry instructions for proteins synthesis. Information from the non-coding introns is removed from messenger RNA, and the exons are stitched together before proteins are made.

These exons can be spliced in many different ways and this alternative splicing means that one gene can specify multiple proteins. Some genes make only parts of proteins and these parts can then be joined in various orders to make a variety of larger proteins. Some proteins are modified after they are made. All these processes result in a human protein population or 'proteome' that is more diverse than the 20 000–24 000 gene genome would imply.

The complexity of gene interactions has also helped clarify common misperceptions that there is 'a gene for a particular disorder'. In fact, there are no genes for diseases. Even if only one gene seems to be involved, such as in haemophilia or colour blindness, the gene is not for the disease. It codes for a protein that is crucial to some aspect of human physiology and when an individual receives mutated alleles of the gene, then that individual makes abnormal proteins resulting in symptoms of the genetic disease. For many aspects of human behaviour and physiology, such as intelligence, autism, schizophrenia and sexual orientation, there are many genes involved and they interact with each other and with environmental factors in extremely subtle ways. Mutations in some alleles can affect the way we each respond to medicines and our understanding of the genome is leading to personalised medicine with tailor made medicines studied by a branch of science called pharmacogenomics.

Our individual genetic profiles can also influence the way we respond to particular foods. People with phenylketonuria have to follow a strict diet to prevent brain damage. In people with no apparent disorders, genetic variations may influence our nutritional needs, and some companies are launching nutrigenomics services, purporting to deliver genetically tailored diets. However, the links between genetics and nutrition are, as yet, poorly understood and most scientists say that eating a balanced diet with plenty of vegetables, low fat, low salt and low in processed food, is the best strategy, as this is good for us all, regardless of our genes.

Source
● Henderson, Mark. (2008) *50 genetics ideas you really need to know*. Quercus Publishing Inc.

Where else will I encounter these themes?

The first question relates to the style of writing.

1. What features of the writing suggest that the source is aimed at an informed general public?

Now let us look at the biological concepts underlying the information in this article.

2. Explain how information from genome sequencing, in particular the Human Genome Project, has showed that the Beadle and Tatum mantra is incorrect.

3. What is meant by the term 'proteome'?

4. Explain why it is misleading to refer to genes for genetic diseases.

5. What is 'pharmacogenomics'?

Activity

In 1982, the evolutionary biologist Richard Dawkins published his book *The Extended Phenotype*. The hypothesis at its centre was that an organism's genotype determines its phenotype, and the phenotype can go beyond what the organism's features are, but include the organism's behaviour or how it influences its environment, including the behaviour of other organisms. If the hypothesis was correct, parasites would have genes that could alter the behaviour of their host in a way that would favour the survival of the parasite. Now, over 30 years later, scientists are finding out more about this mind control of hosts by their parasites. The following are some examples:

A parasitic protoctist, *Toxoplasma gondii*, infects rats and alters their behaviour so they become curious instead of fearful about the smell of cat urine. Rats not fearful of cats are more likely to be caught and eaten. The parasite has to spend some of its life cycle inside cats.

A parasitic wasp, *Hymenoepimecis argyraphaga*, causes its host, the spider *Leucauge argyra*, to build an unusual web that allows the wasp larvae to be suspended from it out of reach of enemies. This allows the wasp larvae to develop into adult wasps.

A parasitic wasp, *Ampulex compressa*, injects neurotoxins into its host's (a cockroach's) brain, making the cockroach fail to escape and instead act as a nursery for the developing wasp larvae inside its body.

A flatworm, *Ribeiroia ondatrae*, infects frog tadpoles and causes them to develop extra hind legs. This makes the frogs easy prey for herons, inside which the parasites reproduce, passing out of the herons' bodies in faeces into water to infect snails and then break out of the snails and infect more tadpoles.

Choose one of the examples above or another example of your choice, and research it using the Internet. Give a short presentation (three minutes maximum) to others in your class, explaining how the ways in which the parasite changes its host's behaviour promotes the survival of the parasite.

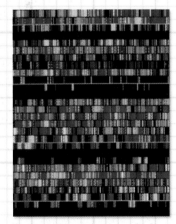

Figure 1 Computer screen display showing DNA sequence of part of the human genome. Each colour represents a specific base. Photographed at the Sanger Centre, Cambridge, UK.

DID YOU KNOW?

Our phenotypes (visible characteristics) are determined by our genomes. An organism's phenotype can extend to the way it manipulates its environment (e.g. a bird that builds a good nest will produce more surviving offspring) or to the way a parasite affects its host behaviour. A mutation in a gene, *egt*, in a virus that infects some types of caterpillars, causes the infected caterpillars, when their bodies are full of viruses, to climb high so that when their bodies turn to 'goo' the viruses particles are showered down onto other caterpillars that become infected. This version of the gene gives the virus an advantage. If this gene is knocked out, viruses can still infect the caterpillars, but do not cause them to climb.

Practice questions

1. What name is given to enzymes obtained from bacteria that can cleave DNA at a specific recognition site? [1]

   A. Ligases

   B. Reverse transcriptases

   C. Restriction endonucleases

   D. DNA polymerases

2. Five molecules of DNA were amplified using the polymerase chain reaction. After 11 cycles, there are about 10 000 molecules of DNA. How many more cycles are needed to produce 80 000 molecules of DNA? [1]

   A. 3

   B. 7

   C. 14

   D. 15

3. Read the following statements.

   (i) The optimum temperature for the enzyme Taq polymerase is 72 °C

   (ii) Primers are needed for the PCR, as DNA polymerase cannot bind to single stranded DNA

   (iii) Nucleotides are added to the newly synthesised strand of DNA in the 3' to 5' direction

   Which statement(s) is/are true? [1]

   A. (i), (ii) and (iii)

   B. (i) and (ii) only

   C. (i) and (iii) only

   D. (ii) and (iii) only

4. Read the following statements.

   (i) Synthetic DNA can be used as databases to store information.

   (ii) In modern DNA profiling, 13 short tandem repeats (STRs) are analysed simultaneously.

   (iii) Pyrosequencing is an example of high throughput DNA sequencing.

   Which statement(s) is/are true? [1]

   A. (i), (ii) and (iii)

   B. (i) and (ii) only

   C. (i) and (iii) only

   D. (ii) only

5. Who developed the polymerase chain reaction? [1]

   A. Alec Jeffreys

   B. James Watson

   C. Kary Mullis

   D. Fred Sanger

   [Total: 5]

6. The following questions relate to recombinant DNA.

   (a) Explain what is meant by the term 'recombinant DNA'. [2]

   (b) Outline the roles of the following enzymes in the process of producing recombinant DNA:
       (i) restriction endonucleases
       (ii) DNA polymerase
       (iii) reverse transcriptase
       (iv) DNA ligase. [6]

   (c) Describe how bacteria such as *E. coli* can be treated to make them take up recombinant plasmids. [3]

   (d) Explain how replica plating can be used to identify the transformed bacteria after the treatment you described in (c). [6]

   [Total: 17]

7. Anti-GM campaigners think that GM crops may escape into the wild and become invasive species. Scientists say that crop plants are not good at competing in the wild, which is why we do not see agricultural crops growing outside of carefully tended fields. A 10-year study was carried out into the possibility of transgenic crops persisting in the wild, in the event of their escaping from cultivation. Four different transgenic crops and their non-transgenic counterparts were grown in 12 different habitats.

   The four different crop plants were:

   • herbicide-tolerant oilseed rape
   • herbicide-tolerant maize
   • herbicide-tolerant sugar beet
   • potato plants that made a toxin poisonous to caterpillars.

   (a) Outline the advantages of growing one of the transgenic crops listed above. [4]

   The mean percentage survival rate, at the end of the first growing season, of all the transgenic crop plants was less than that of their non-transgenic counterparts. The population sizes of all the crops declined sharply after the first year and all the crop populations of maize, rape and sugar beet were extinct in the wild after four years. One crop of non-GM potatoes survived in one habitat for 10 years.

(b) Using the information above and your own knowledge, assess the potential danger to natural habitats of cultivating these transgenic crops. [6]

(c) Discuss the ethical implications of growing GM crops. [6]

[Total: 16]

8. In early trials of gene therapy to treat patients with cystic fibrosis, an engineered adenovirus containing copies of a functioning *CFTR* gene was administered to patients via a nasal aerosol spray. These trials were stopped after one participant developed a severe inflammatory reaction and died. Liposomes may be used as vectors to introduce the functioning alleles.

(a) Describe how (i) viruses and (ii) liposomes may be used as gene vectors in gene therapy. [8]

(b) Outline the potential problems with using viruses as gene vectors. [4]

(c) Explain why the liposome therapy for cystic fibrosis has to be repeated about every 10–14 days. [2]

[Total: 14]

9. The following is from an abstract of an academic research paper: *LMO2*-Associated clonal T-cell proliferation in two patients after gene therapy for SCID-X1. http://www.sciencemag.org/content/302/5644/415.short

We have previously shown correction of X-linked severe combined immunodeficiency in 9 out of 10 patients by retrovirus-mediated gene transfer into CD34 bone marrow T-cells. However, almost three years after gene therapy, uncontrolled exponential clonal proliferation of mature T cells has occurred in the two youngest patients. Both patient's T-cell clones showed retrovirus vector integration in proximity to a proto-oncogene promoter, leading to aberrant transcription and expression of the *LMO2* gene. Thus, retrovirus vector insertion can trigger deregulated premalignant cell proliferation with unexpected frequency.

(a) Explain the following terms:
  (i) X-linked [1]
  (ii) T cells [1]
  (iii) uncontrolled exponential clonal proliferation [1]
  (iv) retrovirus [1]
  (v) proto-oncogene [1]
  (vi) promoter [1]
  (vii) aberrant transcription and expression of *LMO2*. [1]

(b) Discuss the ethical issues surrounding the use of this type of gene therapy to treat SCID. [3]

[Total: 10]

10. (a) Explain the difference between somatic cell line gene therapy and germ line gene therapy. [5]

(b) Describe how you could use restriction enzymes and gel electrophoresis to digest lengths of DNA and separate the resulting fragments. You may use annotated diagrams to help answer the question. [9]

(c) How might small samples of DNA found at a crime scene be amplified for forensic analysis? [5]

[Total: 19]

CLONING AND BIOTECHNOLOGY

Introduction

Cloning is the production of genetically identical copies of cells or organisms. Many plants undergo cloning as a part of their natural vegetative propagation. Farmers and growers often exploit this natural cloning process to produce uniform crops that are consistent in their growth and easy to harvest.

Artificial clones of plants and animals can also be produced. These can be used to increase numbers of carefully selected individuals, which display particular combinations of desired characteristics.

Biotechnology is the use of living organisms (or parts of living organisms) in industrial processes. Biotechnology can be used to produce food, drugs and other products relatively cheaply. Using microorganisms in biotechnology has many advantages, as they reproduce quickly and are relatively easy to keep.

Both cloning and the use of organisms in industrial processes have associated ethical considerations and you should develop a balanced understanding of these issues.

All the maths you need

To unlock the puzzles of this chapter you need the following maths:

- Recognise and make use of appropriate units in calculations
- Recognise and use expressions in decimal form and standard form
- Use ratios, fractions and percentages
- Use calculators to find and use power, exponential and logarithmic functions
- Use an appropriate number of significant figures
- Find arithmetic means
- Construct and interpret frequency tables and diagrams, bar charts and histograms
- Understand simple probability
- Understand the terms mean, median and mode
- Understand measure of dispersion, including standard deviation and range
- Use logarithms in relation to quantities that range over several orders of magnitude
- Translate information between graphical, numerical and algebraic forms
- Plot two variables from experimental data
- Determine the intercept of a graph
- Calculate rate of change from a graph showing a linear relationship
- Draw and use the slope of a tangent to a curve as a measure of rate of change
- Calculate the circumferences, surface areas and volumes of regular shapes

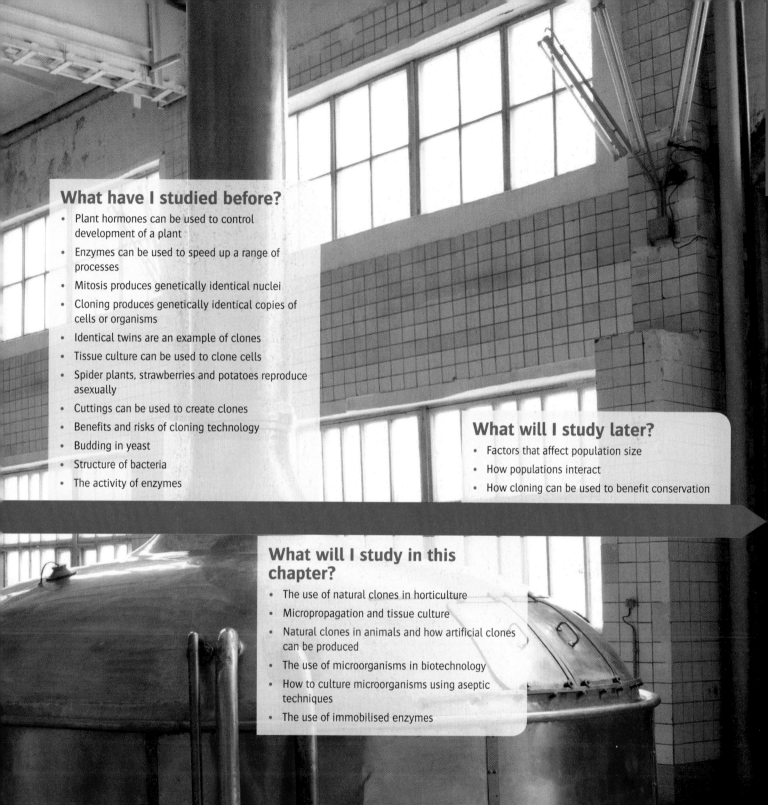

What have I studied before?

- Plant hormones can be used to control development of a plant
- Enzymes can be used to speed up a range of processes
- Mitosis produces genetically identical nuclei
- Cloning produces genetically identical copies of cells or organisms
- Identical twins are an example of clones
- Tissue culture can be used to clone cells
- Spider plants, strawberries and potatoes reproduce asexually
- Cuttings can be used to create clones
- Benefits and risks of cloning technology
- Budding in yeast
- Structure of bacteria
- The activity of enzymes

What will I study later?

- Factors that affect population size
- How populations interact
- How cloning can be used to benefit conservation

What will I study in this chapter?

- The use of natural clones in horticulture
- Micropropagation and tissue culture
- Natural clones in animals and how artificial clones can be produced
- The use of microorganisms in biotechnology
- How to culture microorganisms using aseptic techniques
- The use of immobilised enzymes

(1) Natural clones

By the end of this topic, you should be able to demonstrate and apply your knowledge and understanding of:

* natural clones in plants and the production of natural clones for use in horticulture
* natural clones in animal species

Natural clones

Clones are genetically identical copies. The term can apply to cells or to whole organisms. Clones are produced by asexual reproduction in which the nucleus is divided by mitosis. Mitosis creates two identical copies of the DNA, which are then separated into two genetically identical nuclei before the cell divides to form two genetically identical cells. These cells may not be physically or chemically identical as, after division, they may differentiate to form two different types of cell.

Clones are formed in nature. Any organism that reproduces asexually will produce clones of itself. For example, single-celled yeasts reproduce by budding and bacteria reproduce by binary fission. Both processes involve exact replication of DNA, so the cells produced are genetically identical.

Advantages of natural cloning

The offspring produced by cloning are genetically identical to the parent. The advantages of reproduction by cloning include:

* If the conditions for growth are good for the parent, then they will also be good for the offspring.
* Cloning is relatively rapid – so the population can increase quickly to take advantage of the suitable environmental conditions.
* Reproduction can be carried out, even if there is only one parent and sexual reproduction is not possible.

However, there are disadvantages:

* The offspring may become overcrowded.
* There will be no genetic diversity (except that caused by mutation during DNA replication).
* The population shows little variation.
* Selection is not possible.
* If the environment changes to be less advantageous, the whole population is susceptible.

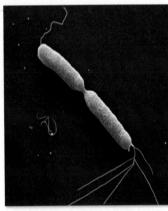

Figure 1 An electronmicrograph of bacteria completing binary fission (×16 500).

Plant cloning by vegetative propagation

The differentiation of many plant cells is not as complete as that in animals. Many parts of a plant contain cells that retain the ability to divide and differentiate into a range of types of cell. This means that plants are able to reproduce by cloning. Natural cloning involves a process called **vegetative propagation**. This is the process of reproduction through vegetative parts of the plant, rather than through specialised reproductive structures.

Runners, stolons, rhizomes and suckers

Many plants grow horizontal stems that can form roots at certain points. These stems are called **runners or stolens** if they grow on the surface of the ground, and **rhizomes** if they are underground. Some rhizomes are adapted as thickened over-wintering organs from which one or more new stems will grow in the spring.

Suckers are new stems that grow from the roots of a plant – these may be close to the base of an older stem or could be some distance away. In all cases, the original horizontal branch may die, leaving the new stem as a separate individual.

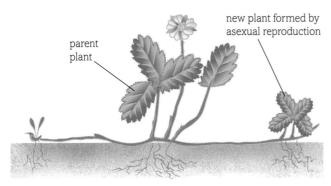

Figure 2 A plant can reproduce by runners producing clones.

Bulbs

Bulbs (for example, onions) are an over-wintering mechanism for many perennial monocotyledonous plants. Bulbs consist of an underground stem from which grow a series of fleshy leaf bases. There is also an apical bud, which will grow into a new plant in the spring. Often a **bulb** contains more than one apical bud and each will grow into a new plant. Figure 3 shows a hyacinth bulb with two apical buds growing into two separate plants.

Corms

Corms are often mistaken for bulbs. However, corms are solid rather than fleshy like a bulb. A corm is an underground stem with scaly leaves and buds. Corms remain in the ground over winter. In the spring the buds grow to produce one or more new plants.

Croci and gladioli reproduce using corms. Figure 4 shows corms of the root vegetable taro.

Figure 3 Hyacinth bulb showing vegetative propagation.

Figure 4 Corms of the root vegetable taro.

Leaves

The Kalanchoe plant reproduces asexually, as clones grow on the leaf margins (see Figure 5). The immature plants drop off the leaf and take root.

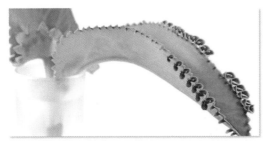

Figure 5 New plants growing on the edge of a Kalanchoe leaf.

Tubers

Tubers are another type of underground stem. Potatoes are tubers. One potato will grow into one or more plants. Each new plant can then produce many new tubers (potatoes) later that year.

> **LEARNING TIP**
> Many types of vegetative propagation are over-wintering mechanisms. They become successful means of reproduction when more than one bud grows from the over-wintering organ.

Cloning in animals

Animals do not clone as often as plants. There are, however, a few examples of natural cloning.

Mammals clone when identical twins are formed. This occurs when a fertilised egg (zygote) divides as normal, but the two daughter cells then split to become two separate cells. Each cell grows and develops into a new individual.

Figure 6 Identical twins are clones.

The water flea (*Daphnia pulex*) and greenfly (*Acyrthosiphon pisum*) are examples of animals that commonly reproduce asexually to produce clones.

Questions

1. Explain why asexual reproduction is quicker than sexual reproduction.

2. Explain why plants are more likely to reproduce asexually than animals.

3. Why is it an advantage to small mobile animals, such as water fleas, to be able to reproduce asexually?

4. Why is it likely that a plant such as Kalanchoe will become overcrowded?

5. Suggest why many plants will grow suckers after the main stem has been damaged.

② Clones in plants

By the end of this topic, you should be able to demonstrate and apply your knowledge and understanding of:

* how to take plant cuttings as an example of a simple cloning technique
* the production of artificial clones of plants by micropropagation and tissue culture
* the arguments for and against artificial cloning in plants

KEY DEFINITIONS

micropropagation: growing large numbers of new plants from meristem tissue taken from a sample plant.
tissue culture: growing new tissues, organs or plants from certain tissues cut from a sample plant.

Figure 1 Roots growing from a cutting.

Using natural clones

For many years gardeners and growers have made use of vegetative propagation. The easiest way to create clones is through making cuttings. To make a cutting, a stem is cut between two leaf joints (**nodes**). The cut end of the stem is then placed in moist soil. New roots will grow from the tissues in the stem – usually from the node, but they may grow from other parts of the buried stem.

Some plants such as pelargonium (geraniums) and blackberry will take root easily. Other plants may need further treatment. Dipping the cut stem in rooting hormone helps to stimulate root growth. It may also be helpful to wound or remove the bark from the cut end of the stem, as this encourages the plant to produce a callus.

This technique can be used to produce large numbers of plants very quickly.

Cuttings can also be made successfully from other parts of a plant:

* root cuttings, in which a section of root is buried just below the soil surface, and produces new shoots
* scion cuttings, which are dormant woody twigs
* leaf cuttings, in which a leaf is placed on moist soil. The leaves develop new stems and new roots. Some leaves may produce many new plants from one cutting.

Tissue culture

Large-scale cloning by taking cuttings can be time-consuming and needs a lot of space. Also some plants do not respond well to taking cuttings. Many commercially grown houseplants are cloned using **tissue culture** techniques.

Tissue culture is a series of techniques used to grow cells, tissues or organs from a small sample of cells or tissue. It is carried out on a nutrient medium under sterile conditions. Application of plant growth substances at the correct time can encourage the cells in the growing tissue to differentiate.

Tissue culture is widely used commercially to increase the number of new plants, in **micropropagation**.

Micropropagation

Micropropagation involves taking a small piece of plant tissue (the **explant**) and using plant growth substances to encourage it to grow and develop into a whole new plant.

Micropropagation involves these steps:

1. Suitable plant material is selected and cut into small pieces. These are called explants. Explants could be tiny pieces of leaf, stem, root or bud. Meristem tissue is often used, as this is always free from virus infection.

2. The explants are sterilised using dilute bleach or alcohol. This is essential to kill any bacteria and fungi, as these would thrive in the conditions supplied to help the plant grow well.

3. The explants are placed on a sterile growth medium (usually agar gel) containing suitable nutrients such as glucose, amino acids and phosphates. The gel also contains high concentrations of the plant growth substances auxin and cytokinin. This stimulates the cells of each explant to divide by mitosis to form a callus (a mass of undifferentiated, **totipotent** cells).

4. Once a callus has formed, it is divided to produce a larger number of small clumps of undifferentiated cells.

5. These small clumps of cells are stimulated to grow, divide and differentiate into different plant tissues. This is achieved by moving the cells to different growth media. Each medium contains different ratios of auxin and cytokinin. The first medium contains the ratio 100 auxin : 1 cytokinin, and this stimulates roots to form. The second medium contains the ratio 4 auxin : 1 cytokinin, which stimulates the shoots to form.

6. Once tiny plantlets have been formed, these are transferred to a greenhouse to be grown in compost or soil and acclimatised to normal growing conditions.

> **DID YOU KNOW?**
> Seed banks use tissue culture techniques to store plants at a growth stage when they are not too large. This technique is important in the conservation of species whose seeds do not remain viable for long periods.

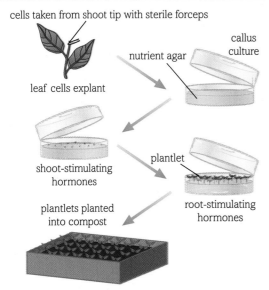

Figure 2 The steps in micropropagation.

Advantages and disadvantages of artificial cloning

Advantages of artificial cloning

- Cloning is a relatively rapid method of producing new plants compared with growing plants from seed.

- Cloning can be carried out where sexual reproduction is not possible. Plants that have lost their ability to breed sexually can be reproduced, for example commercially grown bananas. Similarly, plants that are hard to grow from seed can be reproduced, for example orchids for the horticulture industry.

- The plants selected will all be genetically identical to the parent plant. They will therefore display the same desirable characteristics such as high yield, resistance to a common pest or disease, or a particular colour of flower.

- If the original plant had an unusual combination of characteristics due to selective breeding or genetic

modification, then this combination can be retained without the risk of losing that combination through sexual reproduction.

- The new plants are all uniform in their phenotype, which makes them easier to grow and harvest.

- Using the apical bud (meristem) as an explant for tissue culture ensures the new plants are free from viruses.

Disadvantages of artificial cloning

- Tissue culture is labour intensive.

- It is expensive to set up the facilities to perform tissue culture successfully.

- Tissue culture can fail due to microbial contamination.

- All the cloned offspring are genetically identical and are therefore susceptible to the same pests and/or diseases. Crops grown in monocultures allow rapid spread of a disease or pest between the closely planted crop plants.

- There is no genetic variation, except that introduced by mutation.

> **INVESTIGATION**
>
> **Cloning cauliflower**
> - Cut a mini-floret from a cauliflower. Cut the floret into small 3–5 mm pieces; these are the explants.
>
> - Transfer the explants to a sterilising solution such as sodium dichloroisocyanurate and swirl the solution periodically. Be aware that sodium dichloroisocyanurate releases chlorine; avoid inhalation or skin contact and take particular care if you have asthma.
>
> - After 15 minutes, transfer the explants onto a sterile nutrient agar surface in a specimen tube. It is important to use aseptic techniques to avoid contaminating the explants during this stage. Place the cap on the tube.
>
> - Incubate in a warm lab. Growth and greening of some tissues should be visible within two weeks.

Questions

1. For successful vegetative propagation, the cutting or explant must contain meristem cells. Explain why.

2. Nodes are regions on the stem where side branches grow. Suggest why the new roots on a cutting usually grow from the node.

3. Careful selective breeding can produce a plant that possesses an unusual combination of characteristics. Explain why a grower who wants to produce a large number of these plants must breed them by cloning.

4. Why is it essential to use sterile conditions when carrying out tissue culture?

5. What is a callus?

By the end of this topic, you should be able to demonstrate and apply your knowledge and understanding of:

* how artificial clones in animals can be produced by artificial embryo twinning or by enucleation and somatic cell nuclear transfer (SCNT)

* the arguments for and against artificial cloning in animals

---

KEY DEFINITIONS

embryo twinning: splitting an embryo to create two genetically identical embryos.
enucleation: removal of the cell nucleus.
somatic cell nuclear transfer (SCNT): a technique that involves transferring the nucleus from a somatic cell to an egg cell.

Artificial cloning in animals

Some invertebrate species such as greenfly and water fleas have evolved the ability to clone naturally. In other species it is a rare event. Therefore, most cloning of animals is artificial.

Successful cloning starts with cells that are **totipotent** – such cells can divide and differentiate into all types of cell found in the adult organism. In animals, the only truly totipotent cells are very early embryo cells.

Reproductive cloning

Reproductive cloning can produce large numbers of genetically identical animals. Cloning may be useful for:

* Elite farm animals produced by selective breeding (artificial selection) or genetic modification. For example, a particularly good individual bull whose value is as a stud – supplying sperm for artificial insemination.

* Genetically-modified animals developed with unusual characteristics, for example, goats that produce spider silk in their milk and cows that produce less methane.

The two main techniques to achieve reproductive cloning are **embryo twinning** and **somatic cell nuclear transfer (SCNT)**.

LEARNING TIP

Cloning is the production of genetically identical cells or organisms – it is not sufficient simply to refer to 'identical' cells or organisms.

Embryo splitting

Mammals can produce identical offspring (twins) if an embryo splits very early in development. This process has given rise to an artificial technique that has been used since the 1970s:

1. A zygote (fertilised egg) is created by *in vitro* fertilisation (IVF).

2. The zygote is allowed to divide by mitosis to form a small ball of cells.

3. The cells are separated and allowed to continue dividing.

4. Each small mass of cells is placed into the uterus of a surrogate mother.

This technique has been used to clone elite farm animals or animals for scientific research. However, the precise genotype and phenotype of the offspring produced will depend upon the sperm and egg used. Therefore, the precise phenotype will be unknown until the animals are born.

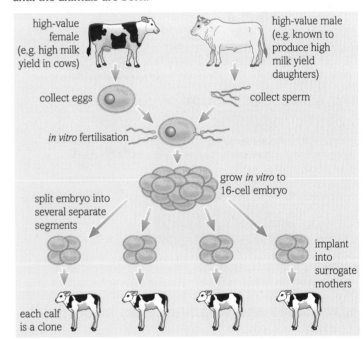

Figure 1 Cloning high-value farm animals by embryo splitting.

Somatic cell nuclear transfer (SCNT)

SCNT is the only way to clone an adult. The advantage is that the phenotype is known before cloning starts. This process was first performed successfully on a mammal in 1996, to produce Dolly the sheep.

1. An egg cell is obtained and its nucleus is removed, known as **enucleation**.

2. A normal body cell (somatic cell) from the adult to be cloned is isolated and may have the nucleus removed.

3. The complete adult somatic cell or its nucleus is fused with the empty egg cell by applying an electric shock.

4. The shock also triggers the egg cell to start developing, as though it had just been fertilised.

5. The cell undergoes mitosis to produce a small ball of cells.

6. The young embryo is placed into the uterus of a surrogate mother.

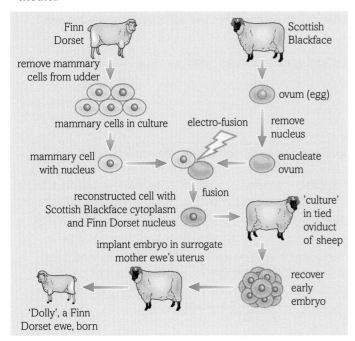

Figure 2 Using somatic cell nuclear transfer to clone an adult mammal.

DID YOU KNOW?

Dolly suffered ill health, including arthritis and tumours in her chest. She apparently aged prematurely. This fuelled speculation that SCNT produced babies who were 'old at birth'. There is still controversy surrounding Dolly's health.

Non-reproductive cloning

Non-reproductive cloning is the production of cloned cells and tissues for purposes other than reproduction.

Therapeutic cloning

New tissues and organs can be grown as replacement parts for people who are not well:

- Skin can be grown *in vitro* to act as a graft over burned areas.
- Cloned cells have been used to repair damage to the spinal cord of a mouse and to restore the capability to produce insulin in the pancreas.
- There is the potential to grow whole new organs to replace diseased organs.

Tissues grown from the patient's own cells will be genetically identical and so avoid rejection, which is a problem when transplanting donated organs.

Cloning for scientific research

Cloned genetically identical embryos can be used for scientific research into the action of genes that control development and differentiation. They can also be used to grow specific tissues or organs for use in tests on the effects of medicinal drugs.

Arguments for and against artificial cloning in animals

Arguments for artificial cloning in animals	Arguments against artificial cloning in animals
Can produce a whole herd of animals with a high yield or showing an unusual combination of characteristics (such as producing silk in their milk).	Lack of genetic variation may expose the herd to certain diseases or pests. Animals may be produced with little regard for their welfare, which may have undesirable side effects such as meat-producing chickens that cannot walk.
Produces genetically identical copies of very high value individuals retaining the same characteristics.	The success rate of adult cell cloning is very poor and the method is a lot more expensive than conventional breeding. Cloned animals may be less healthy and have shorter life spans.
Using genetically identical embryos and tissues for scientific research allows the effects of genes and hormones to be assessed with no interference from different genotypes.	There are ethical issues regarding how long the embryo survives and whether it is right to create a life simply to destroy it.
Testing medicinal drugs on cloned cells and tissues avoids using animals or people for testing.	
Can produce cells and tissues genetically identical to the donor, for use in repairing damage caused by disease or accidents.	
Individuals from an endangered species can be cloned to increase numbers.	This does not help increase genetic diversity.

Table 1 Arguments for and against artificial cloning in animals.

Questions

1. Describe the advantages of repairing a damaged heart by injecting cloned cells compared to a heart transplant.

2. In normal sexual reproduction, DNA from two parents is combined. What are the advantages of sexual reproduction over cloning?

3. Mitochondria contain mitochondrial DNA. Which parent does this come from in sexual reproduction?

By the end of this topic, you should be able to demonstrate and apply your knowledge and understanding of:

* the use of microorganisms in biotechnological processes (continued in topics 6.4.5 and 6.4.6)

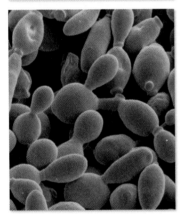

Figure 1 *Saccharomyces cerevisiae* yeast cells used in brewing and baking.

History of biotechnology

The name **biotechnology** was first coined by Karoly Ereky in 1919. He was a Hungarian agricultural engineer who set up a huge industrialised farming unit. He used the term to describe any technological process that made use of living organisms or parts of living organisms to manufacture useful products or provide useful services. This included the domestication of animals, planting of crops, mechanisation of agricultural processes and selective breeding of plants and animals over many generations.

Biotechnology has been in use for a very long time. The oldest documented example of biotechnology is the production of beer or ale 7000 years ago. Brewing makes use of yeast (a single-celled fungus) to ferment maltose sugars in germinating barley. Other early examples of biotechnology include making yoghurt and cheese, as well as baking.

More recent examples of biotechnology include:

* the use of the bacterium *Clostridium acetobutylicum* to produce acetone, which was needed to make explosives during World War I

* the manufacture of penicillin from the fungus *Penicillium notatum* during World War II.

These examples and others shifted the emphasis of biotechnology away from food towards the manufacture of drugs.

However, the new science of DNA technology brought biotechnology to its current position. Our increasing understanding of genetics and genetic engineering, along with the ability to manipulate the living conditions of living organisms, has led to a huge expansion in biotechnology. The biggest expansion is in the use of microorganisms in industrial processes.

Figure 2 Industrial fermenters used for growing microorganisms.

Biotechnology today

The four main areas in which microorganisms are used in biotechnology are summarised in Table 1.

Product	Example	Organism used
Food	• Ethanol in beer and wine • Carbon dioxide used to make bread rise • Lactic acid used to make yoghurt and cheese • Mycoprotein – a filamentous fungus protein used to make vegetarian food • Soya – soya beans are fermented to produce soy sauce	Yeast (*S. cerevisiae*) Yeast (*S. cerevisiae*) *Lactobacillus* bacteria *Fusarium venenatum* fungus Yeast or *Aspergillus*
Pharmaceutical drugs	• Penicillin • Other antibiotics • Insulin, other therapeutic human proteins	*Penicillium* fungus Other fungi and bacteria (mainly bacteria of the genus *Streptomyces*) Genetically-modified bacteria
Enzymes	• Protease and lipase used in washing powders • Pectinase used to extract juice from fruit • Sucrase used to digest sugar to make food sweeter • Amylase to digest starch into sugar to produce syrup used as a sweetener in food production • Protease used to tenderise meat • Lactase to make lactose-free milk • Removing sticky residues from recycled paper	Bacteria, e.g. *Bacillus licheniformis* *Aspergillus niger* *Yeasts* and *Aspergillus* spp *A. oryzae* *Aspergillus* spp *A. niger* and *A. oryzae*
Other products	• Biogas, which is a combination of carbon dioxide and methane • Citric acid, E330, a food preservative • Bioremediation – cleaning waste water	Anaerobic bacteria (decomposers) *A. niger* fungus Variety of bacteria and fungi

Table 1

The advantages of using microorganisms in biotechnology

Using microorganisms in biotechnology has many advantages:

• Microorganisms are relatively cheap and easy to grow.

• In most cases, the production process takes place at lower temperatures than would be required to make the molecules by chemical engineering means. This saves fuel and reduces costs.

• The production process can take place at normal atmospheric pressure, which is safer than using chemical reactions that may require very high pressure for successful manufacture of certain molecules.

• The production process is not dependent on climate – so it can take place anywhere in the world with the resources to build and run suitable equipment.

• The microorganism can be fed by-products from other food industries, e.g. starch, waste water or molasses. (Note, however, that the starting ingredients often have to be pre-treated, which adds to the cost.)

• Microorganisms have a short life cycle and reproduce quickly. Some microorganisms may reproduce as often as every 30 minutes under ideal conditions. Therefore, a large population can grow very quickly inside the reaction vessel (**fermenter**).

- Microorganisms can be genetically modified relatively easily. This allows very specific production processes to be achieved.
- There are fewer ethical considerations to worry about in using microorganisms.
- The products are often released from the microorganism into the surrounding medium. This makes the product easy to harvest.
- The product is often more pure or easier to isolate than in conventional chemical engineering processes. This means lower downstream processing costs.

Using other organisms in biotechnology

Microorganisms are not the only organisms used in biotechnology. Genetically-modified mammals such as sheep, goats and cows can be used to produce useful proteins.

In some mammals the proteins are incorporated into the milk and can be easily harvested. For example, goats that have been genetically modified to possess the gene for spider silk secrete it into their milk. In other cases the protein may be secreted into the blood. For example, cows have been genetically modified to synthesise human antibodies, which can be isolated from their blood.

Other forms of biotechnology

In recent years the term 'biotechnology' has come to mean using organisms in production processes. However, biotechnology also encompasses the following processes:

- gene technology
- genetic modification and gene therapy
- selective breeding
- cloning by embryo-splitting and micropropagation
- the use of enzymes in industrial processes
- immunology.

Questions

1. Suggest two reasons why microorganisms are particularly useful in bioremediation – the treatment of contaminated waste.

2. Antibiotics prevent the growth of bacteria. Suggest why an organism, such as *Penicillium*, would evolve the ability to synthesise such a molecule.

3. State three reasons why using microorganisms in biotechnology enables very rapid production.

4. Suggest what conditions need to be maintained to ensure rapid production from microorganisms.

5. Explain how the antibodies made by a cow can be human antibodies.

⑤ Using biotechnology to make food

By the end of this topic, you should be able to demonstrate and apply your knowledge and understanding of:

* the use of microorganisms in biotechnological processes (continued in topic 6.4.6)
* the advantages and disadvantages of using microorganisms to make food for human consumption

Microorganisms in food manufacture

As described in topic 6.4.4, microorganisms are used in manufacturing food. Microorganisms have been used for many years, and many traditional foods are made with the help of microorganisms.

> **LEARNING TIP**
>
> The biological principles behind these methods of producing food are important. The examination is more likely to test these principles rather than test your memory of the production processes.

Yoghurt

Yoghurt is milk that has undergone **fermentation** by *Lactobacillus bulgaricus* and *Streptococcus thermophilus*. The bacteria convert lactose to lactic acid. The acidity denatures the milk protein, causing it to **coagulate**. The bacteria partially digest the milk, making it easy to digest. Fermentation also produces the flavours characteristic of yoghurt.

Other bacteria, such as *L. acidophilus*, *L.* subsp. *casei* and *Bifidobacterium*, may be added as probiotics – bacteria which may benefit human health by improving digestion of lactose, aiding gastrointestinal function and stimulating the immune system.

Cheese

Milk is usually pre-treated with a culture of bacteria (*Lactobacillus*) that can produce lactic acid from the lactose. Once it is acidified, the milk is mixed with **rennet**. Rennet contains the enzyme **rennin** (**chymosin**), which is found in the stomachs of young mammals. Rennin coagulates the milk protein (**casein**) in the presence of calcium ions:

1. **Kappa-casein**, which keeps the casein in solution, is broken down. This makes the casein insoluble.

2. The casein is precipitated by the action of calcium ions, which bind the molecules together.

The resulting solid, called **curd**, is separated from the liquid component (whey) by cutting, stirring and heating. The bacteria continue to grow, producing more lactic acid. The curd is then pressed into moulds.

Treatment while making and pressing the curd determines the characteristics of the cheese. Flavour is determined during the later ripening and maturing processes. The cheese can be given additional flavour by inoculation with fungi such as *Penicillium* to produce 'blue' cheese.

(a)

(b)

Figure 1 (a) Stirring salts and rennet into milk to make cheese. (b) The finished product.

> **DID YOU KNOW?**
>
> Rennet is traditionally produced from the stomachs of young calves, but the enzyme rennin is now also produced by genetically-modified bacteria.

Baking

Bread is a mixture of flour, water and salt with some yeast (which is a single-celled fungus, *Saccharomyces cerevisiae*). Bread-making processes have three key steps:

1. Mixing – the ingredients are mixed together thoroughly by kneading. This produces dough.

2. Proving/fermenting – the dough is left in a warm place for up to three hours while the yeast respires anaerobically. This produces carbon dioxide bubbles, causing the dough to rise.

3. Cooking – the risen dough is baked. Any alcohol evaporates during the cooking process.

Alcoholic beverages

Alcoholic beverages are also the product of the anaerobic respiration of yeast (*S. cerevisiae*). Wine is made using grapes that naturally have yeasts on their skin. Grapes contain the sugars fructose and glucose. When the grapes are crushed, the yeast uses these sugars to produce carbon dioxide and alcohol.

Ale or beer is brewed using barley grains that are beginning to germinate. This process is called malting. As the grain germinates it converts stored starch to maltose, which is respired by the yeast. Anaerobic respiration again produces carbon dioxide and alcohol. Hops are used to give a bitter taste to the liquid.

Single-cell protein (SCP)

More recently, microorganisms have been used to manufacture protein that is used directly as food. The microorganism used most frequently is the fungus *Fusarium venenatum*. The fungal protein or mycoprotein is also known as single-cell protein (SCP). The best known example of a mycoprotein is Quorn™, which was first produced in the early 1980s. It is marketed as a meat substitute for vegetarians and a healthy option for non-vegetarians, as it contains no animal fat or cholesterol.

There is huge potential in SCP production using such microorganisms as *Kluyveromyces*, *Scytalidium* and *Candida*. These fungi can produce protein with a similar amino acid profile to animal and plant protein. They can grow on almost any organic substrate, including waste materials such as paper and whey (curdled milk from which the curds have been removed).

Figure 2 Quorn™ – a protein produced from fungi.

DID YOU KNOW?

A top-quality bull weighing 1 tonne will produce 1–2 kg of protein in a day. Fungal cells kept under the right conditions can reproduce every 1–3 hours. One tonne of fungal cells could therefore produce at least a tonne of protein in a day!

Advantages and disadvantages of using microorganisms

Advantages

- Production of protein can be many times faster than that of animal or plant protein.

- The biomass produced has very high protein content (45–85%).

- Production can be increased and decreased according to demand.

- There are no animal welfare issues.

- The microorganisms provide a good source of protein.

- The protein contains no animal fat or cholesterol.

- The microorganisms can easily be genetically modified to adjust the amino acid content of the protein.

- SCP production could be combined with removal of waste products.

- Production is independent of seasonal variations.

- Not much land is required.

Disadvantages

- Some people may not want to eat fungal protein or food that has been grown on waste.

- Isolation of the protein – the microorganisms are grown in huge fermenters and need to be isolated from the material on which they grow.

- The protein has to be purified to ensure it is uncontaminated.

- Microbial biomass can have a high proportion of nucleic acids, which must be removed.

- The amino acid profile may be different from traditional animal protein – and particularly it can be deficient in methionine.

- Infection – the conditions needed for the microorganisms to grow are also ideal for pathogenic organisms. Care must be taken to ensure the culture is not infected with the wrong organisms.

- Palatability – the protein does not have the taste or texture of traditional protein sources.

Questions

1. In the production of cheese, milk is acidified before rennet is added. Suggest why.

2. Suggest why extra calcium salt is often added to milk during cheese making.

3. During the brewing process, why must the yeast be encouraged to respire anaerobically?

4. Production of both yoghurt and cheese uses the bacterium *Lactobacillus*. Explain why cheese is so much harder than yoghurt.

5. What are the advantages of producing rennin from bacteria?

(6) Other processes involving biotechnology

By the end of this topic, you should be able to demonstrate and apply your knowledge and understanding of:

* the use of microorganisms in biotechnological processes

* the importance of manipulating the growing conditions in batch and continuous fermentation in order to maximise the yield of product required

Scaling up production of drugs

Commercial drug production uses large stainless steel containers called **fermenters**, in which the growing conditions can be controlled to ensure the best possible yield of the product. The conditions that must be controlled include:

* temperature – too hot and enzymes will be denatured, too cool and growth will be limited

* nutrients available – microorganisms require nutrients to grow and synthesise the product. Sources of carbon, nitrogen, minerals and vitamins are needed

* oxygen availability – most microorganisms respire aerobically

* pH – enzyme activity and hence growth and synthesis are affected by extremes of pH

* concentration of product – if the product is allowed to build up, it may affect the synthesis process.

A fermenter must first be sterilised using superheated steam. It can then be filled with all the components required for growth and supplied with a starter culture of the microorganism to be used. The culture will be left to grow and synthesise the products.

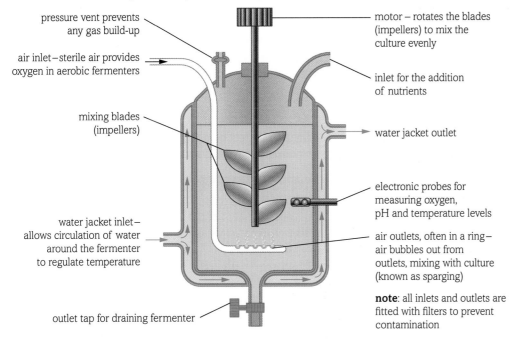

pressure vent prevents any gas build-up

air inlet – sterile air provides oxygen in aerobic fermenters

mixing blades (impellers)

water jacket inlet – allows circulation of water around the fermenter to regulate temperature

outlet tap for draining fermenter

motor – rotates the blades (impellers) to mix the culture evenly

inlet for the addition of nutrients

water jacket outlet

electronic probes for measuring oxygen, pH and temperature levels

air outlets, often in a ring – air bubbles out from outlets, mixing with culture (known as sparging)

note: all inlets and outlets are fitted with filters to prevent contamination

Figure 1 A generalised diagram of an industrial fermenter.

Batch and continuous culture

Some products are synthesised by the microorganism during normal metabolism when they are actively growing. These are called primary metabolites. Such products are continually released from the cells and can be extracted continuously from the fermenting broth. The broth is topped up with nutrients as these are used by the microorganisms. Some of the broth is removed regularly to extract the product and remove cells from the broth – otherwise the population becomes too dense. This is known as continuous culture and keeps the microorganism growing at a specific growth rate.

Other products are produced only when the cells are placed under stress, such as high population density or limited nutrient availability. These are called secondary metabolites and are produced mostly during the stationary phase of growth. Here, the culture is set up with a limited quantity of nutrients and allowed to ferment for a specific time. After this time, the fermenter is emptied and the product can be extracted from the culture. This is known as batch culture.

The importance of asepsis

Asepsis is ensuring that sterile conditions are maintained. The nutrient medium would also support the growth of unwanted microorganisms which would reduce production, because the unwanted microorganisms:

- compete with the cultured microorganisms for nutrients and space
- reduce the yield of useful products
- spoil the product.

They may also:

- produce toxic chemicals
- destroy the cultured microorganisms and their products.

In processes where foods or medicinal chemicals are produced, all products must be discarded if contamination by unwanted organisms occurs.

Production of penicillin

Florey and Chain devised the process to successfully mass produce penicillin through fermentation by the fungus *Penicillium chrysogenum*. Modern strains of the fungus have been selectively bred to be more productive than the early strains.

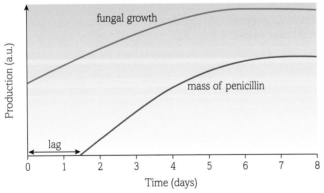

Figure 2 A graph representing the relative production of fungal growth and penicillin in a fermenter.

Penicillin is a secondary metabolite – it is only produced once the population has reached a certain size. Therefore, penicillin is manufactured by batch culture.

1. The fermenter is run for six to eight days. The culture is then filtered to remove the cells.

2. The antibiotic is precipitated as crystals by the addition of potassium compounds. The antibiotic may be modified by the action of other microorganisms or by chemical means.

3. The antibiotic is mixed with inert substances and prepared for administration in tablet form, as a syrup or in a form suitable for injection.

> **LEARNING TIP**
> Microorganisms use enzymes to carry out the required processes. Remember that any condition suitable for enzyme activity is likely to be suitable for bacterial and fungal growth.

Production of insulin

Insulin is widely used to treat Type 1 diabetes. It was previously extracted from the pancreas of animals such as cattle or pigs sent for slaughter. Insulin from slaughtered animals is not identical to human insulin and so is less effective than human insulin, and expensive to extract.

In 1978, synthetic human insulin was developed by genetically modifying a bacterium. The gene for human insulin was combined with a plasmid to act as a vector, so the gene could be inserted into the bacterium *Escherichia coli*. The resulting genetically-modified bacterium enabled the production of vast quantities of human insulin at relatively low cost. Insulin is manufactured by continuous culture.

Bioremediation

Bioremediation is the use of microorganisms to clean the soil and underground water on polluted sites. The organisms convert the toxic pollutants to less harmful substances. The idea really started when Ananda Chakrabarty modified a *Pseudomonas* bacterium in 1971, enabling it to break down crude oil, and he proposed that it could be used in treating oil spills. Solvents and pesticides can also be treated using bioremediation.

> **DID YOU KNOW?**
> Four species of bacteria were known that had some capability to digest oil. Chakrabarty transferred plasmids from these species into one organism and used UV light to activate the genes. This produced an organism that could digest oil much more quickly – it is now known as a separate species, *Pseudomonas putida*.

Bioremediation involves stimulating the growth of suitable microbes that use the contaminants as a source of food. It requires the right conditions for the growth of microorganisms, which are:

- available water
- a suitable temperature
- suitable pH.

Where conditions are not quite suitable, they may be modified by the addition of suitable substances. In some cases, additional nutrients such as molasses may be needed to ensure the microorganisms can grow effectively. It may also be necessary to pump in oxygen for aerobic bacteria.

Where conditions cannot be made suitable *in situ*, the soil may be dug up and moved to be treated *ex situ*.

Figure 3 A pond used for bioremediation on an oil field. Contaminated soil can be treated in the pond.

Advantages of bioremediation

- Uses natural systems
- Less labour/equipment is required
- Treatment *in situ*
- Few waste products
- Less risk of exposure to clean-up personnel

However, bioremediation is only suitable for certain products; heavy metals such as cadmium and lead cannot be treated.

Questions

1. Explain why populations of *Penicillium* needed to be treated to produce mutations.

2. Suggest why the fermenters must be sterilised before use.

3. Explain why the pancreas is the only tissue from which insulin can be extracted.

4. What is the usual role of mRNA in protein synthesis?

5. Suggest what concerns people might have about bioremediation.

(7) Microorganism cultures

By the end of this topic, you should be able to demonstrate and apply your knowledge and understanding of:

* how to culture microorganisms effectively, using aseptic techniques

---

KEY DEFINITIONS

agar: a polysaccharide of galactose obtained from seaweed, which is used to thicken the medium into a gel.
aseptic technique: sterile techniques used in culturing and manipulating microorganisms.

Growing microorganisms

Microorganisms will grow on almost any material that provides the carbon compounds for respiration and a source of nitrogen for protein synthesis. However, in the laboratory, microorganisms are usually grown in one of two types of **growth medium**:

* a soup-like liquid called a broth, kept in bottles or tubes

* a set jelly-like substance called **agar**, which is melted and poured into **Petri dishes**.

Typical nutrient agar contains peptones (from the enzymatic breakdown of gelatine), yeast extract, salts and water; it may also contain glucose or blood.

Aseptic techniques

Aseptic techniques have been developed to reduce the likelihood of contaminating the medium with unwanted bacteria or fungi. The standard procedure includes the following:

1. Wash your hands.

2. Disinfect the working area.

3. Have a Bunsen burner operating nearby to heat the air. This causes the air to rise and prevents air-borne microorganisms settling. It also creates an area around it of sterile air in which the microbiologist can work.

4. As you open a vessel, pass the neck of the bottle over the flame to prevent bacteria in the air entering the bottle. The bottle should also be flamed as it is closed.

5. Do not lift the lid of the Petri dish off completely – just open it enough to allow introduction of the desired microorganism.

6. Any glassware or metal equipment should also be passed through the flame before and after contact with the desired microorganisms.

Techniques used in microbiology

Microorganisms can grow almost anywhere. The key is to ensure that they grow on the medium used and that you grow the desired microorganism rather than others that have infected the medium by mistake.

Growing microorganisms on agar plates involves three main steps:

1. sterilisation

2. **inoculation**

3. **incubation**.

DID YOU KNOW?

Bacteria are everywhere. They are so small that their numbers can be unimaginable – there can be around 500 billion bacteria in one tablespoon of soil!

Sterilisation

The nutrient agar medium and any equipment to be used must be sterilised.

The medium is sterilised by heating in an **autoclave** at 121 °C for 15 minutes (the high temperature is achieved by boiling water under high pressure inside the autoclave). This kills all living organisms, including any bacterial or fungal spores. When the medium has cooled sufficiently to handle, it is poured into sterile Petri dishes and left to set. It is important that the lid is kept on the Petri dish to prevent infection.

Figure 1 An autoclave used to sterilise equipment before and after use.

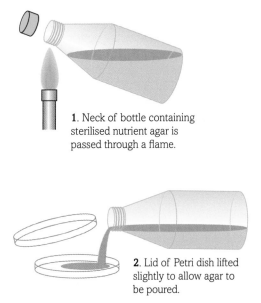

1. Neck of bottle containing sterilised nutrient agar is passed through a flame.

2. Lid of Petri dish lifted slightly to allow agar to be poured.

Figure 2 Pouring an agar plate.

All equipment used from this point onwards must be sterilised by heating.

Inoculation

Inoculation is the introduction of microorganisms to the sterile medium. This can be achieved in a number of ways:

- Streaking – a wire inoculating loop is used to transfer a drop of liquid medium onto the surface of the agar (see Figure 3). The drop is drawn out into a streak by dragging the loop across the surface. Take care not to break the surface of the agar.

- Seeding – a sterile pipette can be used to transfer a small drop of liquid medium to the surface of the agar or to the Petri dish before the agar is poured in.

- Spreading – a sterile glass spreader may be used to spread the inoculated drop over the surface of the agar.

- A small cotton swab or cotton bud can be moistened with distilled water and used to collect microorganisms from a surface and then carefully wiped over the surface of the agar medium.

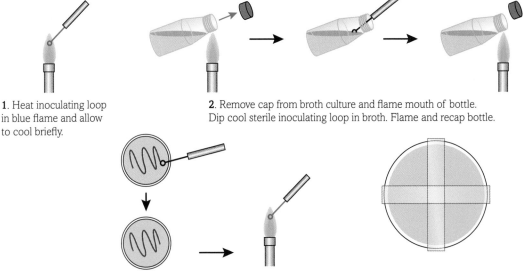

1. Heat inoculating loop in blue flame and allow to cool briefly.

2. Remove cap from broth culture and flame mouth of bottle. Dip cool sterile inoculating loop in broth. Flame and recap bottle.

3. Spread a streak of culture over surface of agar and cover with a lid. Reheat inoculating loop in blue flame.

4. Use adhesive tape to hold lid on Petri dish and incubate at 25 °C.

Figure 3 Inoculating an agar plate.

Incubation

The Petri dish must be labelled and the top taped to the bottom using two strips of adhesive tape – be careful not to seal the Petri dish completely, as this can lead to selection of anaerobic bacteria which may be pathogenic. The Petri dish is then placed in a suitable warm environment such as an incubator. It should be placed upside down as this prevents drops of condensation falling onto the surface of the agar. It also prevents the agar medium from drying out too quickly.

Suitable temperatures will depend on the type of organisms being grown.

Cultures can be examined after 24–36 hours. Do not open the Petri dish! Bacteria grow into visible colonies, which may be shiny or dull. Some colonies are round with entire edges, while others can have crenated edges. The colonies can also be a range of different colours. Each colony results from a single bacterium.

Filamentous fungi grow into a mass of hyphae, which may also be circular, but the mass is not shiny and often looks like cotton wool with fluffy aerial hyphae. Single-celled fungi (yeasts) grow as circular colonies.

All Petri dishes must be completely sterilised after use and before disposal. Thoroughly wash your hands after handling a Petri dish, as any moisture coming out of the dish could be a source of infection.

Using a liquid medium

A liquid broth is initially clear but will turn cloudy when bacteria have grown. A liquid broth can be useful to increase the numbers of microorganisms before transferring to agar plates for counting or identification. Similar aseptic techniques as described above must be applied when using a broth. A liquid broth can be used to investigate population growth, as described in topic 6.4.8.

LEARNING TIP

In schools, it is important to avoid incubation temperatures close to 37 °C, which may encourage the growth of pathogenic microorganisms. Incubation at 37 °C may be used in a medical lab that is trying to identify the organism causing a disease.

Questions

1 Explain why the inoculating loop must be sterilised before and after use.

2 When measuring population growth, why is it important to inoculate the agar plate with a standard sized drop of the broth culture?

3 When an agar plate is inoculated with microorganisms, explain why:
   (a) the lid of the Petri dish should be taped to the bottom
   (b) the dish should not be sealed
   (c) the dish should be inverted.

4 Explain why nutrient agar must contain peptones.

(8) Population growth in a closed culture

By the end of this topic, you should be able to demonstrate and apply your knowledge and understanding of:

* the standard growth curve of a microorganism in a closed culture

* practical investigations into the factors affecting the growth of microorganisms

A liquid broth (see topic 6.4.7) can be used to measure the growth rate of a microorganism population. A sterile broth is inoculated and the population size measured at regular intervals during incubation. The population size can be measured by transferring a small sample to an agar plate and incubating the agar culture – each individual microorganism will produce a visible colony.

Serial dilutions

The numbers of individual microorganisms in a broth can be high. If the broth is used to inoculate an agar plate, there may be too many colonies which merge together making a count impossible. In order to investigate the rate of growth of the population of microorganisms, it will be essential to reduce the population density. This can be achieved by **serial dilution**.

This is a step-wise dilution of the broth culture. At each step the broth is diluted by a factor of 10. Take a $1\,cm^3$ sample of the broth and add $9\,cm^3$ of distilled water – label the diluted broth as '10^{-1}'. Take a $1\,cm^3$ sample of this diluted broth and add $9\,cm^3$ of distilled water – label as '10^{-2}'. Continue this procedure until you have a series of dilutions with suitable labels.

A drop of each dilution can be used to inoculate an agar plate. One of them will produce a culture plate in which the number of colonies can be counted. When recording the population density, do not forget to multiply your count by the dilution factor and also by the volume added to the plate.

> **KEY DEFINITIONS**
>
> **closed culture:** a culture which has no exchange of nutrients or gases with the external environment.
> **serial dilution:** a sequence of dilutions used to reduce the concentration of a solution or suspension.

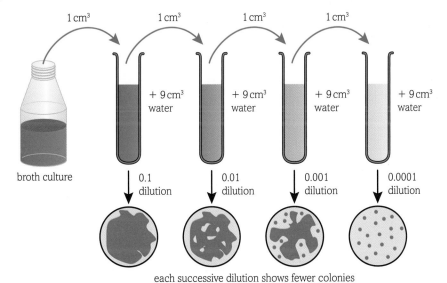

Figure 1 Serial dilution and plating to count population density.

LEARNING TIP

A 10^{-1} strength dilution of broth has been diluted to form a 1 in 10 ratio of broth to water. The number of colonies observed on the agar plate must be multiplied by 10 to calculate the population density in the broth. An accurate population density can only be calculated if the size of drop used to inoculate the agar plate is known precisely. The volume of this drop must also then be used to calculate a colony forming unit (cfu) per cm^3.

Investigating population growth

Use aseptic techniques throughout and avoid skin contact with the bacteria or the inoculated broth.

- Use a commercially supplied colony of bacteria (e.g. *Bacillus subtilis*) to inoculate a nutrient broth in a bottle.
- Thoroughly shake the broth and incubate the broth at 25 °C.
- At time zero take a small sample of the broth using a sterile pipette that gives a standard size drop. Inoculate the agar plate and use a sterile glass spreader to spread the drop over the agar surface.
- At intervals of 2 hours, shake the broth and take another sample to inoculate another nutrient agar plate.
- When the broth starts to look cloudy, carry out a serial dilution and use each dilution to inoculate a sterile agar plate.
- Incubate the plates at 25 °C.
- After 24 hours, record the number of bacterial colonies visible on each plate and multiply by any dilution factor used.
- Use the volume of the drop added to the plate to calculate the density of cells per cm^3.
- Plot the population density calculated (in cfu/cm^3) against time. Typically the *y*-axis should be as a log10 scale.

This method can be used to measure population growth under different conditions by changing factors such as temperature, type of sugar supplied, pH, etc. in the broth.

The growth curve

A small population of microorganisms in a **closed culture** that contains all the nutrients required for growth will undergo population growth. A 'closed culture' refers to a population in which all the conditions are set at the start and there is no exchange with the external environment. The population growth will follow a predictable pattern, as shown in Figure 2.

These are similar to the conditions set up for batch production in a fermenter. In batch production, however, certain substances such as oxygen may be added to keep the population growing until the nutrients are used up.

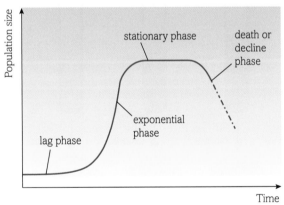

Figure 2 Growth of a population of microorganisms in a closed culture.

Lag phase

In the early part of population growth, the population does not grow quickly. This is partly because the population is still small, but also because the organisms are adjusting to their new environment. This may involve:

- taking up water
- cell growth
- switching on (activating) certain genes
- synthesising specific proteins (enzymes).

Log (exponential) phase

In the log (exponential) phase, the organisms have adjusted to their environment. They each have the enzymes needed to survive. Each individual has sufficient nutrients and space to grow rapidly and reproduce. The population doubles in size with each generation. In some microorganisms, this can be as frequently as once every 20–30 minutes.

Stationary phase

Eventually the increasing numbers of organisms use up the nutrients and produce increasing amounts of waste products such as carbon dioxide and other metabolites. The rate of population growth declines and the number of individuals dying increases until the reproduction rate equals the death rate. This is the **stationary phase**, where there is no population growth.

Death (decline) phase

The nutrients run out and the concentration of waste products may become lethal. More individuals die than are produced and the population begins to fall. Eventually all the organisms will die.

Primary and secondary metabolites

Primary metabolites produced during the normal activities of the microorganism will be collected from a fermenter during the log phase. In a fermenter, the population is not kept in a closed culture, but conditions are maintained for optimal growth.

Secondary metabolites are produced in the stationary phase. The population must be kept in a closed culture and the metabolites can be collected at the end of the stationary phase or during the decline phase.

Questions

1 Addition of nutrients will affect the population growth. Suggest how population growth will be affected if nutrients are added at:
(a) the lag phase
(b) the stationary phase.

2 Suggest why antibiotics (which are a secondary metabolite) are not produced until the stationary phase.

3 Suggest how the population growth curve may be different in an open system.

LEARNING TIP

The factors that limit the population growth of a microorganism are the same as those that will limit a population of organisms in any environment.

DID YOU KNOW?

Antibiotics produced by some fungi are a mechanism to reduce competition. When the population is dense they release antibiotics which kill competitor species, leaving more food for the fungus.

⑨ Immobilised enzymes

By the end of this topic, you should be able to demonstrate and apply your knowledge and understanding of:

* the uses of immobilised enzymes in biotechnology and the different methods of immobilisation

immobilised enzyme: an enzyme that is held in place and not free to diffuse through the solution.

Immobilised enzymes in biotechnology

Some biotechnological processes can be simplified by taking the enzymes out of the microorganisms. Enzymes are large proteins that act on substrates to generate the product. Enzymes are not used up in the reaction and remain in suspension when the reaction has been completed. In an industrial process, this means that the product must be isolated from the enzymes before use. This could be expensive.

Immobilised enzymes are taken out of suspension and held so that they do not mix freely with the substrate. This has the following advantages:

* Enzymes do not mix with the product, so extraction costs are lower.

* The enzymes can easily be reused.

* A continuous process is made easier, as there are no cells requiring nutrients, reproducing and releasing waste products.

* The enzymes are surrounded by the immobilising matrix, which protects them from extreme conditions – so higher temperatures or a wider pH range can be used without causing denaturing.

However, setting up the immobilised enzyme process is more expensive, and immobilised enzymes are usually less active than free enzymes, making the process slower.

Methods used to immobilise enzymes

Enzymes can be immobilised by binding them to a surface or by simply trapping them so that they cannot enter the substrate solution.

Adsorption

Enzyme molecules are bound to a supporting surface by a combination of hydrophobic interactions and ionic links (Figure 1a). Suitable surfaces include clay, porous carbon, glass beads and resins. The enzyme molecules are bound with the active site exposed and accessible to the substrate. However, the active site may be slightly distorted by the additional interactions affecting enzyme activity.

The bonding forces are not always strong, and enzymes can become detached and leak into the reaction mixture.

Covalent bonding

Enzyme molecules are bonded to a supporting surface such as clay using strong covalent bonds (Figure 1b). The enzymes are bonded using a cross-linking agent, which may also link them in a chain.

The production of covalent bonding can be expensive and can distort the enzyme active site, reducing activity. However, the enzymes are much less likely to become detached and leak into the reaction mixture.

Entrapment

Enzyme molecules are trapped in a matrix that does not allow free movement (Figure 1c). The enzyme molecules are unaffected by **entrapment** and remain fully active. However, the substrate molecules must diffuse into the entrapment matrix, and the product molecules must be able to diffuse out. The method is therefore suitable only for processes where the substrate and product molecules are relatively small. Calcium alginate beads are often used in schools to immobilise enzymes by entrapment. Industrial processes may also use a cellulose mesh.

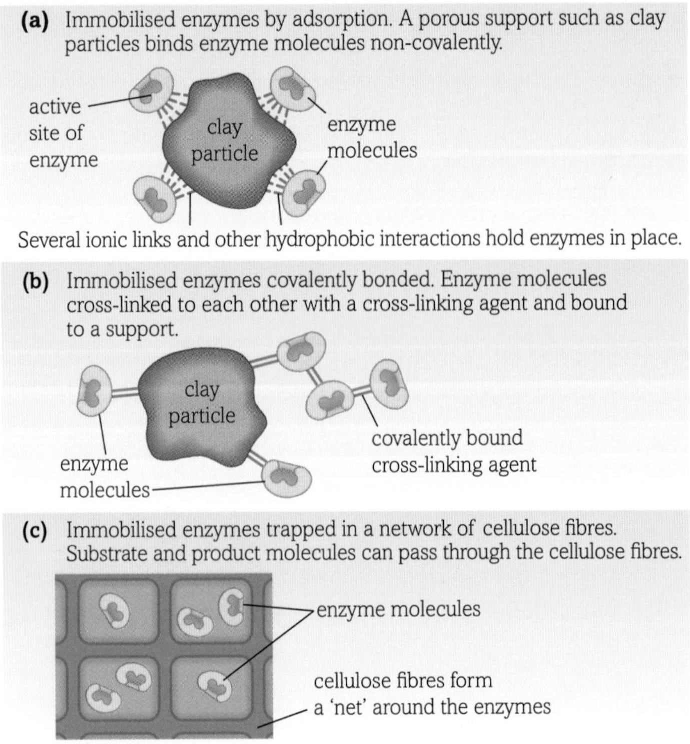

(a) Immobilised enzymes by adsorption. A porous support such as clay particles binds enzyme molecules non-covalently.

active site of enzyme — clay particle — enzyme molecules

Several ionic links and other hydrophobic interactions hold enzymes in place.

(b) Immobilised enzymes covalently bonded. Enzyme molecules cross-linked to each other with a cross-linking agent and bound to a support.

clay particle

enzyme molecules — covalently bound cross-linking agent

(c) Immobilised enzymes trapped in a network of cellulose fibres. Substrate and product molecules can pass through the cellulose fibres.

enzyme molecules

cellulose fibres form a 'net' around the enzymes

Figure 1 Methods used to immobilise enzymes: (a) adsorption; (b) covalent bonding; (c) entrapment.

Membrane separation

Enzyme molecules are separated from the reaction mixture by a partially permeable membrane. As in entrapment, the substrate and product molecules must be small enough to pass through the partially permeable membrane by diffusion. This access to the enzymes may limit the reaction rate.

Industrial use of immobilised enzymes

Glucose isomerase (also known as xylose isomerase)

- Converts glucose to fructose.

- Probably the most widely used enzyme, because of the number of applications of the syrup produced.

- Used to produce high fructose corn syrup (HFCS), which is much sweeter than sucrose. HFCS is often used in 'diet foods', as less sugar needs to be added for the equivalent sweetness. It may also be used as a sweetener in foods for diabetics. HFCS is cheaper than sucrose and so is widely used in the food industry to replace sucrose – especially in soft drinks, but also in many processed foods such as breakfast cereals, jam, ice cream, yoghurt and even sliced ham.

Penicillin acylase (also known as penicillin amidase)

Formation of semi-synthetic penicillins, such as amoxicillin and ampicillin, which were first developed during the 1960s. Some penicillin-resistant microorganisms are not resistant to these semi-synthetic penicillins.

Lactase

- Converts lactose to glucose and galactose by hydrolysis. Used to produce lactose-free milk.

- Milk is an important source of calcium, which is needed for strong bones and teeth. People with insufficient calcium in their diet are more likely to develop weak bones or osteoporosis. It is therefore important that people who are lactose intolerant (unable to digest and absorb the lactose in milk) are given calcium supplements or lactose-free milk.

Aminoacylase

- A hydrolase used to produce pure samples of L-amino acids by removing the acyl group from the nitrogen of an N-acyl-amino acid.

- L-amino acids are used as the building blocks for synthesis of a number of pharmaceutical and agrochemical compounds. They may also be used as additives for human food and animal feedstuffs.

Glucoamylase

- Converts dextrins to glucose. During the hydrolysis of starch, short polymers of glucose (dextrins) are formed. Hydrolysis by glucoamylase can convert these dextrins to glucose. Glucoamylase can be immobilised on a variety of surfaces and used to digest sources of starch such as corn and cassava.

- The enzyme is used in a wide range of fermentation processes, including the conversion of starch pulp to alcohol used to produce gasohol – an alternative fuel for motor vehicles. It is also used within the food industry to make high fructose corn syrup.

Nitrile hydratase

- Converts nitriles to amides, including acrylonitrile to acrylamide. Acrylamide can be polymerised to form polyacrylamide, which is a plastic used as a thickener.

- The most common use of polyacrylamide is in treatment of water. It helps to stick many small contaminants together, so that they are precipitated or are easy to filter out of the water.

- Polyacrylamide is also used in paper-making and to make gel for electrophoresis.

DID YOU KNOW?

Early biological washing powders left enzymes in the clothes, thus causing skin irritation. Modern biological powders contain enzymes immobilised on tiny beads, which rinse out more easily.

Questions

1. Explain why immobilising enzymes may reduce the rate of productivity.

2. Explain why bonding enzymes to a surface may alter the active site.

3. Explain why some immobilised enzyme processes can be carried out at temperatures well above the normal optimum for that enzyme.

4. What are the advantages of using immobilised enzymes over whole microorganisms?

COULD CLONING CURE DIABETES?

Scientists have used human cloning techniques to create healthy stem cells from a diabetic patient which they hope will produce insulin when transplanted into the body.

HOW HUMAN CLONING COULD CURE DIABETES

Human cloning has been used to create stem cells which could cure diabetes by triggering insulin production.

For the first time, scientists have successfully replaced the damaged DNA of a Type 1 diabetes sufferer with the healthy genetic material of an infant donor.

It is hoped that when these cells are injected back into the diabetic patient they will begin to produce insulin of their own accord.

Using the cloning technique which produced Dolly the sheep in 1996, the procedure would prevent the need for daily insulin injections and effectively 'cure' the disease.

"We are now one step closer to being able to treat diabetic patients with their own insulin-producing cells," said Dr Dieter Egli, the New York Stem Cell Foundation scientist who led the research.

"From the start, the goal of this work has been to make patient-specific stem cells from an adult human subject with Type 1 diabetes that can give rise to the cells lost in the disease."

Patients with Type 1 diabetes lack insulin-producing beta cells, resulting in insulin deficiency and high blood-sugar levels.

Because the stem cells are made using a patient's own skin cells, the engineered cells for replacement therapy would match the patient's DNA and so would not be rejected.

It is hoped that in future the stem cell therapy could be used for a wide range of conditions including Parkinson's disease, macular degeneration, multiple sclerosis, and liver diseases and for replacing or repairing damaged bones.

"I am thrilled to say we have accomplished our goal of creating patient-specific stem cells from diabetic patients using somatic cell nuclear transfer," said Susan Solomon, CEO and co-founder of NYSCF whose own son is Type 1 diabetic.

"Seeing today's results gives me hope that we will one day have a cure for this debilitating disease."

The technique works by removing the nucleus from an adult oocyte – an early stage egg – and replacing it with the nucleus of a healthy infant skin cell.

An electric shock causes the cells to begin dividing until they form a 'blastocyst' – a small ball of a few hundred cells which can be harvested.

Dr Rudolph Leibel, a co-author and co-director with Dr Robin Goland of the Naomi Berrie Diabetes Center, where aspects of these studies were conducted, said: "The resulting technical and scientific insights bring closer the promise of cell replacement for a wide range of human disease."

In 2011, the team reported creating the first embryonic cell line from human skin using nuclear transfer when they made stem cells and insulin-producing beta cells from patients with Type 1 diabetes.

However, those stem cells were triploid, meaning they had three sets of chromosomes, and therefore could not be used for new therapies.

Earlier this month a separate team reported that they had used human cloning to create stem cells for adults for the first time in a breakthrough which could lead to tissue and organs being regrown.

Researchers were able for the first time to turn adult human skin cells into stem cells, which can grow into any type of tissue in the body.

Last year a team managed to create stem cells from the skin cells of babies but it was unclear whether it would work in adults.

A team at the Research Institute for Stem Cell Research at CHA Health Systems in Los Angeles and the University of Seoul said they had achieved the same result with two men, one aged 35 and one 75.

However, both breakthroughs are likely to reignite the debate about the ethics of creating human embryos for medical purposes and the possible use of the same technique to produce cloned babies – which is illegal in Britain.

Dr Solomon said: "This research is strictly for therapeutic purposes. Under no circumstances do we or any other responsible scientific group have any intention to use this technique for human cloning, nor would it be possible."

The study was published in the journal *Nature*.

Source
- http://www.telegraph.co.uk/news/science/science-news/10794029/How-human-cloning-could-cure-diabetes.html

Where else will I encounter these themes?

Book 1	5.1	5.2	5.3	5.4	5.5	5.6

Let's start by considering the nature of the writing in the article.

1. Suggest why the writer states: "the procedure would ….effectively 'cure' the disease" rather than being more definite about having found a cure.
2. The article states that the research is strictly for therapeutic purposes and not for human cloning. Explain the difference between therapeutic cloning and reproductive cloning.

Now we will look at the biology in, or connected to, this article.

3. Suggest why researchers who successfully created stem cells from the skin of a baby were unsure that the procedure would work in adults.
4. Cells produced by somatic cell nuclear transfer show mutations in some of their genes. Suggest how producing stem cells by SCNT could cause mutations.
5. The technique has been used for reproductive cloning in some animals. Explain how the introduction of mutations may cause the cloned animal to be unhealthy.
6. Stem cells can be totipotent, pluripotent, multipotent, oligopotent or unipotent. Explain the difference between these terms.
7. Explain why embryonic cells are the only truly totipotent cells.
8. How can unipotent and pluripotent cells be useful in research?
9. What ethical issues are associated with use of stem cells in research?

> Don't worry if you are not ready to give answers to these questions yet. You may like to return to the questions once you have covered other topics later in the book. Use the timeline at the bottom of the page to help you to put this work in context with what you have already learned and what is ahead in your course.

Activity

Produce a leaflet designed to explain to a year 11 Biology student how a mammal can be cloned using the somatic cell nuclear transfer procedure.

Your leaflet should be A4 size folded in two or three.

It should include information about the following:

- The main steps in the procedure
- An explanation of why each step is required
- Ethical issues associated with reproductive cloning
- An explanation of the differences between reproductive cloning and non-reproductive cloning
- Potential therapeutic benefits of cloning

Figure 1 Cloned animals may look the same but the clone may not be as healthy as the original.

1. Which of the following is not an advantage of natural cloning? [1]

   A. Requires only one parent

   B. Rapid means of increasing numbers

   C. All the offspring are the same

   D. Occurs when conditions are ideal

2. Which row correctly describes the sequence of processes in tissue culture and micropropagation? [1]

A	Cut plant into small pieces (explants)	Grow callus on sterile medium	Separate into small clumps of cells (callus)	Grow on medium with specific ratios of auxin and cytokinin
B	Separate into small clumps of cells (callus)	Grow callus on sterile medium	Cut plant into small pieces (explants)	Grow on medium with specific ratios of auxin and cytokinin
C	Cut plant into small pieces (explants)	Grow on medium with specific ratios of auxin and cytokinin	Separate into small clumps of cells (callus)	Grow callus on sterile medium
D	Separate into small clumps of cells (callus)	Grow callus on sterile medium	Grow on medium with specific ratios of auxin and cytokinin	Cut plant into small pieces (explants)

3. The following statements are about biotechnology:
   (i) Biotechnology always uses living organisms.
   (ii) Immobilised enzymes are less active than free enzymes.
   (iii) Biotechnology makes use of cloning.
   (iv) All foods produced by biotechnology contain genetically-modified products.
   (v) Immobilised enzymes can be used for continuous fermentation processes.

Which row correctly identifies the statements that are **correct**? [1]

   A. All the statements are incorrect.

   B. Statements (i), (iii) and (v) are correct.

   C. All the statements except (iv) are correct.

   D. Statements (ii), (iii) and (v) are correct.

4. The following statements are about growth of microorganisms:
   (i) Bacteria can grow on any medium.
   (ii) Bacteria in a closed system follow an S-shaped growth curve.
   (iii) Secondary metabolites are formed during the log phase.
   (iv) Primary metabolites are produced as a part of normal metabolism.
   (v) During the stationary phase, more bacteria die than are produced.

Which row correctly identifies the statements that are **incorrect**? [1]

   A. All the statements are incorrect.

   B. Statements (i), (iii) and (v) are incorrect.

   C. All the statements except (ii) are incorrect.

   D. All the statements except (ii) and (v) are correct.

5. Which statement identifies an advantage of continuous fermentation over batch fermentation? [1]

   A. The fermenter does not need to be sterilised.

   B. Primary metabolites can be produced continuously.

   C. The fermenter must be stopped and cleaned after each production session.

   D. Nothing needs to be added to the fermenter during fermentation.

   [Total: 5]

6. (a) Potatoes are an important food crop. Each year, some of the potatoes harvested are kept to replant the next year.
   (i) State two advantages of reproducing potatoes in this way. [2]
   (ii) In 1845–1850, most of the potato crop in Ireland was destroyed by the pathogen *Phytophthora infestans*. Explain why it was able to cause such devastation. [2]

   (b) Today, new stocks of potato plants are guaranteed to be free of pathogens. This is achieved through tissue culture and micropropagation.
   (i) Explain why these plants can be guaranteed disease free. [2]
   (ii) Describe how micropropagation is carried out. [7]

   [Total: 13]

7. Modern techniques can produce large numbers of genetically identical animals.

    (a) State three advantages of having a herd of genetically identical animals. [3]

    (b) Cloning animals can be achieved by somatic cell nuclear transfer. Describe one other way in which an animal selected for its good characteristics can be used to make more animals with the same characteristics. [8]

    (c) Explain why the precise characteristics of the cloned animals are unknown until birth. [4]

    [Total: 15]

8. Figure 1 shows the typical growth of a population of microorganisms in a closed culture.

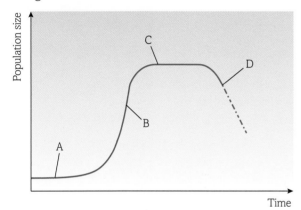

Figure 1

    (a) Identify the phases labelled A, B, C and D. [4]

    (b) Explain what is meant by a closed culture. [2]

    (c) Add a line to the graph to show how the concentration of a primary metabolite would rise in the culture. Label this line 'primary metabolites'. [1]

    (d) Add a second line to show how the concentration of a secondary metabolite would rise. Label this line 'secondary metabolites'. [1]

    (e) Explain the shape of the curve at point A and at point B. [4]

    (f) Suggest how the shape of the curve would change in an open system. Explain your answer. [4]

    [Total: 16]

9. High fructose corn syrups are produced using immobilised glucose isomerase.

    (a) To what category of molecules does glucose isomerase belong? [1]

    (b) State three advantages of immobilising this type of molecule. [3]

    (c) Describe two ways in which these molecules can be immobilised. [4]

    (d) A student mixed 1 cm³ of lactase solution with milk at three different temperatures. After 2 minutes, she measured the concentration of glucose in the milk. The student then immobilised the same volume of lactase solution in calcium alginate balls. She filled a syringe barrel with these alginate balls and poured milk into the syringe barrel. She then held the milk there for 2 minutes before allowing it to run out into a beaker beneath the syringe. The milk was tested for the concentration of glucose present. This was repeated with milk at the same three temperatures. Table 1 shows the results:

Temperature (°C)	Concentration of glucose (a.u.)	
	Free enzymes	**Immobilised enzymes**
10	7	5
30	23	17
70	5	42

Table 1

Explain the results shown. [5]

[Total: 13]

MODULE **6**

CHAPTER **6.5**

Genetics and ecosystems

ECOSYSTEMS

Introduction

In any ecosystem, organisms interact with other living organisms and with physical components of their environment. It is these interactions which this module focuses upon. You will learn how ecosystems work, and how to study them.

You may think that ecosystems tend to stay the same all the time. However, many ecosystems are dynamic – they change all the time. Because all the populations of living organisms in an ecosystem interact with each other and with their physical environment, any small changes in one can affect the others. You will learn about the process of succession and how a more stable climax community is eventually reached.

You already know that energy is passed from one member of a food chain to another. The original source of energy for most food chains is light energy from the Sun, which is converted by plants during photosynthesis and stored as biomass. However, at each stage of the food chain, some biomass is lost. We say that transfers between members of the food chain are inefficient. When biomass is lost, it is often respired, releasing energy (eventually as heat), which is lost from the ecosystem. When this happens, or when it decomposes, materials are released, which are commonly recycled within the ecosystem. You will learn about how nitrogen and carbon are recycled, and the role of microorganisms in such recycling. Humans have an interest in reducing the biomass which is lost during food production. You will learn how humans can manipulate such flow of energy through ecosystems.

Ecologists are scientists who study ecosystems, and they use techniques and tools to allow them to understand the interactions described above. By the end of this chapter, you will be familiar with such approaches, and you will be able to study ecosystems in just the same way.

All the maths you need

To unlock the puzzles of this chapter you need the following maths:

- Substitute numerical values into equations
- Use ratios and percentages
- Estimate results
- Find arithmetic means
- Plot data in an appropriate format
- Understand the principles of sampling
- Use random numbers

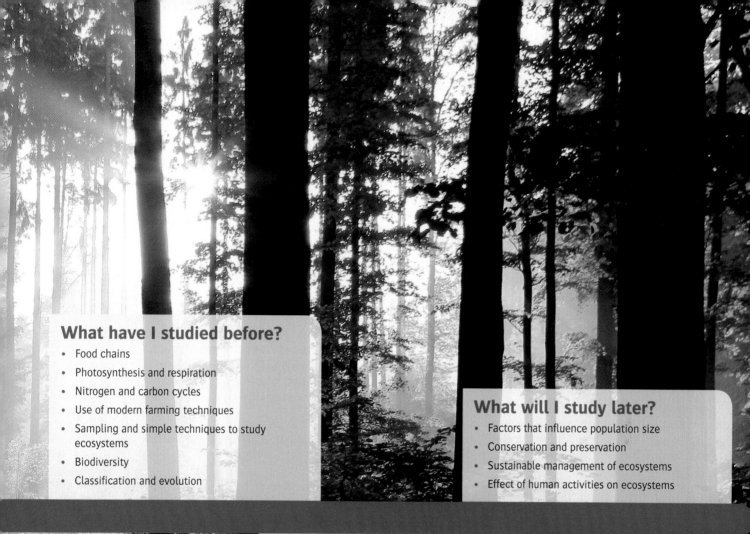

What have I studied before?

- Food chains
- Photosynthesis and respiration
- Nitrogen and carbon cycles
- Use of modern farming techniques
- Sampling and simple techniques to study ecosystems
- Biodiversity
- Classification and evolution

What will I study later?

- Factors that influence population size
- Conservation and preservation
- Sustainable management of ecosystems
- Effect of human activities on ecosystems

What will I study in this chapter?

- Introduction to ecosystems
- The flow of biomass through ecosystems
- Manipulating biomass flow through ecosystems
- Recycling of matter within an ecosystem
- Succession
- Techniques to study ecosystems

① Ecosystems

By the end of this topic, you should be able to demonstrate and apply your knowledge and understanding of:

* ecosystems, which range in size, are dynamic and are influenced by both biotic and abiotic factors

---

KEY DEFINITIONS

abiotic factors: non-living components of an ecosystem that affect other living organisms.
biotic factors: environmental factors associated with living organisms in an ecosystem that affect each other. e.g. predation, disease.
ecosystem: a community of animals, plants and bacteria interrelated with the physical and chemical environment.

What is an ecosystem?

Any group of living, and non-living things, and the interrelationships between them, can be thought of as an **ecosystem**. Ecosystems can be on a large scale (like an African grassland), a medium scale (like a playing field) or on a smaller scale (like a rock pool or a large tree) (Figure 1).

Figure 1 Different ecosystems: African grassland, an oak tree, a playing field and a rock pool.

The components of an ecosystem include:

* **Habitat** – the place where an organism lives.
* **Population** – all of the organisms of one species, who live in the same place at the same time, and who can breed together.
* **Community** – all the populations of different species, who live in the same place at the same time, and who can interact with each other.

The role of each species in an ecosystem is its **niche**. Because each organism interacts with both living and non-living things, it is almost impossible to define its niche entirely. A description of its niche could include things like how and what it feeds on, what it excretes, and how it reproduces. It is impossible for two species to occupy *exactly* the same niche in the same ecosystem.

Ecosystems do not have clear edges. It is impossible to draw a line around a group of living things and say that they only interact with each other rather than with organisms outside that ecosystem. However, it is often useful to think of ecosystems as being 'closed', as it makes them easier to understand.

Factors affecting ecosystems

Biotic factors

Depending on their niche, the living organisms in an ecosystem can affect each other:

* **Producers.** Plants (and some photosynthetic bacteria), which supply chemical energy to all other organisms.
* **Consumers. Primary consumers** are herbivores, which feed on plants, and which are eaten by carnivorous **secondary consumers**. These in turn are eaten by carnivorous **tertiary consumers**.
* **Decomposers.** Decomposers (bacteria, fungi and some animals) feed on waste material or dead organisms.

Because these components of the ecosystem require their own source of materials and energy, they can affect other organisms' food supply. They can also be responsible for predation and disease.

Abiotic factors

Abiotic factors describe the effects of the non-living components of an ecosystem: pH, relative humidity, temperature and the concentration of pollutants are all examples. These can vary in space and time. Such factors could also include disturbance to the ecosystem by other factors such as turbulence and storms. Abiotic factors may also be influenced by the biotic components of the ecosystem. For example, in a rainforest, the forest canopy influences the temperature and humidity of the ecosystem.

At extreme values of an abiotic factor, a species may perform better or worse, or even die. A generalised curve depicting the effect of an abiotic factor on an organism's activity is shown in Figure 2. The abiotic factor could be pH, temperature or any variable that has some sort of optimum level, and where there are lethal levels at both extremes.

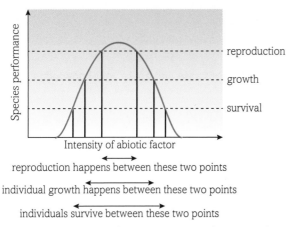

Figure 2 The effect of an abiotic factor on an organism's activity, when the abiotic factor is lethal at both extremes.

reproduction happens between these two points

individual growth happens between these two points

individuals survive between these two points

When there is not a lethal level at both extremes, an organism's response can be plotted differently. For example, at low levels of pollutants, an organism may survive without any detrimental effect. But at high levels, the pollutants may be lethal. Such a situation is shown in Figure 3.

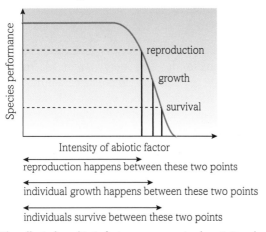

reproduction happens between these two points

individual growth happens between these two points

individuals survive between these two points

Figure 3 The effect of an abiotic factor on an organism's activity, when the abiotic factor is lethal only at one extreme.

Ecosystems are dynamic

Because ecosystems change, we refer to them as dynamic. The non-living elements change, and the living elements grow and die, with populations of particular species rising and falling. In most ecosystems, population sizes rise and fall very slightly or very noticeably. Living things in an ecosystem interact with each other and with their physical environment. Any small changes in one can affect the other. For example, if a predator's population size goes up, the population size of the prey will go down, because more are eaten more quickly. Nitrogen levels in soil affect the population sizes of plants growing there. Nitrogen-fixing plants grow successfully in nitrogen-deficient soil, but affect their environment by increasing the soil nitrogen levels. This change then helps other plants to grow there as well.

Three types of change in ecosystems affect population size.

- **Cyclic changes**. These changes repeat themselves in a rhythm. For example, movement of tides and changes in day length are cyclic. The way in which predator and prey species fluctuate is cyclic.

- **Directional changes**. These changes are not cyclic. They go in one direction, and tend to last longer than the lifetime of organisms within the ecosystem. Within such change, particular variables continue to increase or decrease. Examples include the deposition of silt in an estuary, or the erosion of coastline.

Figure 4 House in danger of coastal erosion.

DID YOU KNOW?

The East coast of England has eroded to such an extent that villages have been decimated by falling into the sea. Dunwich, which was a huge trading city on the East coast, is now almost completely under water. Local legend says you can hear the church bells ring sometimes during rough weather!

- **Unpredictable/erratic changes**. These have no rhythm and no constant direction. For example, such changes may include the effects of lightning or hurricanes.

Living things respond to changes in ecosystems. For example, small mammals may hibernate on a rhythmical basis to avoid the cold temperatures of winter. Likewise, deciduous trees may shed their leaves. A mammal may change the thickness and/or colour of its fur between summer and winter (Figure 5).

Figure 5 An Arctic fox changes its coat colour as winter begins, so it is camouflaged against the snow.

Questions

1 Look at the list below. Which of these are (a) biotic, (b) abiotic? parasitism; water; pH; light intensity; competition

2 What is the difference between:
(a) a consumer and a producer
(b) a habitat and a niche?

3 Suggest why two species never occupy *exactly* the same niche in an ecosystem.

4 Thermal oceanic vents are examples of unusual ecosystems. There is no light on the ocean bed. What are the producers in this ecosystem and what form of energy do they use?

(2) Transfer of biomass

By the end of this topic, you should be able to demonstrate and apply your knowledge and understanding of:

* biomass transfers through ecosystems

Energy and materials in an ecosystem

Materials are constantly recycled within an ecosystem (Figure 1) – nutrient cycles, such as the nitrogen cycle and carbon cycle, are good examples. Energy is not recycled – it flows through the ecosystem (see Figure 1).

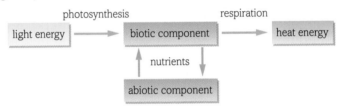

Figure 1 Energy flow and nutrient cycling in an ecosystem.

All living things need energy and materials. Energy is captured by plants in photosynthesis to produce organic molecules like glucose from water and carbon dioxide; such energy is released from glucose during respiration. The products of photosynthesis are not only used immediately for respiration, but incorporated into tissues and organs (e.g. cellulose – the building block of plant cell walls – is made up of large numbers of glucose molecules). Mineral ions are also absorbed through plant roots.

Together, the organic components (such as glucose molecules) and inorganic components (such as mineral ions, but excluding water) of the plant make up its biomass. So when a plant is eaten, its biomass is consumed by a primary consumer. The flow of biomass through one food chain is represented in Figure 2.

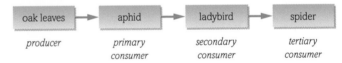

Figure 2 Flow of materials and energy in a food chain. The arrows represent the direction of flow from producer (plant) to primary consumer to secondary consumer.

Each level of the food chain is a **trophic level**. Tracking how biomass changes in a food chain helps us to track the movement of materials and energy through the food chain. We can do that for one food chain or for a whole food web in an ecosystem. An example of a food web is shown in Figure 3.

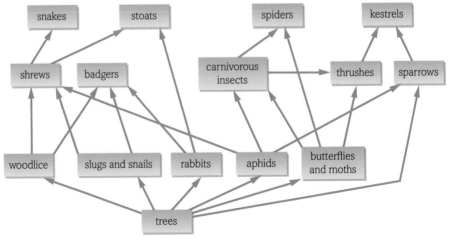

Figure 3 Flow of materials and energy in a food web. The arrows represent the direction of flow.

Biomass transfers through ecosystems

At each trophic level, some biomass is lost from a food chain and is therefore unavailable to the organism at the next trophic level (see Figure 4):

- At each trophic level, living organisms need energy to carry out life processes. Respiration releases energy from organic molecules like glucose. Some of this energy is eventually converted to heat, and materials are lost in carbon dioxide and water.

- Biomass is also lost from a food chain in dead organisms and waste material, which is then only available to decomposers such as fungi and bacteria. This waste material also includes parts of animals and plants that cannot be digested by consumers, such as bones and hair.

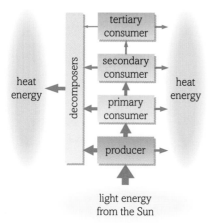

Figure 4 Biomass and energy loss through a food chain (the size of the arrows denotes the relative amounts of biomass).

Therefore, biomass is less at higher levels of the food chain. When the organisms in a food chain are about the same size, this means there will be fewer consumers at the higher levels. Ecologists draw a **pyramid of numbers** to represent this idea (see Figure 5). The area of each bar in the pyramid is proportional to the number of individuals, as an approximation for the total biomass at that level. Pyramids can be drawn for individual food chains or for an ecosystem as a whole.

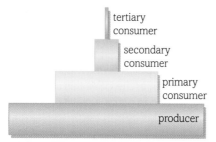

Figure 5 Pyramid of numbers.

Calculating the efficiency of biomass transfer

Counting the number of organisms does not always provide an accurate picture of how much biomass exists at each level. A better approach is to draw a **pyramid of biomass**, where the area of each bar is proportional to the **dry mass** of all the organisms at that trophic level. To do this properly, an ecologist collects all the organisms and puts them into an oven at 80 °C until all the water in them has been evaporated. They check this by periodically finding the mass of the organisms. Once the mass stops reducing, they can be certain that all the water has been removed. Unfortunately, doing this is rather destructive to the ecosystem being studied, so ecologists often just measure the **wet mass** of the organisms and calculate the dry mass on the basis of previously published data.

To calculate the efficiency of **biomass transfer** between trophic levels ecologists make the following calculation:

$$\text{Ecological efficiency} = \left(\frac{\text{Biomass at the higher trophic level}}{\text{Biomass at the lower trophic level}}\right) \times 100$$

For example:

$$\text{Ecological efficiency} = \left(\frac{\text{Biomass of primary consumer}}{\text{Biomass of producer}}\right) \times 100$$

LEARNING TIP

If you see a pyramid of numbers which is not pyramid shaped, it is likely that the individuals involved have very different masses.

Questions

1. In what ways are energy and materials lost from a food chain?

2. Explain why ecologists may prefer to draw:
   (a) a food web instead of a food chain
   (b) a pyramid of biomass instead of a pyramid of numbers.

3. Suggest why:
   (a) there are fewer individuals at higher trophic levels in a food chain
   (b) most food chains have no more than five stages.

4. Explain why a pyramid of biomass helps to show how energy is lost from a food chain between trophic levels.

5. What is the difference between dry mass and wet mass? Suggest why ecologists are interested in dry mass, rather than wet mass.

By the end of this topic, you should be able to demonstrate and apply your knowledge and understanding of:

* biomass transfers through ecosystems

Increasing primary productivity – the entry of biomass into a food chain

The rate at which energy passes through each trophic level in a food chain is a measure of its **productivity**. **Gross primary productivity** is the rate at which plants convert light energy into chemical energy through photosynthesis. Even at the start of a food chain, this is inefficient. Because photosynthesis produces glucose, entry of biomass into the food chain is also inefficient. In optimal conditions, only 40% of light energy from the Sun enters the light reaction of photosynthesis, and only half of this is involved in glucose production. Only two-thirds of the glucose is then used for production of starch, cellulose, lipids and proteins, contributing to growth. The rest is respired. Hence, only a small proportion (between 1 and 8%) of the energy from the Sun remains to enter the food chain: the **net primary productivity** (NPP) is 8%. By manipulating environmental factors, humans make energy conversion more efficient, reduce energy loss and increase the amount of biomass which is incorporated into plants.

* Light levels limit the rate of photosynthesis and hence production of biomass. Some crops are planted early to provide a longer growing season to harvest more light. Others are grown under light banks (Figure 1).

Figure 1 High value crop plants are grown under light banks.

* As well as irrigating crops, drought-resistant strains have been bred, for example drought-resistant barley in North Africa, wheat in Australia and sugar beet in the UK. Water is a reactant in photosynthesis when glucose is produced.

* Growing plants in greenhouses provides a warmer temperature, increases the rate of photosynthesis, and increases the rate of production of biomass (Figure 2). Planting field crops early to provide a longer growing season also helps to avoid the impact of temperature on final yield. For example, winter wheat has a longer growing season than spring wheat.

Figure 2 Greenhouse cultivation of lettuce plants.

* Lack of available nutrients slows the rate of production of biomass through photosynthesis. Crop rotation can help – growing a different crop in each field on a rotational cycle. This stops the reduction in soil levels of inorganic materials such as nitrate or potassium. Including a nitrogen-fixing crop like peas or beans in that cycle replenishes nitrogen levels. Many crops have been bred to respond to high levels of fertiliser which provides ammonium, nitrate, potassium and phosphorus.

* Pests like insects or nematodes eat crop plants, removing biomass from the food chain and lowering yield. Spraying with pesticides may help. Some plants have also been bred to be pest-resistant or have been genetically modified with a bacterial gene (*Bt* gene) from *Bacillus thuringiensis* (see topic 6.3.7). In Bt cotton (Figure 3) in the USA, this confers resistance against bollworm, and in maize against corn-borers.

Figure 3 Genetically-modified cotton is resistant to bollworm.

- Fungal disease reduces biomass. Fungi cause root rot (reducing water absorption), damage xylem vessels (interfering with water transport), damage foliage through wilt, blight or spotting (interfering with photosynthesis directly), damage phloem tubes (interfering with translocation of sugars), or damage flowers and fruit (interfering with reproduction). Farmers spray crops with fungicides. Many crops have been bred to resist fungal infections (e.g. *Rhizomania* resistance in sugar beet). Potatoes have been genetically modified to resist potato blight.

> **DID YOU KNOW?**
> Before sugar beet became resistant, wherever *Rhizomania* was discovered in a sugar beet crop, the crop had to be destroyed, and the field never be used for sugar beet again.

- Competition from weeds for light, water and nutrients reduces a crop's NPP. Farmers use herbicides to kill weeds. The herbicide usually binds to an enzyme, stopping it from working, and frequently leading to a toxic build-up of the enzyme's substrate.

Improving secondary productivity

Transfer of biomass between trophic levels is inefficient.

Primary consumers do not make full use of plants' biomass – some plants die, consumers do not eat every part of the plant, and they do not digest everything they eat (such as cellulose), egesting a lot of it in their faeces.

Even when food is digested and absorbed, much of it is respired, with only a small amount contributing to an increase in biomass and being available to the next consumer in the food chain.

Humans can manipulate energy transfer:

- A young animal invests a larger proportion of its energy into growth than an adult. Harvesting animals just before adulthood minimises loss of energy from the food chain.

- Selective breeding has been used to produce improved animal breeds with faster growth rates, increased egg production and increased milk production.

- Animals may be treated with antibiotics to avoid unnecessary loss of energy to pathogens and parasites.

- Mammals and birds waste a lot of energy finding food and keeping their body temperature stable. Zero grazing for pig and cattle farming maximises energy allocated to muscle (meat) by stopping the animals from moving about, by supplying food to them, and by keeping the environmental temperature constant (Figure 4).

Figure 4 Keeping cows indoors.

Although transfer of energy from producers to consumers is inefficient, and grain could be used to feed humans directly as opposed to feeding cattle or pigs first, in some infertile areas grain cannot be grown but animals can survive; for example, sheep often live on mountainsides, producing food for humans.

Many people have serious concerns about modern farming practices and animal welfare. Deciding where the balance lies between welfare and efficient food production is a contentious topic that is constantly kept under review and should include informed public debate.

> **LEARNING TIP**
> Remember that biomass and energy are inextricably linked. When biomass is respired, it releases energy. Some people refer to biomass as a store of energy.

Questions

1. Explain why modern agricultural practices may involve:
   (a) selective breeding
   (b) genetic modification
   (c) greenhouses
   (d) fertilisers
   (e) pesticides.

2. Why does harvesting animals just before adulthood minimise energy loss from the food chain?

3. Describe how selective breeding and genetic modification can increase primary and secondary productivity.

4. Suggest which animals would be most efficient to farm: endotherms like birds and mammals (with a constant body temperature), or ectotherms like worms, fish and reptiles (whose body temperature varies). Explain your answer.

5. Rainforests are an important source of biodiversity. Some rainforests are being cleared to grow crops for animal feed. Because of this, some people have chosen to adopt a vegetarian diet. Explain their reasoning in terms of energy loss from the food chain.

By the end of this topic, you should be able to demonstrate and apply your knowledge and understanding of:

* recycling within ecosystems

Recycling within ecosystems

Topic 6.5.2 indicates how energy and materials are lost from a food chain when living things excrete waste or die. This dead and waste organic material can be broken down by decomposers – microorganisms such as bacteria and fungi.

Bacteria and fungi involved in decomposition feed in a different way from animals. They feed saprotrophically, so they are described as **saprotrophs**. The steps in saprotrophic decomposition are:

1. Saprotrophs secrete enzymes onto dead and waste material.

2. Enzymes digest the material into small molecules, which are then absorbed into the saprotroph's body (Figure 1).

3. Having been absorbed, the molecules are stored or respired to release energy.

If bacteria and fungi did not break down dead organisms, energy and valuable nutrients would remain trapped within the dead organisms. By digesting dead and waste material, microorganisms obtain a supply of energy to stay alive, and the trapped nutrients are recycled.

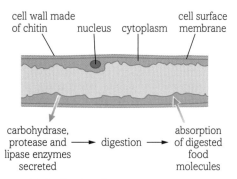

Figure 1 Saprotrophic feeding by a fungus.

Microorganisms have a particularly important role to play in cycling carbon and nitrogen within ecosystems.

Recycling nitrogen

Living things need nitrogen to make proteins and nucleic acids. Figure 2 shows how nitrogen atoms are cycled between the biotic and abiotic components of an ecosystem. Bacteria are involved in **ammonification**, nitrogen fixation, nitrification and denitrification.

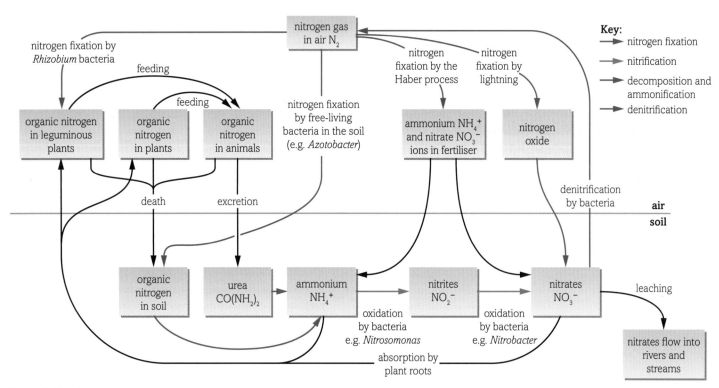

Figure 2 The nitrogen cycle.

Nitrogen fixation

Although nitrogen gas makes up 79% of the Earth's atmosphere, it is very unreactive. This means it is impossible for plants to use it directly (even though it is so abundant). Instead, plants need a supply of 'fixed' nitrogen such as ammonium ions (NH_4^+) or nitrate ions (NO_3^-). Nitrogen fixation can occur when lightning strikes or through the Haber process in making fertiliser. However, these processes only account for about 10% of nitrogen fixation around the world.

Nitrogen-fixing bacteria supply the rest of the fixed nitrogen. *Azotobacter* are bacteria that live freely in the soil and fix nitrogen gas, which is in the air within soil, using it to manufacture amino acids. Nitrogen-fixing bacteria such as *Rhizobium* also live inside the root nodules (Figure 3) of plants such as peas, beans and clover, which are all members of the bean family. These nitrogen-fixing bacteria have a mutualistic relationship with the plant: the bacteria provide the plant with fixed nitrogen and receive carbon compounds such as glucose in return.

Proteins such as leghaemoglobin in the nodules absorb oxygen and keep the conditions anaerobic. Under these conditions, the bacteria use an enzyme, nitrogen reductase, to reduce nitrogen gas to ammonium ions that can be used by the host plants.

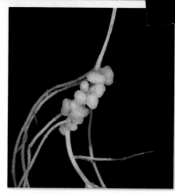

Figure 3 Root nodules of a plant from the legume family. Nitrogen-fixing bacteria live inside the nodules, which grow in response to chemical signals from the bacteria (×1.25).

> **DID YOU KNOW?**
>
> Thomas Coke (1754–1842) of Holkham Hall in Norfolk developed a four-crop-rotation (the Norfolk four-course rotation), in which crops were rotated around different fields year on year. By growing nitrogen-fixing clover within that cycle, he was able to replenish the soil's fixed nitrogen levels.

Ammonification and nitrification

Ammonium ions are released through ammonification by bacteria involved in putrefaction of proteins found in dead or waste organic matter. Rather than getting their energy from sunlight (like photoautotrophic bacteria, algae and plants), some **chemoautotrophic bacteria** in the soil (*Nitrosomonas* bacteria) obtain it by oxidising ammonium ions to nitrites, while others (*Nitrobacter* bacteria) obtain it by oxidising nitrites to nitrates. These processes are called nitrification.

Oxidation requires oxygen; therefore these reactions only happen in well-aerated soils.

Nitrates can be absorbed from the soil by plants and used to make nucleotide bases (for nucleic acids) and amino acids (for proteins).

Denitrification

Other bacteria convert nitrates back to nitrogen gas. When the bacteria involved are growing under anaerobic conditions, such as in waterlogged soils, they use nitrates as a source of oxygen for their respiration and produce nitrogen gas (N_2) and nitrous oxide (N_2O).

> **DID YOU KNOW?**
>
> Venus flytraps grow in acidic, anaerobic, waterlogged conditions. Because there is no fixed nitrogen available in the soil, they have evolved to acquire nitrogen by digesting animal tissues, from the prey they catch in their traps.

> **LEARNING TIP**
>
> Ensure you distinguish between the processes of nitrogen fixation (conversion of atmospheric nitrogen into nitrate or ammonium), nitrification (oxidation of ammonium to nitrite and nitrite to nitrate) and denitrification (conversion of nitrates to nitrogen gas).

Recycling carbon

Carbon is also cycled between the biotic and abiotic components of an ecosystem.

The carbon cycle (see Figure 4) is driven by the processes of respiration and photosynthesis, with carbon dioxide being the main vehicle for the cycling of carbon between biotic and abiotic components of the cycle.

Animals, plants and microorganisms respire to release carbon dioxide. Microorganisms are particularly important in decomposition of dead organisms and waste.

Terrestrial plants use gaseous carbon dioxide in photosynthesis, whereas aquatic plants use dissolved carbonates.

Carbon is exchanged between the air and water when carbon dioxide dissolves in water and then reacts to form carbonic acid. Carbon also enters rivers and lakes from weathering of limestone and chalk in the form of hydrogen carbonate.

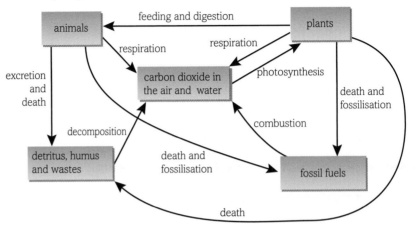

Figure 4 The carbon cycle.

Combustion of fossil fuels has increased across the last century, so that the balance of the carbon cycle has changed and atmospheric carbon dioxide levels are higher. This change is responsible for global warming.

Questions

1 Why are bacteria called saprotrophs?

2 Write down the names of three species of bacteria that are involved in the nitrogen cycle, together with their roles.

3 Explain the difference between *nitrogen fixation* and *nitrification*.

4 What type of respiration do denitrifying bacteria carry out?

5 Why is it important for materials to be recycled in an ecosystem?

6 What is the role of each organism in the carbon cycle?

7 The carbon cycle can be disrupted by human activity. List the ways in which humans may disrupt the carbon cycle, and explain the effect of such disruption.

8 List the differences and similarities between the carbon and nitrogen cycles.

⑤ Succession

By the end of this topic, you should be able to demonstrate and apply your knowledge and understanding of:

* the process of primary succession in the development of an ecosystem

---

KEY DEFINITIONS

climax community: the final stable community that exists after the process of succession has occurred.
deflected succession: happens when succession is stopped or interfered with, such as by grazing or when a lawn is mowed.
pioneer species: the species that begin the process of succession, often colonising an area as the first living things there.
succession: progressive change in a community of organisms over time.

Changing ecosystems

Any change in a community of organisms can cause a change in their habitat. Any change in a habitat can also cause a change in the make-up of the community. These ideas can help explain why gradual directional changes happen in a community over time. Such a process of directional change is called **succession**.

How does succession happen?

The island of Surtsey in Iceland was created by a volcanic eruption in the 1960s, but is now home to a community of plants. Development of such a community from bare ground is known as primary succession, and comes about as follows:

1. Algae and lichens begin to live on the bare rock. This is called a pioneer community.

2. Erosion of the rock and a build-up of dead and rotting organic material produce enough soil for larger plants like mosses and ferns to grow (see Figure 1). These replace, or succeed, the algae and lichens.

3. In a similar way, larger plants succeed these small plants, until a final, stable community is reached. This is called a **climax community**. In the UK, climax communities are often woodland communities.

Succession does not always start from bare ground. Secondary succession takes place on a previously colonised but disturbed or damaged habitat.

Figure 1 Moss plants on the island of Surtsey.

Figure 2 Sand dunes.

Succession on sand dunes

It can be difficult to understand how a habitat has changed and how it will change. Sand dunes are interesting, because they display all the stages of succession in the same place at the same time.

Look at the beach and sand dunes in Figure 2. Because the sea deposits sand on the beach, the sand nearest to the sea is deposited more recently than the sand further away. This means that the sand just above the high water mark is at the start of the process of succession, whereas the sand much further away already hosts its climax community. By walking up the beach and through the dunes, it is possible to see each stage in the process of succession.

The stages of succession are outlined below (see Figure 3). Eventually, a dunes community will develop into grassland and then woodland.

The stages of succession are:

1. **Pioneer species** like sea rocket (*Cakile maritima*) and prickly sandwort (*Salsola kali*) colonise the sand just above the high water mark. These can tolerate being sprayed with salty water, lack of fresh water and unstable sand.

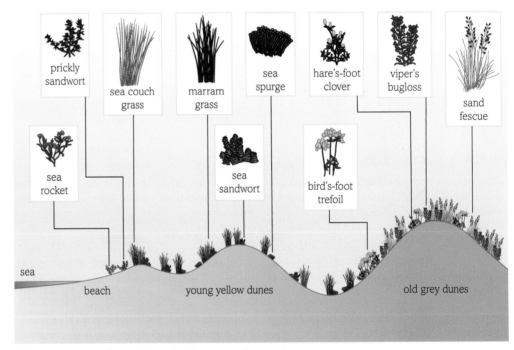

Figure 3 Cross-section of a sand dune showing stages of succession.

2. Wind-blown sand builds up around the base of these plants, forming a mini sand dune. As plants die and decay, nutrients accumulate in this mini dune. As the dune gets bigger, plants like sea sandwort (*Honkenya peploides*) and sea couch grass (*Agropyron junceiforme*) colonise it. Because sea couch grass has underground stems, it helps to stabilise the sand.

3. With more stability, and accumulation of more nutrients, plants like sea spurge (*Euphorbia paralias*) and marram grass (*Ammophila arenaria*) start to grow. Marram grass is special: its shoots trap wind-blown sand, and as the sand accumulates the shoots grow taller to stay above the growing dune, trapping more sand in the process.

4. As the sand dune and nutrients build up, other plants colonise the sand. Many are leguminous, such as hare's foot clover (*Trifolium arvense*) and bird's-foot trefoil (*Lotus corniculatus*), which convert nitrogen into nitrate. With nitrate available, more species colonise the dunes, like sand fescue (*Festuca rubra*) and viper's bugloss (*Echium vulgare*), which stabilise them further.

LEARNING TIP

Remember that succession happens over time. A sand dune often shows you the stages of succession in order of occurrence. However, usually you will not be able to see the previous stage in a succession.

Deflected succession

The landscape in the UK is heavily influenced by agriculture; therefore it can be difficult to work out whether a particular location has reached its climax community.

For example, when a groundsman cuts the grass on a golf course, he is keeping that particular area at one stage in a succession. If the grass were left unmown over a number of years, succession would continue without interference and the course would most likely reach a climax community of woodland. When succession is stopped or interfered with in this way, we refer to it as **deflected succession**. The sub-climax community that results is called a plagioclimax.

Other ways in which succession can be deflected include grazing, burning, application of fertiliser, application of herbicide, and exposure to excessive amounts of wind.

Succession in many locations is deflected by human activity, and has been for centuries, which can make it hard for preservationists and conservationists to decide which habitats warrant preservation or conservation (see topic 6.6.5).

> **DID YOU KNOW?**
>
> If you don't manage a pond well, succession will happen. Plants will colonise the water's edge, trapping organic matter, until eventually the banks will have encroached so much, the pond 'dries up'!

Questions

1. Explain the meaning of the terms:
   (a) primary succession
   (b) pioneer community
   (c) climax community.

2. Which of the following are most likely to be examples of deflected succession?
   (a) A golf course
   (b) A rock pool
   (c) A sand dune
   (d) A garden
   (e) A garden pond.

3. Compare and contrast the process of succession on bare rock and on a sand dune.

4. Explain why all stages of succession are visible in a sand dune.

5. Many conservationists argue that we should conserve particular habitats, some of which may not have reached their climax community. Suggest reasons why it can be useful to conserve such habitats.

6. The remnants of the walls of an old monastery in Norfolk are under a field which was used for grazing of cattle. The farm went out of business and the grazing stopped. Suggest how succession is a problem for conservation of the walls remaining.

6 Studying ecosystems

By the end of this topic, you should be able to demonstrate and apply your knowledge and understanding of:

* how the distribution and abundance of organisms in an ecosystem can be measured

* the use of sampling and recording methods to determine the distribution and abundance of organisms in a variety of ecosystems

Sampling

Ecologists usually study ecosystems to find out whether the abundance and distribution of a species is related to that of other species, or to environmental factors such as light intensity or soil pH. It would be ideal to count every individual of every species, but in most habitats this is impossible. Ecologists get around this problem by taking samples from the habitat – selecting small portions of the habitat and studying them carefully.

Quadrats

> **LEARNING TIP**
>
> A quadrat and a quadrant are different. A quadrat is a square frame used for studying ecosystems, but a quadrant is an instrument used to measure angles.

Imagine that an ecologist wants to compare the abundance and distribution of plant species in two different fields. It is impossible to count all the individuals in the fields, so they sample small parts of the habitat using a **quadrat** to define the sample areas (Figure 1). It is often 1 m square, and can have strings across every 10 cm, separating it into 100 smaller squares.

You can collect two types of data using a quadrat:

* Presence or absence of each species (distribution). Usually at least 50% of the plant needs to be inside the quadrat to count.

* Number of individuals (the abundance) of each species – either estimated or counted. For some plants, like grass and moss, it is difficult to count individuals, so ecologists tend to estimate percentage cover.

Figure 1 Using a quadrat and point frame.

Estimating percentage cover is difficult. Using a point frame can help make that estimate more accurate (Figure 1). You lower the frame into the quadrat and record any plants touching the needles. If the frame has 10 needles and you lower it 10 times in each quadrat, you will have 100 readings. So every time an individual touches a needle, this represents 1% cover. Do not forget to record bare ground!

Before starting to sample, decide:

* **Where to place the quadrats**: If you take samples only from one corner of the field, it may be that the soil in this corner is particularly rich in nitrogen and the species growing there are different to those in the rest of the field. To avoid biasing the sample, and to provide a sample which is representative of the whole habitat, either:

  ○ Randomly position the quadrats across the habitat, using random numbers to plot coordinates for each one (random sampling). Lay out two tape measures on two edges of the study site, so that they look like axes on a graph. Use your calculator, a pair of dice or a random number table to generate pairs of random numbers. Use these pairs as coordinates to place your quadrats. Lay the bottom left hand corner of the quadrat at the coordinate.

  or

  ○ Take samples at regular distances across the habitat, so you sample every part of the habitat to the same extent (systematic sampling).

* **How many samples to take**: Just looking at one quadrat will not accurately represent the whole habitat, but using 20 000 would take forever! In a pilot study looking at species distribution, take random samples from across the habitat, making a cumulative frequency table like that shown in Table 1. Plot cumulative frequency against quadrat number. The point where the curve levels off tells you how many quadrats to use.

Quadrat number	Number of new species found in this quadrat	Total number of different species found
1	5	5
2	4	9
3	4	13
4	3	16
5	3	19
6	2	21
7	2	23
8	2	25
9	2	27
10	1	28
11	0	28
12	0	28

Table 1 Cumulative frequency table of quadrat number against number of species found.

LEARNING TIP

If you used a quadrat, always state the size of the quadrat you used.

Having collected the data, the equation below shows you how to estimate the size of the population of each species in the whole habitat:

population size of a species

$$= \frac{\text{mean number of individuals of the species in each quadrat}}{\text{fraction of the total habitat area covered by a single quadrat}}$$

Transects

You can also look for changes in vegetation across a habitat. For example, you may want to look at the changes in abundance and distribution of species as you walk up a beach and through the sand dunes, measuring the environmental conditions at the same time. This is done using a **transect**, which is a line taken across a habitat.

It is easiest to stretch out a tape measure and then take samples at regular intervals along the tape. The distance between samples will depend on the length of the line you want to look at, and the density of plants in the habitat.

There are two approaches to using a transect:

- **Line transect** – at regular intervals, make a note of which species is touching the tape (Figure 2).

- **Belt transect** – at regular intervals, place a quadrat next to the line (interrupted belt transect), studying each, as described above. Alternatively, place a quadrat next to the line, moving it along the line after looking at each quadrat (continuous belt transect) (Figure 3).

To understand how living things vary along the transect line, you can compare the number of organisms of each species at each sampling point along the transect line (Figure 4). You can also

Figure 2 Using a line transect.

Figure 3 Using a belt transect.

plot abiotic factors (e.g. soil temperature) on the same scale, with distance on the x-axis and temperature on the y-axis. By lining up your graph with your kite diagram, you can start to see how temperature affects distribution and abundance.

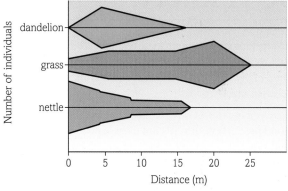

Figure 4 A kite diagram, which represents distribution and abundance of different species along a transect line.

Questions

1. Give reasons why ecologists take samples from habitats, rather than studying the whole habitat.

2. Suggest the advantages of taking those samples (a) at random, and (b) at regular intervals through the habitat.

3. Explain:
   (a) how you would estimate the abundance and distribution of species through a salt marsh.
   (b) how you would compare the abundance and distribution of species between a school playing field and a meadow.

4. Suggest the advantages and disadvantages of using a continuous belt transect compared to (a) a line transect and (b) an interrupted belt transect.

5. Suggest how you would select the specific location for a line or belt transect on a rocky shore.

6. Explain why you calculate population size according to the equation above. You need to justify in words why the equation allows you to calculate population size.

AN ENVIRONMENTALLY-FRIENDLY DIET

Many people are now trying to live in an environmentally-friendly way. This article considers the impact of our diet on the environment and how changes to our diet may influence our overall carbon footprint.

DOES GOING VEGETARIAN REDUCE YOUR CARBON FOOTPRINT?

If you stop eating meat, your food-related carbon footprint could plummet to less than half of what it was. That is a much bigger drop than many previous estimates, and it comes from a study of people's real diets.

As much as a quarter of our greenhouse gas emissions come from food production. But it's not clear how much would really be saved if people swapped their beef steaks for tofu burgers. On some estimates, going vegetarian could cut out 25 per cent of your diet-related emissions. But it all depends on what you eat instead of the meat. With some substitutions, emissions could even rise.

So Peter Scarborough and his colleagues at the University of Oxford took data on the real diets of more than 50 000 people in the UK, and calculated their diet-related carbon footprints.

"This is the first paper to confirm and quantify the difference," says Scarborough.

Stop those emissions

They found that the benefits could be huge. If those eating more than 100 grams of meat a day – a fairly small rump steak – went vegan, their food-related carbon footprint would shrink by 60 per cent, saving the equivalent of 1.5 tonnes of carbon dioxide a year.

Perhaps more realistically, if someone eating more than 100 grams of meat a day simply cut down to less than 50 grams a day, their food-related emissions would fall by a third. That would save almost a tonne of CO_2 each year, about as much as an economy return flight between London and New York.

Pescatarians, who eat fish but not other meat, are almost as carbon-friendly as vegetarians, creating only about 2.5 per cent more food-related emissions. But vegans can feel the most superior, pumping out 25 per cent less emissions than vegetarians, who still eat eggs and dairy.

"In general there is a clear and strong trend with reduced greenhouse gas emissions in diets that contain less meat," says Scarborough.

Where to cut?

There are other ways of reducing emissions, such as driving and flying less, but changing food habits will be easier for many, says Scarborough. "I think it is easier to change your diet than to change your travel behaviour, but others may not agree."

"This research presents a strong case for the greenhouse gas benefits of a low-meat diet," says Christopher Jones of the University of California, Berkeley.

In 2011, Jones compared all the ways US households can cut their emissions. Although food was not the biggest source of emissions, it was where people could make the biggest and most cost-effective savings, by wasting less food and eating less meat. Jones calculated that saving each tonne of CO_2 emissions would also save the household $600 to $700.

"Americans waste about a third of the food they buy, and eat about 30 per cent more calories than recommended, on average," says Jones. "Reducing food purchases and physical consumption would have even greater greenhouse gas benefits than reducing meat consumption in the American case."

Source
● *New Scientist* 26th June 2014 (www.newscientist.com/article/dn25795-going-vegetarian-halves-co2-emissions-from-your-food.html#.VNZv8EfXerU)

Let's look at the biology in, or connected to, this article. Use the timeline at the bottom of the page to help place this work in context with what you have already learned and what is ahead in your course.

1. What is your 'carbon footprint' and why is it important to think about it?
2. If everybody in the UK became pescatarian:
   a. suggest what would happen to the population sizes of fish in the North Sea
   b. suggest what effects that could have on the food webs with which they are involved
   c. suggest what would happen to a person's carbon footprint if fish had to be imported.
3. In parts of the world where over-fishing has taken place, some species have become endangered. Governments then impose strict fishing quotas, which must not be exceeded.
   a. Draw and annotate a population growth curve for a recovering population of a species of fish. Justify the shape of the curve you have chosen.
   b. Explain how fishing quotas are used in Antarctica to protect the environment.
4. Explain, using what you know about the carbon cycle:
   a. why eating meat involves an inefficient transfer of biomass from producers to humans
   b. whether fish or farm animals provide the most efficient energy transfer from producer to humans. Use evidence from the article to support your answer.
5. Explain why Peter Scarborough conducted his study with such a large sample of 50 000 people.
6. According to this article, going vegan means the 'food-related carbon footprint would shrink by 60 per cent, saving the equivalent of 1.5 tonnes of carbon dioxide a year.'
   a. Calculate the food-related carbon footprint for a non-vegan and for a vegan.
   b. How much money would a typical American save by becoming vegan?
7. Animals and plants produce carbon dioxide in respiration. Describe how the process occurs in the cell and compare and contrast gaseous exchange in animals and plants.
8. The article presents a number of headline figures, calculated from large data sets, averaged across either the UK or the USA. Governments can sometimes base policy on studies of this type. Suggest why it may be inappropriate to encourage a vegan lifestyle amongst cattle farmers in Namibia, on the edge of the Kalahari Desert.

Activity

Research and explain why eating meat can increase an individual's carbon footprint. It is impossible to grow arable crops in some habitats, including moorland. In a world with a rapidly increasing population, it is important to maximise food production, for example by using such areas for grazing by sheep. Grazing can also be important for conservation. Research and evaluate the arguments for and against meat consumption under such circumstances.

1. Which is the process by which nitrogen is converted from nitrogen gas into a form which is usable by living things? [1]

   A. Nitrification

   B. Denitrification

   C. Nitrogen fixation

   D. Oxidation

2. The graphs in Figure 1 show the density of two different plant species as proximity to the coast changes.

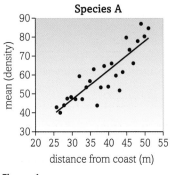

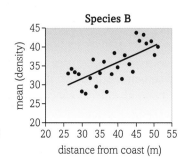

Figure 1

   Which of the following statements correctly describes one aspect of the technique used to collect these data? [1]

   A. Quadrats were randomly placed using a random number generator and coordinates.

   B. Larger quadrats were required for **species A** because their mean density was higher.

   C. A belt transect has been used to allow calculation of density.

   D. Abiotic factors were measured at every point of quadrat sampling.

   [Q7, sample paper H420/02, 2014]

3. Which of the following best describes a community?

   A. The place where an organism lives

   B. A group of organisms of the same species living in the same place at the same time

   C. A group of organisms of different species living in the same place at the same time

   D. A group of living, and non-living, things and the interrelationships between them [1]

   [Total: 3]

4. Knowledge of the nitrogen cycle can be used to make decisions about the management of farmland. A farmer uses her grass meadow to raise sheep. In a separate field she grows cabbages.

   (a) Figure 2 shows part of the nitrogen cycle. The four boxes on the bottom line of the diagram refer to substances in the soil.

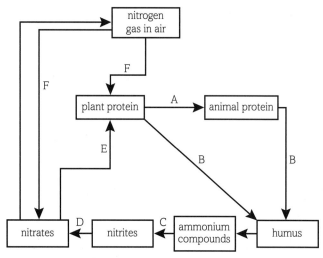

Figure 2 Diagram showing part of the nitrogen cycle.

   (i) Briefly summarise the steps that must occur for plant protein to be converted into animal protein in the farmer's sheep, as shown by arrow A on Figure 2. [4]

   (ii) List the processes which contribute to B in the meadow where sheep are raised. [2]

   (iii) Name the bacteria that carry out processes C and D, and explain the significance of these groups of bacteria for the growth of plants. [3]

   (iv) Use the letters in Figure 2 to explain why the soil nitrate concentration will decrease in the cabbage field, if it is used to grow repeated crops of cabbages year after year. [3]

   (v) The farmer does not want to use inorganic fertiliser to replace the nitrate in the soil of the cabbage field. She wants to make use of process F. Suggest a crop she could plant that would allow process F to occur and explain how this would add nitrate to the soil. [3]

   [Total: 15]

   [Q1, F215 June 2011]

5. (a) Organisms do not live in isolation, but interact with other organisms and their physical environment.

State the word used to describe:

(i) the study of interactions between an organism and its environment [1]

(ii) the physical (non-living) factors in the environment [1]

(iii) a physical area that includes all the organisms present and their interactions with each other and with the physical environment. [1]

[Total: 3]

[From Q3, F215 January 2012]

6. (a) Explain the meaning of term 'primary succession'. [2]

(b) Figure 3 shows a primary succession in a temperate climate. X represents an example of deflected succession.

grassland
X
dry initial conditions (bare rock or sand dune)
lichens, mosses
grasses, herbaceous plants
small shrubs
mixed woodland

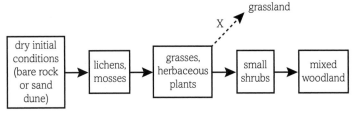

Figure 3

Explain the role of pioneer plants in succession on a bare rock or sand dune. [3]

(c) Suggest two ways in which deflected succession at X could be caused. [2]

(d) Explain how biomass changes during a primary succession. [2]

[Total: 9]

[Q8, sample paper F215, 2007]

7. State and describe **two types** of ecological interaction that can occur between different species in a habitat. As part of each description, you should **name** the two species involved in your chosen example. [6]

[Total: 6]

[From Q3, F215 January 2012]

8. The distribution and abundance of plants can show how a physical factor varies across the habitat. Describe how you would measure the distribution and abundance of plants over a distance of 100 metres. *In your answer, you should make clear the sequence of procedures you would follow.* [6]

[Total: 6]

[From Q3, F215 January 2012]

9. Peat bogs are large areas of waterlogged land that support a specialised community of plants. Peat bogs take thousands of years to form. Figure 4 lists the main stages in the formation of a peat bog.

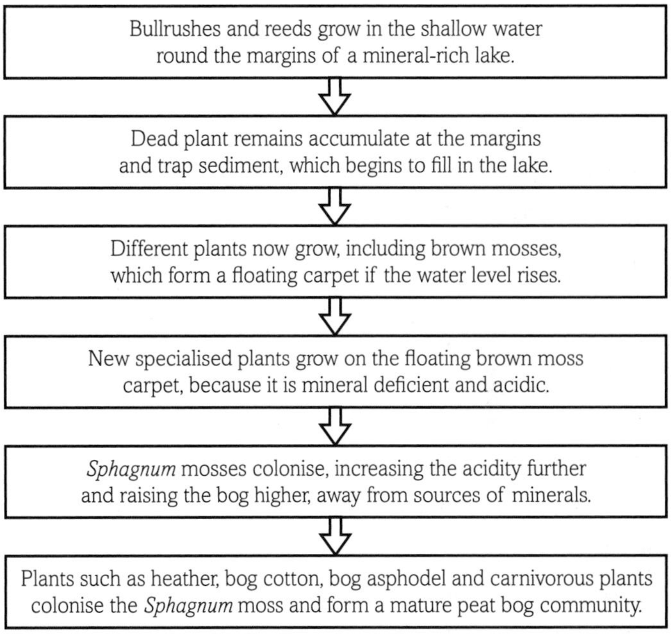

Bullrushes and reeds grow in the shallow water round the margins of a mineral-rich lake.
Dead plant remains accumulate at the margins and trap sediment, which begins to fill in the lake.
Different plants now grow, including brown mosses, which form a floating carpet if the water level rises.
New specialised plants grow on the floating brown moss carpet, because it is mineral deficient and acidic.
Sphagnum mosses colonise, increasing the acidity further and raising the bog higher, away from sources of minerals.
Plants such as heather, bog cotton, bog asphodel and carnivorous plants colonise the Sphagnum moss and form a mature peat bog community.

Figure 4

(a) (i) Name the process summarised in Figure 4 that changes a lake community into a peat bog community. [1]

(ii) Using Figure 4, list **two abiotic** factors that play a role in determining what species of plant can grow in an area. [2]

(b) Most of the minerals in a peat bog are held within the living plants at all times, **not** in the soil.

- Plants like bog cotton and bog asphodel recycle the minerals they contain.
- The leaves of these plants turn orange as the chlorophyll within them is broken down.
- Minerals such as magnesium ions are transported from the leaves to the plants' roots for storage.

Describe **one** similarity and **two** differences in mineral recycling in a peat bog and in a **deciduous forest**. [3]

[Total: 6]

[From Q5, F215 June 2012]

CHAPTER 6.6

POPULATIONS AND SUSTAINABILITY

Introduction

If you are a conservationist, trying to conserve a particular species, your aim is to keep their population size at a high enough level for the species to survive and reproduce. To conserve it effectively, you have to know the factors that affect its population size. The effect of those factors can be positive or negative, and they determine the size of the population that the environment can support. The factors can be biological or non-biological in origin. For example, predators and prey populations affect each other, and competition within and between species can affect population size.

Conservation is important, because the effect of human activity on the environment can cause drastic reductions in population size. Having said that, conservation can involve artificially deflecting succession – deliberately having an effect – in order to maintain biodiversity. Being able to exploit resources in an environmentally sustainable way is essential to maintaining habitats and conserving particular species. You will learn about sustainable management of forests and fisheries, to minimise impact on the environment. Balancing the conflict between human needs and conservation is essential, and you will learn about efforts to reduce the impacts of human activity on environmentally sensitive habitats.

All the maths you need

To unlock the puzzles of this chapter you need the following maths:

*   Represent and interpret data using graphs

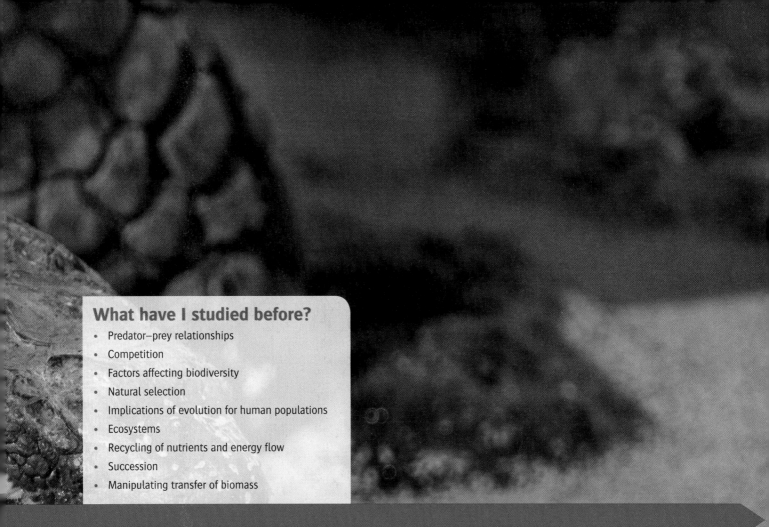

What have I studied before?

- Predator–prey relationships
- Competition
- Factors affecting biodiversity
- Natural selection
- Implications of evolution for human populations
- Ecosystems
- Recycling of nutrients and energy flow
- Succession
- Manipulating transfer of biomass

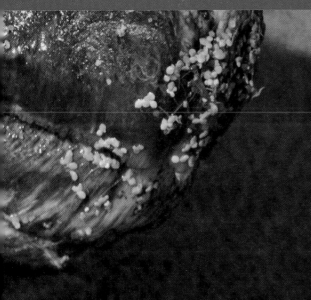

What will I study in this chapter?

- Factors which determine population size
- How populations interact
- Conservation
- Sustainable management of ecosystems
- Balancing the needs of humans with conservation of species
- How humans affect their environment and how to control their effects

By the end of this topic, you should be able to demonstrate and apply your knowledge and understanding of:

* the factors that determine size of a population

Population size and carrying capacity

The size of population of a species may remain stable, it may rise or fall quite suddenly, or oscillate up and down with a regular pattern. The balance between the death rate (mortality) and the rate of reproduction determines the size of a population. Figure 1 shows how many populations grow.

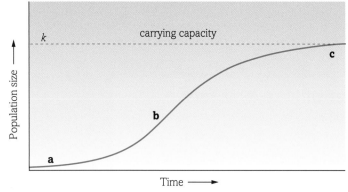

Figure 1 Population growth.

In Figure 1:

* At point **a** (the **lag phase**), there may only be a few individuals, which are still acclimatising to their habitat. At this point, the rate of reproduction is low, and the growth in population size is slow.

* At point **b** (the **log phase**), resources are plentiful, and conditions are good. Reproduction can happen quickly, with the rate of reproduction exceeding mortality. The population size increases rapidly.

* At point **c** (the stationary phase), the population size has levelled out at the **carrying capacity** of the habitat – the habitat cannot support a larger population. In this phase, the rates of reproduction and mortality are equal. The population size therefore stays stable, or fluctuates very slightly up and down in response to small variations in environmental conditions each year.

Limiting factors

A habitat that has reached carrying capacity cannot support a larger population, because factors limit the growth in population size. These are called **limiting factors**.

Some limiting factors are density independent. These act just as strongly, irrespective of the size of the population. For example, particularly low temperatures may kill the same proportion of individuals in a population, irrespective of its size.

Other limiting factors are density dependent, where the factor influences population more strongly as population size increases. For example, the availability of resources like food, water, light, oxygen, nesting sites or shelter may decrease. Similarly, as population size increases, levels of parasitism and predation from other species may increase, as do the intensity of competition for resources, both with individuals of the same species and individuals of other species. The carrying capacity is the upper limit that these factors place on the population size.

Types of strategist

r-Strategies and *k*-strategies represent two ends of a continuum of strategies adopted by living things.

k-Strategists

Species whose population size is determined by the carrying capacity are often called ***k*-strategists**. For these populations, limiting factors exert a more and more significant effect as the population size gets closer to the carrying capacity, causing the population size to gradually level out.

k-Strategists, such as birds, larger mammals, like humans, elephants and lions, and larger plants, often exhibit many of the following characteristics:

* low reproductive rate

* slow development

* late reproductive age

* long lifespan

* large body mass.

r-Strategists

Some species adopt a different type of population growth. In these species, the population size increases so quickly that it can exceed the carrying capacity of the habitat before the limiting factors start to have an effect. Once the carrying capacity has been exceeded, there are no longer enough resources to allow individuals to reproduce or even to survive. Likewise, an excessive build-up of waste products may start to poison the species, and they begin to die, entering a death phase (Figure 2). This type of population growth is known as boom and bust.

r-Strategists, such as mice, insects, spiders and weeds, tend to exhibit many of the following characteristics:

- high reproductive rate
- quick development
- young reproductive age
- short life span
- small body mass.

The most important influence on population growth is the physical rate (*r*) at which individuals can reproduce. This type of growth is characteristic of species with short generation times (such as bacteria) and of pioneer species (see topic 6.5.5). Quick population growth means pioneer *r*-strategist species colonise a disturbed habitat before *k*-strategists, dispersing to other habitats once limiting factors start to have an effect.

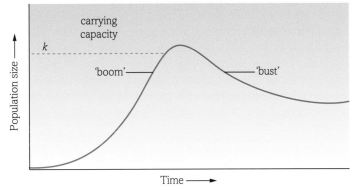

Figure 2 Population growth in *r*-strategist species.

DID YOU KNOW?

Malthus predicted that human population size would be limited by periodic catastrophes, such as famine or war. These are known as Malthusian catastrophes, and he predicted they would arise because population size grows exponentially, whereas resources only grow in a linear way. Advances in agriculture have helped to limit the occurrence of Malthusian catastrophes.

Questions

1. List three biotic and two abiotic factors which could act as limiting factors.

2. During each of the following, indicate which is higher – death rate or reproduction rate.
   (a) lag phase
   (b) log phase
   (c) stationary phase.

3. Explain the connection between limiting factors and carrying capacity.

4. Explain the difference between *r*- and *k*-strategists.

5. Look back at topic 6.5.5.
   (a) What type of succession begins from a disturbed habitat?
   (b) Why are *r*-strategists good at colonising disturbed habitats?
   (c) Are *k*-strategists or *r*-strategists more likely to be members of a climax community? Explain your answer.

By the end of this topic, you should be able to demonstrate and apply your knowledge and understanding of:

* interactions between populations

interspecific competition: competition between individuals of different species.
intraspecific competition: competition between individuals of the same species.

Predators and prey

A predator is an animal that hunts other animals (prey) for food. Predation can act as a limiting factor on a prey's population size, which in its turn can affect the predator's population size. Figure 1 shows how this happens:

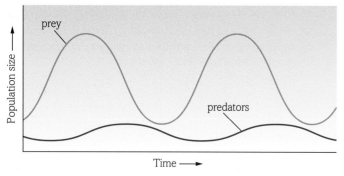

Figure 1 Relationship between a population of a predator and its prey.

1. When the predator population gets bigger, more prey are eaten.

2. The prey population then gets smaller, leaving less food for the predators.

3. With less food, fewer predators can survive and their population size reduces.

4. With fewer predators, fewer prey are eaten, and their population size increases.

5. With more prey, the predator population gets bigger, and the cycle starts again.

Figure 1 comes from an experiment conducted in a laboratory, where the predators only ate one type of prey, and predation was the main limiting factor on the prey's population. However, in the wild, predators often eat more than one type of prey, and there are a number of other limiting factors. Because of this, studies of predators and prey in the wild yield graphs of a similar, but not so well-defined, shape (Figure 2).

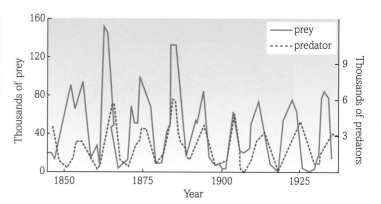

Figure 2 Relationship between the number of predators and prey over a number of years.

Competition

Competition happens when resources (like food or water) are not present in adequate amounts to satisfy the needs of all the individuals who depend on those resources. If a resource is in short supply in an ecosystem, there will be competition between organisms for that resource. As the intensity of competition increases, the rate of reproduction decreases (because fewer organisms have enough resources to reproduce), whilst the death rate increases (because fewer organisms have enough resources to survive).

There are two types of competition: **intraspecific competition** and **interspecific competition**.

Intraspecific competition

Intraspecific competition happens between individuals of the same species. As factors such as food supplies become limiting, individuals compete for food. Those individuals best adapted to obtaining food survive and reproduce, while those not so well adapted fail to reproduce, or die. As explained in topic 6.6.1, this slows down population growth and the population enters the stationary phase.

Although there are slight fluctuations in population size during the stationary phase, intraspecific competition keeps the population size relatively stable.

* If the population size drops, competition reduces and the population size increases.

* If the population size increases, competition increases and the population size drops.

Interspecific competition

Interspecific competition happens between individuals of different species, and can affect both the population size of a species, and the distribution of species in an ecosystem.

Some of the classic work on interspecific competition was carried out in 1934 by a Russian scientist called Gause. He grew two species of *Paramecium* (Figure 3), both separately and together. When together, there was competition for food, with *P. aurelia* obtaining food more effectively than *P. caudatum*. Over 20 days, the population of *P. caudatum* reduced and died out, whereas the population of *P. aurelia* increased, eventually being the only species remaining (see Figure 4).

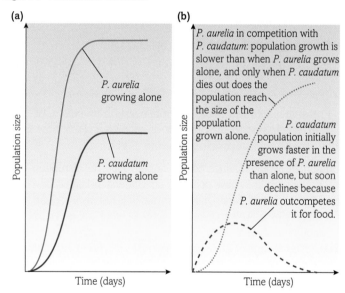

Figure 3 *Paramecium caudatum.*

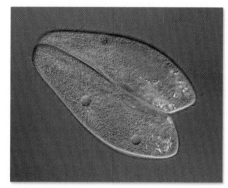

Figure 4 Population growth of two species of *Paramecium* grown (a) separately and (b) together.

(b) *P. aurelia* in competition with *P. caudatum*: population growth is slower than when *P. aurelia* grows alone, and only when *P. caudatum* dies out does the population reach the size of the population grown alone. *P. caudatum* population initially grows faster in the presence of *P. aurelia* than alone, but soon declines because *P. aurelia* outcompetes it for food.

Gause concluded that more overlap between two species' niches results in more intense competition. If two species have exactly the same niche, one is out-competed by the other and dies out or becomes extinct in that habitat; two species cannot occupy the same niche. This idea became known as the *competitive exclusion principle*, and can be used to explain why particular species only grow in particular places.

Often, though, it is not quite that simple. Other observations and experiments suggest that extinction is not necessarily inevitable. Sometimes, interspecific competition simply results in one population being much smaller than the other, with both population sizes remaining relatively constant.

It is also important to realise that in the laboratory it is easy to exclude the effects of other variables, so the habitat of the two species remains stable. In the wild, however, a wide range of variables may act as limiting factors for the growth of different populations, and may change on a daily basis or over the course of a year. For example, experiments on competition between flour beetles *Tribolium confusum* and *T. castaneum* initially confirmed the competitive exclusion principle – the *T. castaneum* population size increased, whilst the *T. confusum* population died out (Figure 5) – but even a small change in the temperature could change the outcome, so *T. confusum* would survive instead.

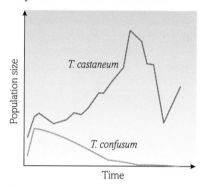

Figure 5 Changes in population size of *T. castaneum* and *T. confusum* over time.

DID YOU KNOW?

Some plants have a mechanism whereby they can interfere with neighbouring plants' physiology by releasing chemicals into their habitat. This phenomenon is called allelopathy, and is a form of competition because it stops their neighbours from using the resources in the habitat. The chemicals may inhibit growth, germination, or nutrient uptake itself. Allelopathic chemicals can be found in almost any part of a plant. They can be released into the soil directly by the roots, or they may leach out of leaves and fruit when a plant sheds them.

LEARNING TIP

Make sure you refer to inter- and intraspecific competition correctly! They are easy to mix up!

Questions

1. Figure 1 derives from a laboratory experiment, and shows clearly how the populations of predators and prey can be dependent upon each other. Figure 2 comes from studies on wild populations. Explain why the relationship shown in Figure 2 is not as clear as that shown in Figure 1.

2. What is the difference between intraspecific and interspecific competition?

3. Explain how a species' population size can be affected by:
   (a) intraspecific competition
   (b) interspecific competition.

4. Explain why the competitive exclusion principle does not always apply in natural ecosystems.

③ Conservation and preservation

By the end of this topic, you should be able to demonstrate and apply your knowledge and understanding of:

* the reasons for, and differences between, conservation and preservation

---

KEY DEFINITIONS

conservation: maintenance of biodiversity, including diversity between species, genetic diversity within species, and maintenance of a variety of habitats and ecosystems.

preservation: maintenance of habitats and ecosystems in their present condition, minimising human impact.

Conservation and preservation

Preservation

Preservation is keeping species and habitats as they are now. The approach known as preservation focuses on keeping things 'natural' and eliminating any human effects on ecosystems that exist today. The most extreme preservationists even query whether fishing or logging should take place.

Conservation

Conservation is a more active management process involving human intervention. Change in many ecosystems is almost inevitable, both through natural succession and human activity. As a result of human activity in the past, very few habitats in the UK are truly 'natural'. Therefore, adopting a preservation strategy would not preserve a natural habitat. Instead, conservation programmes (rather than preservation) focus on maintaining or improving biodiversity. This includes maintaining not just diversity between species, but also genetic diversity within species, and also maintaining a range of habitats and ecosystems.

Threats to biodiversity

Unfortunately, a steadily increasing human population can threaten biodiversity through:

* over-exploitation of wild populations for food (e.g. cod in the North Sea), for sport (e.g. sharks), and for commerce (e.g. pearls collected from mussels); 'over-exploitation' means species are harvested at a faster rate than they can replenish themselves

* habitat disruption and fragmentation as a result of more intensive agricultural practices, increased pollution, or widespread building

* species introduced to an ecosystem by humans that out-compete other native species, leading to their extinction.

Conservation strategies

Successful conservation requires consideration of the social and economic costs to the local community, and effective education and liaison with the community. Conservation can involve establishing protected areas like National Parks, green belt land or sites of special scientific interest (SSSIs). It can also involve giving legal protection to endangered species, or conserving them *ex situ* in zoos or botanic gardens.

Maintaining biodiversity in dynamic ecosystems requires careful management to maintain a stable community, or even reclamation of an ecosystem to reverse the effects of human activity.

Some management strategies are outlined below. Which strategies are adopted depends on the specific characteristics of the ecosystem and species involved.

* Raise carrying capacity by providing extra food.

* Move individuals to enlarge populations, or encourage natural dispersal of individuals between fragmented habitats by developing dispersal corridors of appropriate habitat.

* Restrict dispersal of individuals by fencing.

* Control predators and poachers.

* Vaccinate individuals against disease.

* Preserve habitats by preventing pollution or disruption, or intervene to restrict the progress of succession (see topic 6.5.5) by for example, **coppicing**, mowing or grazing.

Sometimes, simple management is inappropriate, as disruption of a community may have gone too far. Understanding which species was part of the original community is not always clear, and succession is likely to take a long time before it allows such a community to survive again. Short-cutting that process requires detailed knowledge of all the species involved. Where environmental conditions have remained fairly stable, it is possible to clean up pollution, remove unwanted species (see Figure 1), or recolonise with the original species, for example from captive breeding programmes. However, it is often easier, and more successful, to 'replace' a disrupted community with a slightly different community rather than to rehabilitate the original community.

Figure 1 South American coypus were inadvertently released from fur farms in southern England, causing significant damage to local plant communities. A capture programme was established to eradicate them.

Why conserve?

Ethics

Many conservationists argue that the kinds of problem listed above make conservation essential. They believe that every species has value, and that humans have an ethical responsibility to look after them.

Although these arguments are laudable, they are subjective. The arguments in favour of human activities which work against conservation (e.g. burning fossil fuels, open-cast mining, etc.) are objective, and driven by economics. Expressing the value of conservation in economic terms is more effective in driving governments to prioritise conservation.

Economic and social reasons

Many species already have direct economic value when harvested. This value is the easiest to measure. Others may also have direct value that is as yet unrecognised, and may provide benefit in the future. Such value can be hard to estimate.

- Many plant and animal species provide a valuable food source, and were originally domesticated from wild species. Genetic diversity in wild strains may be needed in future to breed for disease resistance and improved yield in animals and plants, and drought tolerance in plants (see Figure 2). Likewise, new plant species may be domesticated for food use.

- Natural environments are a valuable source of potentially beneficial organisms. Many of the drugs we use today were discovered in wild plant species (see Figure 3).

- Natural predators of pests can act as biological control agents. This is preferable to causing pollution with artificial chemicals, but few such species are yet being used.

Figure 2 Sugar beet researchers are studying wild sea beet (*Beta vulgaris* ssp. *maritima*) as a source of disease resistance for the commercial sugar beet crop.

Figure 3 Aspirin was originally derived from the bark of willow trees (*Salix alba*).

Many species also have indirect economic value. This is also quite difficult to quantify. For example, insect species are responsible for pollinating crop plants. Without the insects, a harvest may fail and farmers would go out of business. Likewise, other communities are important in maintaining water quality, protecting soil, and breaking down waste products. There is even evidence that a reduction in biodiversity may reduce climatic stability, with such loss of diversity resulting in drought or flooding and an associated economic cost.

Ecotourism and recreation in the countryside also have significant social and financial value, which derives from the aesthetic value of living things (that is, they look nice). Ecotourism in particular depends on maintenance of biodiversity, and there is even a sizeable industry in natural history books, films and other media.

LEARNING TIP

When considering the economic value of conservation, remember to consider direct value, indirect value and the potential future value of conservation.

DID YOU KNOW?

A UK government report estimated the commercial value of pollinating insects at £430 million per year.

Questions

1. Conservation involves maintenance of biodiversity.
   (a) Explain the meaning of the term biodiversity.
   (b) List the reasons why it is important to maintain biodiversity.
   (c) List the human threats to biodiversity.

2. Explain the differences between the direct and indirect financial value of a species.

3. Explain why:
   (a) it is impossible to preserve an ecosystem simply by putting a fence around it
   (b) reclamation of a habitat is so difficult.

4. Suggest why it is easier to place a value on the direct benefits of conservation, as opposed to the indirect benefits.

(4) Sustainable management

By the end of this topic, you should be able to demonstrate and apply your knowledge and understanding of:

* how the management of an ecosystem can provide resources in a sustainable way

Sustainable management of ecosystems

The human population is getting larger increasingly quickly, and we have had to use more intensive methods to exploit our environment for resources. Such approaches can disrupt or destroy ecosystems, reduce biodiversity, and even completely remove the resource we originally wanted to harvest: our use of natural resources is not sustainable.

There is potential conflict between our need for resources and conservation in, for example, wood and timber production, and in fish production. However, sustainable management and exploitation of these resources is possible, and can mean that biodiversity is maintained, whilst ensuring supplies and maintaining their economic benefits.

Managing timber production

Small-scale timber production

Coppicing provides a sustainable supply of wood. The stem of a deciduous tree (one that loses its leaves in the winter) is cut close to the ground. Once cut, new shoots grow from the cut surface and mature into narrow stems (Figure 1). These can be used for fencing, firewood or furniture. After cutting them off, new shoots start to grow again, and the cycle continues.

Figure 1 A coppiced woodland.

Pollarding involves cutting the stem higher up (see Figure 2), to prevent deer eating the emerging shoots.

Figure 2 A pollarded tree. Pollarding is useful when the population size of deer is high, as deer will eat the emerging shoots from a low coppiced stem, but cannot reach the pollarded stems.

To provide a consistent supply of wood, woodland managers divide a wood into sections and cut one section each year. This is rotational coppicing. By the time they want to coppice the first section again, the new stems have matured and are ready to be cut. In each section, some trees are left to grow larger without being coppiced. These trees are called standards, and are eventually harvested to supply larger pieces of timber.

Rotational coppicing is good for biodiversity. Left unmanaged, woodland goes through a process of succession, blocking out light to the woodland floor and reducing the number of species growing there. In rotational coppicing, different areas of woodland provide different types of habitat, letting more light in, and increasing the number and diversity of species.

DID YOU KNOW?

Bradfield woods in Suffolk has been managed sustainably since 1252. It is one of the oldest managed woodlands in the UK, and was owned by monks from a nearby abbey. They would still recognise the woods today, as the woodland is managed in the same way as it was originally. It is home to over 370 species of plants and has thriving populations of badger and dormouse.

Large-scale timber production

In the past, large-scale production of wood for timber often involved clear felling all the trees in one area. This could destroy habitats on a large scale, reduce soil mineral levels and leave soil susceptible to erosion. Trees usually remove water from soil and stop soil being washed away by rain. Soil may run off into waterways, polluting them. Trees also maintain soil nutrient levels through their role in the carbon and nitrogen cycles.

Clear felling is now rarely practised in the UK. Leaving each section of woodland to mature for 50–100 years before felling allows biodiversity to increase. However, such a time-scale is not cost-effective. Instead, modern sustainable forestry avoids this by working on the following principles:

- Any tree which is harvested is replaced by another tree, either grown naturally or planted.

- The forest as a whole must maintain its ecological function regarding biodiversity, climate, and mineral and water cycles.

- Local people should benefit from the forest.

Selective cutting involves removing only the largest, most valuable trees, leaving the habitat broadly unaffected.

Sustainably managing forests involves balancing conservation against the need to harvest wood, both to maintain biodiversity and to make the woodland pay for itself. If each tree supplies more wood, fewer trees need to be harvested. To achieve this, foresters:

- control pests and pathogens

- only plant particular tree species where they know they will grow well

- position trees an optimal distance apart. If trees are too close, this causes too much competition for light, and they grow tall and thin, producing poor-quality timber.

Managing fish stocks

Fisheries

The fishing industry has high economic value, not just from sales of fish, but in providing livelihoods for millions of people.

Figure 3 Trawlers can catch huge numbers of fish at once.

The Marine Stewardship Council has proposed three princi[?] sustainable management of fisheries:

- Fishing must take place at a level which allows it to continue indefinitely. Over-fishing must be avoided, because it can reduce fish populations to zero. If over-fishing happens, reducing fishing to let stocks recover can rapidly increase productivity and is good for profitability, given high stock values can support a more efficient harvest. The optimum is to maintain fish populations at the carrying capacity of their environment, while fishing continues to harvest fish in excess of that capacity.

- Fishing must be managed to maintain the structure, productivity, function and diversity of the ecosystem. This means there should not be permanent damage to the local habitat, and any effect on dependent species is minimised.

- A fishery must adapt to changes in circumstances and comply with local, national and international regulations.

Aquaculture

Aquaculture can also provide sustainable fish stocks. Raising stocks of fish in aquaculture restricts the impact on oceanic fish stocks. Aquaculture is expanding rapidly, particularly in the developing world, and is expected to feed more people than traditional 'capture fisheries' in the near future.

LEARNING TIP

When considering sustainable management of resources, remember that maintaining biodiversity is just as important as maintaining resources at a steady level.

Questions

1. What is meant by the phrase 'sustainable management of wood production'?

2. Write down the differences and similarities between small-scale and large-scale production of wood.

3. Explain what would happen to biodiversity in Bradfield Woods if the Suffolk Wildlife Trust stopped managing it.

4. Explain why over-fishing is a problem to sustainable fish supplies.

5. Why does sustainable fishing aim to maintain fish stocks at maximum reproductive capacity?

6. What are the benefits and disadvantages of aquaculture, rather than 'capture fisheries'?

⑤ Balancing the conflict between conservation and human needs

By the end of this topic, you should be able to demonstrate and apply your knowledge and understanding of:

* the management of environmental resources and the effects of human activities

Balancing the competing requirements for natural resources by humans and other living things is essential to help secure sustainable use of natural resources, which is compatible with conservation. Three successful examples of conservation being implemented alongside development are the Terai region of Nepal, the Maasai Mara in Kenya and peat bogs in the UK.

The Terai region

In the south of Nepal is the Terai region (Figure 1), made up of marshy grasslands, savannah and forests. It is densely populated, and home to endangered species including the Bengal tiger and the greater one-horned rhinoceros. There are many national parks in the region.

Figure 1 Grassland in the Terai region of south Nepal.

For over 10 years, the forests in the Terai region have been under pressure from expansion of agriculture into forested areas, grazing from farm animals, over-exploitation of forest resources, and replacement of traditional agricultural crop varieties with modern ones. In response, the World Wide Fund for Nature (WWF) found that rural livelihoods are heavily dependent on the forests, which are also home to many of the region's endangered species. The forests can provide local people with a sustainable source of fuel, animal feed, food, building materials, agricultural and household tools, as well as medicines.

Because local people have such a high stake in the forest, the WWF with the Nepalese government in the Terai-arc landscape programme focused on conservation of the forest landscape as a whole. To ensure conservation with development, they introduced community forestry initiatives in which local people had rights to exploit the forest as well as responsibilities to look after it. These community groups helped to create forest corridors between national parks, which are essential to the dispersal and survival of tigers, as well as taking the initiative in counteracting poachers and illegal felling. Forestry work also developed and diversified on- and off-farm activity, built entrepreneurial skills, and stimulated small credit and marketing schemes. The WWF scheme also introduced biogas plants and wood-efficient stoves to reduce demand for firewood.

Other contributions to the Terai-arc project included constructing waterholes, monitoring endangered species, and eradicating invasive species.

Community involvement combined with governmental and non-governmental (WWF) leadership appears to have been successful. Recent data from southern Nepal suggests tigers are using the corridors between national parks, and their population size is steadily growing.

DID YOU KNOW?

Community forest user groups in Nepal have earned hundreds of thousands of dollars from tourism activities in the buffer zone, which protects the edges of national parks in the Terai region.

Maasai Mara

The Maasai Mara in Kenya is a famous destination for wildlife watchers, with large populations of antelope and other large mammals (Figure 2). Because the Maasai Mara combines high endemic poverty with abundant wildlife populations that attract tourism, there has been scope to develop conservation-compatible land use that rewards local people financially, whilst conserving habitats and species that are the basis for tourism.

Figure 2 The Maasai Mara in Kenya.

After creation of national parks in 1945, remaining Maasai land was held in trust until 1968, when the lands were designated as 'group ranches'. Worried about their tenure on the land, many Maasai took individual title over smaller portions of land. This triggered land-use change, including intensification of agriculture. This limited wildlife to increasingly small islands and constrained the mobility of livestock. For example, wheat farms occupy 40 000 acres of the previous wet-season-range of migratory wildlife. The population of wildebeest that use the wet-season-range shrank from 150 000 in 1977 to 40 000 in 2010.

Figure 3 Wildebeest migration in Kenya.

The density of other wildlife has dropped 65% over the last 30 years, while the density of sheep and goats increased. In 2005, several land-owners to the north of the Maasai Mara Reserve consolidated their land to form conservancies, in order to generate tourism income. Partnerships between conservancies and tourism operators have developed payment for wildlife conservation (PWC) schemes. Conservancies are paid PWC revenue proportional to the area of land set aside for conservation. These conservancies are successful, because they have positive social outcomes, as well as positive conservation outcomes.

There are some negative consequences of the conservancies. Land-owners must move their livestock out during the tourist season, which leads to increased stocking densities outside the reserve, where no one receives the PWC money. Likewise, land-owners are often forced to settle elsewhere, and there are constraints on how they use their land.

Livestock have been seen as a problem for conservation. However, there is evidence that limited livestock grazing can have positive impacts on diversity. Given how important livestock are in Maasai culture, it would make sense to continue to integrate conservation and livestock management more directly.

Peat bogs

Peat forms where a lack of oxygen prevents complete decomposition of organic matter. This usually occurs where the area is waterlogged. Accumulation over thousands of years means that undisturbed peat bogs contain a lot of historical data. Controlled archaeological digs can reveal much information about the landscape and vegetation in the past, as well as indicating what weather conditions were like. Over thousands of years rainfall has washed nutrients from the soil, leaving conditions ideal for the growth of *Sphagnum* moss. The peat bog and moss retain moisture, forming some of the UK's most scarce wetland habitats. These support a high biodiversity and provide an important feeding and stopping-off point for migrating birds.

The ability of peat to retain moisture has led to its widespread use as compost for improving soils in gardens. Pressure from expansion of agriculture, forestry, landfill and peat extraction mean that lowland peat bog now covers less than one-tenth of its original area in England. The UK Biodiversity Action Plan (UKBAP) aims to conserve and enhance biodiversity through local-level schemes. Working with partners such as the RSPB and English Nature, UKBAP has identified certain peat bogs for restoration and aims to end the commercial use of peat in the UK.

> **LEARNING TIP**
>
> When revising these two initiatives, try using a graphic organiser to help remember the key points. A simple graphic organiser is a spider diagram. Choose your own structure, but try organising the information above under the headings: Why? How? Positive consequences, Negative consequences.

Questions

1. Describe how both:
   (a) the Terai project
   (b) the Maasai Mara project
      made resource use more sustainable in each region.

2. State the similarities between the strategies adopted in the Terai region and in the Maasai Mara.

3. Explain why both projects tried to involve the local community in conservation efforts.

4. Write down:
   (a) the positive consequences of each project
   (b) the negative consequences of each project.

5. Explain how people working as a group has helped:
   (a) the species in the Terai region
   (b) the species in the Maasai Mara.

6. Suggest why conservation initiatives that do not account for the needs of local people tend to fail.

7. Suggest how local-level schemes can conserve important habitats such as peat bogs.

6 Controlling the effects of human activities

By the end of this topic, you should be able to demonstrate and apply your knowledge and understanding of:

* the management of environmental resources and the effects of human activities

Human activities can affect populations in a number of ways, including habitat destruction, competition for natural resources, hunting and pollution. To overcome these problems, some areas are protected, such as national parks and reserves, green belt land, world heritage sites, marine protected areas, and areas of outstanding natural beauty. There is also legal protection of endangered species, and eradication of invasive species. In this topic, we look at four locations where humans have strived to control their effects on the environment.

Figure 1 The Galapagos Islands.

The Galapagos Islands

The Galapagos Islands (Figure 1) have high numbers of native species, including Darwin's finches, giant tortoises (Figure 2) and marine iguanas. Unfortunately, 50% of vertebrate species and 25% of plant species are endangered. The Galapagos human population has grown in response to increased demand for marine products and increased tourism.

Habitat disturbance

The population size increase has placed huge demands on water, energy and sanitation services. More waste and pollution have been produced, and the demand for oil has increased – an oil spill in 2001 had a dramatic effect on marine and coastal ecosystems. Building and conversion of land for agriculture has caused destruction and fragmentation of habitats. Forests of *Scalesia* trees and shrubs (a species unique to the islands) have been almost eradicated on Santa Cruz and San Christobal, to make way for agricultural land.

Figure 2 The giant tortoise.

DID YOU KNOW?

Fur traders and whalers used to use giant tortoises for food. Because they could be kept alive in the hold of ships with little or no food, the ships could stay out at sea for longer, and still have fresh meat.

Over-exploitation of resources

In the nineteenth century, whaling boats and fur traders killed 200 000 tortoises in less than half a century. The Charles Darwin Research Station has a captive breeding programme to supplement tortoise numbers. The more recent boom in fishing for exotic species has depleted populations. Depletion of sea cucumber populations has a drastic effect on underwater ecology, and the international market for shark fin has led to the deaths of 150 000 shark each year around the islands, including 14 endangered species.

Effects of introduced species

As well as out-competing local species, alien species can eat native species, destroy native species' habitats, or bring disease onto the islands. For example, cats hunt a number of species, including the lava lizard and young iguanas. Goats feed on Galapagos rock-purslane, a species unique to the islands, and trample and feed upon giant tortoises' food supply and disrupt their nesting sites. On northern Isabela Island, the goat has also transformed forest into grassland, leading to soil erosion. The red quinine is an aggressively invasive species on Santa Cruz Island. It occupies the highlands, and spreads rapidly – it has wind-dispersed seeds. The ecosystem in the highlands has changed from low scrub and grassland, to being a closed forest canopy. Because of this, the native *Cacaotillo* shrub has been almost eradicated from Santa Cruz, and the Galapagos petrel has lost its nesting sites. The red quinine also successfully out-competes native *Scalesia* trees.

Managing the effects of human activity

In 1999, the Charles Darwin Research Station adopted two strategies – to prevent the introduction and dispersion of introduced species and to treat the problems caused by such species. They search arriving boats and tourists for foreign species. Natural predators have also been exploited to reduce the damage caused to ecosystems by pest populations – controlled release of a ladybird wiped out a scale insect, which was damaging plant communities. Culling has also been successful against feral goats on Isabela Island and pigs on Santiago Island.

Because most residents were not born on the islands, fostering a culture of conservation and educating new arrivals about the islands is a challenge. The Galapagos marine reserve provides a model of how local stakeholders can work together to sustainably manage a resource. The reserve is managed by the National Park Service, the Charles Darwin Research Station, and representatives of local fishermen, the tourist industry and naturalist guides. At least 36% of coastal zones have been designated 'No-Take' areas, where no extraction of resources is allowed, and communities are left undisturbed.

The Antarctic

Antarctica is not governed by any one country, but countries have research stations there under the terms of the Antarctic Treaty. As such, the continent is relatively untouched by human influence, but with an increasing number of tourists and scientists, and with increased interest from the fishing industry, it remains important to actively protect the Antarctic ecosystem and biodiversity within it.

Krill

Krill are tiny shrimp-like organisms which provide food for whales, seals, penguins, albatrosses and squid. Krill are used to make nutritional supplements and for animal feed. Recent changes in technology mean that large amounts of krill can be harvested very quickly and easily. Fishing boats tend to congregate in those areas with the largest numbers of krill. However, natural predators of krill cannot adapt as easily to find krill elsewhere. For example, penguins do not migrate very far when raising young. To avoid over-exploitation, there is a trigger level catch size in particular areas. When reached, fishing must be conducted equally across all areas, up to the total catch limit, to avoid a catastrophic impact on predators. However, given the size of the krill fishery, there are new recommendations to force the industry to fish evenly across all areas anyway.

Figure 3 Krill are tiny marine organisms, which are a key component of a large number of important food chains in the Antarctic.

Protected areas

To protect whales and the marine environment, a series of protected areas have been established. The Southern Ocean Whale Sanctuary was established in 1994, covering the summer feeding grounds of 80–90% of the world's whales. This followed the 1982 moratorium on whaling established by the International Whaling Commission. Within the sanctuary, it is illegal to hunt and kill whales, although monitoring of whaling activity still needs to be maintained to ensure the sanctuary is effective. Currently there is an initiative to expand a network of marine protected areas, such as in the Ross Sea (with its high biodiversity levels), which is already attracting the attention of the fishing industry.

Albatrosses and petrels

These majestic birds are threatened by human activities, including pollution, hunting and poaching for eggs, habitat destruction and introduction of non-native predators. However, the biggest threat is long-line fishing. This is when fishermen trail a long fishing line behind their boat (up to 130 km). Attached to this line are hundreds of baited hooks. When behind the boat, the birds try to eat the prey and swallow the hooks. To reduce the number of deaths, boats can use bird-scaring lines and streamers, weighted lines which sink more quickly out of reach of the birds, use lines at night to avoid albatross and petrel feeding times, and avoid breeding and nesting time. By implementing these methods, one Chilean fishery reduced its sea bird catch to zero.

DID YOU KNOW?
Birdlife International estimates 100 000 albatrosses are killed by long-line fishing each year.

The Lake District

The Lake District contains an exceptionally rich diversity of species and habitats. The whole National Park is designated an Environmentally Sensitive Area. However, without grazing causing deflected succession, much of the land would revert to a climax community of oak woodland. Financial incentives are available for farmers to reduce chemical use, to safeguard hedges, and to care for hay meadows, heather moor, wetland, chalk downland and native woodland.

Threat to biodiversity	Solution
Spruce and pine in conifer plantations support limited biodiversity.	Recent initiatives have generated more varied planting and felling patterns, giving a mosaic of smaller stands of different aged trees.
Invasive species, including rhododendron and laurel, have escaped from gardens and spread into woodland outcompeting native species. Their dense canopy reduces the light reaching the woodland floor and their roots produce toxic chemicals, which stop other plants growing (allelopathy).	They are physically removed by conservation workers.
Limestone pavement is a unique habitat characterised by solid blocks with fissures between them. Rare ferns grow well in the fissures and rare butterflies thrive in this habitat.	The pavement is legally protected through the Limestone Pavement Orders.
Hay meadows occur in neutral grassland and support a rich diversity of flowers and grasses. They are under threat due to a preference away from haymaking and in favour of silage production, which involves use of artificial fertiliser and an earlier cut. This has caused loss of species diversity on grasslands and pastures.	Farmers are paid to maintain hay meadows.
Heathland is an open habitat with small shrubs like heather, which are important for butterflies, moths, spiders, beetles, birds and reptiles.	By burning strips of vegetation, new shoot growth is promoted, maintaining areas of different ages, which fosters a bigger variety of animals. This ensures a constant food supply for red grouse and merlin. Such areas are also managed by grazing, but overgrazing is a problem as the shoots of heather can be eaten more quickly than they regenerate. Financial incentives are provided to farmers to prevent overgrazing.
Mires are nutrient-poor, waterlogged ecosystems, within which mosses and liverworts, lichens and sedges flourish. The habitat is internationally scarce, provides a breeding ground for moorland birds, but is under threat from burning, grazing and drainage for more intensive agriculture. Peat extraction for gardeners has also threatened the habitat.	Mires are now managed more sympathetically, with some being rewetted with artificially controlled water levels. In areas with rare plants, like the bog orchid, grazing has been controlled.
Cliff, rock and scree communities support a rich diversity of plant life, providing a habitat for the stonechat and wheatear, and nesting sites for the peregrine falcon and golden eagle. These communities are easily damaged by climbers and walkers.	To protect them, there are seasonal restrictions on walking when birds are nesting, walkers are educated to be more aware, and paths are well maintained to prevent people walking off the path.

Table 1 Threats to biodiversity and solutions.

Figure 4 The Lake District.

Snowdonia National Park

Snowdonia in North Wales attracts walkers and climbers. On Mount Snowdon itself, good footpaths are maintained to ensure that rare plants are not trodden on. Gutters take water from the paths, but these can get blocked by rubbish dropped by walkers. To prevent the path eroding, teams of workers clear the rubbish. National Park employees work with farmers to reduce sheep grazing on the mountain. Because they graze very low to the ground, they can leave the landscape very barren. Reducing sheep grazing gives the rare plants a better chance of survival. Feral goats are a problem for grazing, and their numbers and locations are monitored each year. Farmers are encouraged to plant hedges and conserve ancient woodland.

As well as mountain habitat, there is also moorland and bog in Snowdonia. These habitats provide nesting sites for rare birds like the hen harrier, merlin and kestrel, and are home to a number of rare butterflies. Use of such wetland leads to humans having an impact upon it, which needs to be controlled.

Inpact by humans	Solution
Farmers dig open drainage ditches to dry the land. However, this causes poor water quality in rivers, and rain flows quickly through the habitat, increasing flood risk.	To mitigate the effect, drainage ditches can be blocked by hay bales.
Conifers are planted as cash crops. However, this dries out the moorland as the trees absorb water, and roads have to be built to carry the wood away, compacting the land.	When the trees are cut down, branches are used to block drainage ditches to slow water flow and hence keep the land moist.
The moorland was burnt to provide a varied habitat for grouse, but when sheep grazed the land, burning was stopped. Old heather burns easily, and so accidental fires can be a high risk. If the peat sets on fire, that can damage the habitat on a large scale.	Burning controlled fires before heather gets too old and dry is important to prevent such damage.

Table 2 Threats to habitats in Snowdonia.

Questions

1 List the main threats to native species on the Galapagos Islands.

2 Explain why goats pose such a problem to conservation on the Galapagos Islands.

3 Explain why management of the marine reserve around the Galapagos Islands provides a model for effective conservation.

4 Explain why protection of krill is so important to the Antarctic ecosystem.

5 Make a list of:
   (a) factors which work against conservation in the Galapagos, Antarctica, the Lake District and Snowdonia
   (b) strategies which can help conservation in these places.

6 Create a table which has rows depicting human strategies for dealing with human impact, and columns for each of the ecosystems. Place a tick in the correct box to indicate that a particular strategy is in use in an ecosystem.

LIFE IN THE ANTARCTIC OCEAN

Until relatively recently, it was believed that conditions under Antarctica's ice sheets were too inhospitable to sustain life. However, this belief has now been proved wrong.

WHAT LIVES UNDER ANTARCTIC ICE?

After drilling through Antarctica's Ross Ice Shelf, scientists have discovered microbes, crustaceans, and even several kinds of strange fish living in water buried under nearly half a mile (740 metres) of ice. As recently as a decade ago, it was thought that nothing could survive beneath Antarctica's massive ice sheets. The water under the ice sheet is around 33 feet (10 metres) deep, and temperatures hover below freezing.

The new finds include several kinds of fish that have big eyes, maybe because the animals live in darkness. Some were orange, others black, but the biggest fish of all had translucent skin through which the animal's internal organs could be seen. "From a biological perspective, we got the first glimpse of life beneath the ice on the fifth largest continent on our planet – a continent that was previously thought to be nothing more than a benign body of ice," says study team member John Priscu, professor of ecology at Montana State University. The new discoveries come courtesy of the Whillans Ice Stream Subglacial Access Research Drilling (WISSARD) project, an interdisciplinary collaboration of more than 40 scientists.

A fish of a different colour

After using a hot-water drill to punch through 2400 feet (740 metres) of ice, the scientists lowered a remote-controlled submersible down the hole. This robot sent back images and video of life below the shelf. It's not clear yet if the new see-through fish represents a new species, but it's likely that they belong to the suborder Notothenioidei.

These fish, called notothenioids, don't have a lot of company from other animals in Antarctic waters, as they make up 91 per cent of the total animals by weight (or biomass) and 77 per cent of the species, said Reinhold Hanel, a biologist at the Johann Heinrich von Thünen Institute in Germany. Thanks to a combination of geothermal heat and the pressure and movement created by the ice sheets above, these fish live in water that's perpetually 28 °F (−2 °C). That means the fish have had to develop numerous adaptations to survive.

"Their evolutionary success is related to key adaptations, such as antifreeze glycoproteins, which prevent their body fluids from freezing at subzero temperatures," said Hanel, who is not affiliated with the WISSARD Project. As for being able to see their guts, Hanel said the fish are probably translucent as a result of the evolutionary loss of haemoglobin, the protein that makes blood red.

Of microbes and men

Finding bug-eyed fish was certainly a happy surprise, but Priscu is even more interested in the microbes. Last August, Priscu and his WISSARD project colleagues published an article in *Nature* proving for the first time that microbial life existed beneath the West Antarctic Ice Sheet in Lake Whillans. The samples from this latest expedition have not yet been analysed, but Priscu said he's excited to see how the biodiversity of microbes beneath the Ross Ice Shelf compares to samples taken at Lake Whillans, and with others taken in the Arctic.

He's also curious as to whether the microorganisms found living in the mud might produce greenhouse gases such as methane and carbon dioxide. "If so, we could expect to see a large release of these gases as the ice sheets melt," he said.

But climate change isn't the only reason to be interested in microbes. The team's research could also influence the search for life in the cold, dark recesses of space and the way we understand ecosystems here on Earth.

Source
● www.nationalgeographic.com/news/2015/01/150127-antarctica-translucent-fish-microbes-ice/

Where else will I encounter these themes?

Let's start by considering the nature of the writing in the article.

1. This article comes from *National Geographic* and uses many quotations from the scientists involved. What effect do the quotations have on the reader's impression of the story? Make a list of the key points which you think a reader will remember from the article.

Now we will look at the biology in, or connected to, this article.

2. The notothenioids make up 91% of the animal biomass. The remaining biomass is made up of 77 species. What is the mean biomass of these 77 species?

3. What is a sub-order?

4. Translucent fish are likely to have lost haemoglobin as part of their evolution.

   a. Explain how haemoglobin takes up and releases oxygen in the human body.

   b. Haemoglobin is primarily composed of protein. Suggest why the cold temperature may prevent haemoglobin from functioning properly.

5. The article suggests that large eyes have evolved in fish living under the ice sheet. Explain how this process may have happened, using the principles of natural selection to help you. Suggest how it provides evidence that light does penetrate the ice sheet.

6. The research was conducted by an interdisciplinary team of scientists. What is the advantage of this kind of interdisciplinary research?

7. Release of unfamiliar microbes from under the ice is unlikely to pose a threat to human health.

   a. By considering the environment in which they live, explain why this is the case?

   b. If they did survive to enter the body of one of the researchers, create a flow chart to show how the body would defend itself.

8. Priscu is concerned that microbes may release large quantities of carbon dioxide if the ice melts. However, it is not clear from the article whether such microbes undertake aerobic or anaerobic respiration. Compare and contrast the two forms of respiration.

9. One reason for reduced amounts of haemoglobin could be a reduced basal metabolic rate in the cold conditions. Devise an investigation to test this hypothesis.

10. The article makes reference to the biodiversity of populations under the ice shelf. Suggest why drilling down into this community of organisms may pose a threat to the biodiversity of its microbial populations.

Activity

In aquatic environments with little light, the producer role in a food web is often taken by chemosynthetic bacteria. Create a wiki page which describes the role of chemosynthetic bacteria as producers in one marine environment. You should explain how they provide energy and biomass to primary consumers in the absence of light. Your wiki page should make reference to the adaptations of the members of the food web to environmental conditions in the location you have chosen. You may embed video and still images into your wiki page.

Practice questions

1. What is the name for factors which limit the growth in population size? [1]
   - A. Abiotic factors
   - B. Biotic factors
   - C. Limiting factors
   - D. Predation

2. Which of the following is typical of species whose population size grows in excess of the carrying capacity of the environment before dropping back? [1]
   - A. High reproductive rate
   - B. Slow development
   - C. Late reproductive age
   - D. Long lifespan

3. What happens to a population as competition for food increases? [1]
   - A. Death rate decreases
   - B. Birth rate increases
   - C. Birth rate decreases
   - D. Population size increases more quickly

   [Total: 3]

4. This question is about ecosystems in the Southern (Antarctic) Ocean.

   Observe the food chain:

   phytoplankton (producers) → krill (shrimps etc.) → small fish → large fish → seal

   Table 1 shows the transfers of energy and the quantities of energy stored as biomasses for the food chain. Magnitudes are given in kilojoules per square metre of sea surface per year.

(a) The biomass of large fish in the Southern Ocean is a food resource for humans. It is increasingly harvested by powerful, long-distance trawlers. If over-exploited, the Southern Ocean ecosystem may be permanently altered.
   (i) Suggest two measures that an international treaty might impose, to prevent fishing from causing permanent damage to the Southern Ocean.
   (ii) Identify the practical difficulties that might prevent your two measures from being effective. [4]

(b) Krill can also be harvested as a human food source. The fishing industry aims to harvest large fish.

   Some environmentalists say that krill harvesting should be increased. Use this information and Table 1 to put forward arguments for and against harvesting krill instead of large fish as a human food source. [2]

   [Total: 6]

   [From Q21, sample paper H420/02, 2014]

5. A small, permanent pond is the habitat for a climax community of producers (aquatic plants and algae) and consumers (bacteria, protoctista, worms, snails, arthropods and small vertebrates like newts and fish).

(a) Why might ecologists call this a 'climax community'? [1]

(b) The protoctist *Paramecium caudatum* is usually between 200 and 300 µm in length. An accurate measurement would help in the correct identification of a specimen from this pond.

   What laboratory equipment would you select to make an accurate measurement of the length of *Paramecium caudatum*? [2]

	Phytoplankton	Krill	Small fish	Large fish	Seals
Energy input, by photosynthesis or feeding (kJ m^{-1} y^{-1})	900	80	11	1.4	
Energy lost to surroundings by respiration (kJ m^{-1} y^{-1})	180	64	8.8	1.2	1.05
Energy input converted to biomass (kJ m^{-1} y^{-1})	720	16	2.2	0.2	0.05
Biomass energy lost to other consumers or decomposers (kJ m^{-1} y^{-1})	640	5	0.8	0.09	0.05

Table 1

(c) An animal fell into the pond. It drowned and decayed. Within a year the biological compounds in its body had been completely recycled.

(i) What nitrogenous excretory molecule from the decomposers would pass to the next stage of the nitrogen cycle? [1]

(ii) Complete the flow chart to show what happens to this nitrogenous compound, and name the groups of bacteria involved at steps 1 and 2, as it is converted to a form that plants can take up and use. [4]

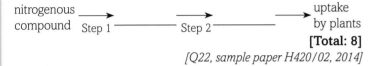

[Total: 8]

[Q22, sample paper H420/02, 2014]

6. Organisms do not live in isolation, but interact with other organisms and with their physical environment. State and describe **two types** of ecological interaction that can occur between different species in a habitat. As part of each description, you should **name** the two species involved in your chosen example. [6]

[Total: 6]

[From Q3, F215 Jan 2012]

7. Sarawak is an area of tropical rainforest in south-east Asia. Logging has been allowed in 60% of the forest.

A study was carried out into the effects of logging on the diversity of mammal species living in the forest. An area of rainforest was sampled before logging, immediately after logging and then again two years and four years after logging.

Before logging began, there were 29 mammal species and four years after logging there were 26 mammal species.

(a) Outline **three** reasons for conserving biological resources, such as the rainforest in Sarawak. [3]

(b) Timber is produced sustainably in the United Kingdom.

Describe **and** explain the benefits of **two** management practices used in sustainable timber production in a temperate country. [4]

[Total: 7]

[From Q5, F215 June 2010]

8. In 1978, the United Nations (UN) declared the Galapagos Islands a World Heritage Site. This led to a rise in the resident human population and the number of visitors to the island.

Table 2 shows how the number of people living on and visiting the Galapagos Islands changed between 1980 and 2005.

Year	Resident population	Number of visitors
1980	5500	16 000
1985	7000	19 000
1990	9500	42 000
1995	12 500	58 000
2000	17 500	68 000
2005	27 500	125 000

Table 2

(a) (i) Calculate the percentage increase in the number of visitors to the Galapagos Islands between 1980 and 2005. Show your working. **Give your answer to the nearest whole number.** [2]

(ii) Outline the main ways in which increased human presence and activity have put endemic species on the Galapagos Islands, and in the sea around them, at risk of extinction.

In your answer you should link the ecological pressures imposed by human activity to examples of Galapagos Island species that have been affected. [7]

(b) In 2007, the United Nations (UN) put the Galapagos Island on its Red List of endangered sites. The Galapagos government's response to this action included making new laws and placing restrictions on human activity, issuing eviction orders and culling introduced species of animals.

Suggest **one** economic and **one** ethical problem that might have arisen from this 2007 UN decision. [2]

[Total: 11]

[From Q2, F215 Jan 2013]

Maths skills

In order to be able to develop your skills, knowledge and understanding in Biology, you will need to have developed your mathematical skills in a number of key areas. This section gives more explanation and examples of some key mathematical concepts you need to understand. Further examples relevant to your A level Biology studies are given throughout this book and in Book 1.

Using logarithms

Calculating logarithms

Many formulae in science and mathematics involve powers. Consider the equation:

$10^x = 62$

We know that the value of x lies between 1 and 2, but how can we find a precise answer? The term logarithm means index or power, and logarithms allow us to solve such equations. We can take the 'logarithm base 10' of each side using the **log**$_\square(\square)$ button of a calculator.

> **EXAMPLE**
> $10^x = 62$
> $\log_{10}(10^x) = \log_{10}(62)$
> $x = 1.792392...$

We can calculate the logarithm using any number as the base by using the **log** button.

> **EXAMPLE**
> $2^x = 7$
> $\log_2(2^x) = \log_2(7)$
> $x = 2.807355...$

Many equations relating to the natural world involve powers of e. We call these exponentials. The logarithm base e is referred to as the natural logarithm and denoted as **ln**.

Using logarithmic plots

An earthquake measuring 8.0 on the Richter scale is more than twice as powerful as an earthquake measuring 4.0 on the Richter scale. This is because the units involved in measuring earthquakes use the concept of logarithm scales in charts and graphs. This helps us to accommodate enormous increases (or decreases) in one variable as another variable changes.

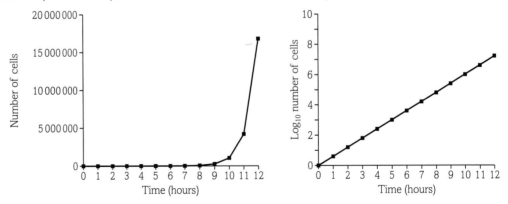

Figure 1 Logarithmic scales are useful when representing a very large range of values, such as in the case of bacterial growth.

Selecting and using a statistical test

In Book 1, you learned how to use measures of average (mean, median and mode) and measures of dispersion (range and standard deviation). We will now look at some statistical tests that can be used to analyse data.

Deciding on a hypothesis

When you use a statistical test, you need to be clear what you are testing for. Scientists do this by creating a hypothesis, then a null hypothesis which we try to disprove. You should clearly state your null hypothesis before you begin a test.

After running your chosen statistical test, you will be left with a number known as the observed value. To know whether or not you should accept your null hypothesis, you need to compare the observed value with a table of critical values. To find the correct value to compare it with, you will also need to calculate the 'degree of freedom' for your results.

In essence, the critical values tell you whether or not you can accept your null hypothesis. If your observed result is less than the 5% significance level given in the table, you can be 95% certain that your null hypothesis is true – and you should therefore accept it. If not, reject it!

Choosing a test

Your choice of the most appropriate statistical test will depend on the type of data your investigation is going to produce and how these data will be organised. This is known as the level of measurement. For each variable, the level of measurement can be nominal, ordinal or interval/ratio.

- **Nominal** data are those which can be categorised. For example, keeping a record of the sex of each participant in an experiment is recording data at the nominal level; males could be put into Category 1 and females into Category 2.

- **Ordinal** data are categorised data that can be placed into rank order. For example, a list of the first, second and third order runners in a race to cross the finish line. The time differences between the runners is not known, so ordinal scales do not have equal intervals.

- **Interval** data are measured on interval scales, for example, the centigrade scale of temperature measurement where the intervals between adjacent values in the centigrade scale are equal. **Ratio** data are special types of interval data that meet certain conditions.

There are many different statistical tests that can be used. The tests that you may meet in your A Level Biology studies are shown in the table below.

Name	Variables	Purpose	Example	Null hypothesis
Spearman's rank correlation coefficient	Two ordinal variables (or interval variables that have been converted to ordinal)	To test whether two ordinal variables display correlation	Test whether there is a correlation between finishing positions in a race and age of runners	There is no correlation between the two variables. (Spearman's rank correlation coefficient equals zero.)
Chi-squared test	One nominal variable for which you have expected and observed results	To test how likely it is that any differences between observed and expected results are due to chance	Test whether a ratio of phenotypes from mating supports a particular inheritance model	The observed results are consistent with the expected distribution. (The differences between observed and expected results are due to chance.)
Student's t-test	One nominal variable with two categories One interval variable	To test whether the difference in the mean value of the interval variable for the two categories is significant	Test whether a drug treatment has been effective compared with a placebo	The means of the interval variable for the two categories are equal.

Using the Spearman's rank correlation coefficient to test for correlation

Step 1: State the null hypothesis

The null hypothesis for this test is: 'r_s is equal to zero, meaning that there is no correlation between the two variables.'

Step 2: Calculate the observed value

To calculate the Spearman's rank correlation coefficient, you first need to ensure that you have ordinal data. If you have one ordinal variable and one interval variable, you can convert interval data to ordinal by placing the values in rank order.

The formula for the Spearman's rank correlation coefficient is:

$$r_s = 1 - \frac{6\Sigma d^2}{n(n^2 - 1)}$$

where d is the difference in rank between each pair of variables, n is the number of pairs and the symbol Σ means 'the sum of'. To use the Spearman's rank correlation coefficient, find the difference in rank between each pair, square these differences and add them all together. Now substitute this value into the formula in place of Σd^2.

Step 3: Decide whether to accept the null hypothesis

To decide whether you can accept the null hypothesis, you need to compare your observed value of r_s with a critical values table. To do this you need to know the degree of freedom, which for correlation tests is the number of pairs minus 2: $df = n - 2$.

Find the critical value that corresponds to the 5% significance level for your calculated degree of freedom. If your value of r_s (ignoring whether it is positive or negative) is less than the critical value, accept the null hypothesis. If not, reject it.

Using the chi-squared test

Step 1: State the null hypothesis

The null hypothesis for this test is: 'The observed results are consistent with the expected distribution, meaning that differences between observed and expected results are due to chance.'

Step 2: Calculate the observed value

The formula for the chi-squared test is:

$$\chi^2 = \Sigma \frac{(O - E)^2}{E}$$

where O is your observed result, E is the expected result and the symbol Σ means 'the sum of'. For each result, you need to find the difference between the observed and expected values, square this difference then divide by the expected value. You then add together the results of all these calculations to get the observed value for χ^2.

Step 3: Decide whether to accept the null hypothesis

To decide whether you can accept the null hypothesis, you need to compare your observed value of χ^2 with a critical values table. To do this you need to know the degree of freedom, which for the chi-squared test is the number of categories minus 1: $df = n - 1$.

Find the critical value that corresponds to the 5% significance level for your calculated degree of freedom. If your value of χ^2 is less than the critical value, accept the null hypothesis. If not, reject it.

Using the Student's t-test

It might seem obvious that you could easily compare the means of the two categories by simply subtracting one from the other. However, can you be sure that the means calculated from the two sample data sets are representative of the whole population? The Student's t-test takes into account the degree of overlap between the two sets of data and allows you to judge whether any difference between the means is statistically significant or just due to chance.

Step 1: State the hypothesis

The null hypothesis for this test is: 'The means of the interval variable for the two categories are equal.'

Step 2: Calculate your observed value

In order to calculate t, you first need to find the mean $\bar{x}$ and the variance s^2 for both of the categories:

$$\bar{x} = \frac{\Sigma x}{n}$$

$$s^2 = \frac{\Sigma x^2 - \frac{(\Sigma x)^2}{n}}{n - 1}$$

You can now use the following formula to calculate the value of t:

$$t = \frac{\bar{x}_1 - \bar{x}_2}{\sqrt{\dfrac{s_1^2}{n_1} + \dfrac{s_2^2}{n_2}}}$$

where:

$\bar{x}_1$ = mean of the first set of data

$\bar{x}_2$ = mean of the second set of data

s_1^2 = variance of the first set of data

s_2^2 = variance of the second set of data

n_1 = number of items in first set of data

n_2 = number of items in second set of data.

Step 3: Decide whether to accept the null hypothesis

To decide whether you can accept the null hypothesis, you need to compare your observed value of t with a critical values table. To do this you need to know the degree of freedom, which for the Student's t-test is the total number of data values minus 2: $df = n_1 + n_2 - 2$.

Find the critical value that corresponds to the 5% significance level for your calculated degree of freedom. If your value of t is less than the critical value, accept the hypothesis. If not, reject it.

Applying your skills

You will often find that you need to use more than one maths technique to answer a question. In this section, we will look at three example questions and consider which maths skills are required and how to apply them.

EXAMPLE

Microorganisms can be grown quickly using a fermenter. If the conditions of the fermenter are set correctly (for example, there are enough nutrients and sufficient space to grow) the growth of the bacterial colony can become exponential. One such microorganism, a species of yeast called S. cerevisiae, *can divide every 100 minutes.*

a) *Assuming there is no cell death, and two yeast cells are introduced into the fermenter, how many cells will there be after 300 minutes?*

b) *Write an equation to show how the population of yeast over a period of time can be calculated, using* d *to represent the number of divisions.*

c) *Using your equation from (b), calculate the population of yeast generated after 32 cell divisions in a fermenter.*

d) *How many cell divisions would have taken place if the population of yeast was only allowed to reach 65 536 cells?*

a) In 300 minutes, the cells will divide three times. After the first division, the population will have doubled to four from the two original yeast cells. The second division will double the population again, to eight cells. The third division will double the population yet again, resulting in 16 cells.

b) If we started with just one cell, the number of cells after time d would be given by 2^d. However, because we are starting with two cells, we need to double this, so $N_d = 2 \times 2^d$ where N_d is the number of cells after d divisions. Using the rules of indices, we could also write this as $N_d = 2^{(d+1)}$.

c) To find the number of cells after 32 divisions, we can just substitute $d = 32$ into our equation:

$N_{32} = 2^{(32+1)} = 2^{33} = 8\,589\,934\,592$ cells.

d) To find the number of divisions necessary to produce 65 536 cells, we can substitute $N_d = 65\,536$ into our equation:

$2^{(d+1)} = 65\,536$

Now we take logarithm base 2 of both sides to simplify the equation:

$\log_2(2^{(d+1)}) = \log_2(65\,536)$

$d + 1 = 16$

$d = 15$

So there will be 65 536 cells after 15 divisions.

A student investigates the effect of two newly developed fertilisers (A and B) on the growth of potato crops. Fourteen plots of 10 m² were sectioned off and treated with either fertiliser A or B. The table below shows the yields of potatoes from the test areas following harvest.

a) *Explain why a Student's t-test would be the most appropriate statistical test to see whether there is a difference in efficacy between fertilisers A and B.*

b) *Is there a significant difference between potato yields for plots treated with fertiliser A or B?*

Fertiliser	Test plot yield (kg)							Mean
	Plot 1	Plot 2	Plot 3	Plot 4	Plot 5	Plot 6	Plot 7	
A	25	27	34	18	21	26	28	25.6
B	17	35	42	19	35	22	44	30.6

a) The Student's t-test can be used to determine whether there is a statistically significant difference between the means of two categories (fertiliser A and B).

b) To answer this question, we need to perform a Student's t-test.

Step 1: State the null hypothesis

Our null hypothesis is: 'There is no difference in potato yield between plots treated with fertiliser A or B.'

Step 2: Calculate the observed value

We have already been given the means of the two data sets:

$$\bar{x}_A = 25.6$$
$$\bar{x}_B = 30.6$$

In order to calculate the variance of the data sets, we need to calculate the sum of squared values, $\sum x^2$, and the sum of the values squared, $(\sum x)^2$.

$$\sum x^2 \text{ for A} = (25)^2 + (27)^2 + (34)^2 + (18)^2 + (21)^2 + (26)^2 + (28)^2 = 4735$$

$$\sum x^2 \text{ for B} = (17)^2 + (35)^2 + (42)^2 + (19)^2 + (35)^2 + (22)^2 + (44)^2 = 7284$$

$$\left(\sum x\right)^2 \text{ for A} = (25 + 27 + 34 + 18 + 21 + 26 + 28)^2 = 32\,041$$

$$\left(\sum x\right)^2 \text{ for B} = (17 + 35 + 42 + 19 + 35 + 22 + 44)^2 = 45\,796$$

We can now calculate the variance (s^2) of each data set:

$$s_A^2 = \frac{[4735 - (32\,041/7)]}{6} = 26.29$$

$$s_B^2 = \frac{[7284 - (45\,796/7)]}{6} = 123.61$$

And so we can now calculate t:

$$t = \frac{\bar{x}_A - \bar{x}_B}{\sqrt{\frac{s_A^2}{n_A} + \frac{s_B^2}{n_B}}}$$

$$t = \frac{(25.6 - 30.6)}{\sqrt{\left[\left(\frac{26.29}{7}\right) + \left(\frac{123.61}{7}\right)\right]}}$$

$$t = \frac{5}{\sqrt{[(3.76) + (17.66)]}}$$

$$t = \frac{5}{4.63}$$

$$t = 1.08$$

Step 3: Decide whether to accept the hypothesis

The number of degrees of freedom is $n_1 + n_2 - 2 = 12$.

The critical value for a 5% significance level with 12 degrees of freedom is 2.18.

The observed value of 1.08 is less than the critical value. Therefore we accept the hypothesis that there is no significant difference in crop yield between plots treated with fertiliser A or B.

EXAMPLE

The inheritance of a recessive genetic disease, A, is being investigated.

a) *If the faulty gene that causes disease A is not sex-linked (i.e. the gene is not found on the X or Y chromosome), out of 300 people affected by the disease, how many would you expect to be female?*

b) *In a sample of 300 patients affected by the disease, 295 were male and 5 were female. Use an appropriate statistical test to show that the difference between the observed and expected results is not due to chance.*

c) *What can you conclude about the inheritance of disease A?*

a) If the disease is not found on either the X or Y chromosome, it would be reasonable to assume that the proportion of males and females inheriting the disorder is equal. Therefore, we would expect 150 females in a sample of 300 patients.

b) Since the question is asking to test whether an observed frequency is different to an expected frequency, we should use a chi-squared test.

Step 1: State the hypothesis

The hypothesis is: 'The observed number of male and female patients with disease A is consistent with the inheritance model.'

Step 2: Calculate the observed value

The formula for the chi-squared test is:

$$\chi^2 = \sum \frac{(O - E)^2}{E}$$

where O is your observed result, E is the expected result and the symbol ($\sum$ means 'the sum of'.

$$\chi^2 = \frac{(295 - 150)^2}{150} + \frac{(5 - 150)^2}{150} = 140.2 + 140.2 = 280.3$$

Step 3: Decide whether to accept the hypothesis

The number of degrees of freedom is $n - 1 = 2 - 1 = 1$.

The critical value for a 5% significance level with 1 degree of freedom is 3.841.

The observed value is greater than the critical value. Therefore we must reject the hypothesis. The difference between the observed and expected numbers of male and female patients is not down to chance.

c) A patient with disease A is much more likely to be male than female. This suggests that the disease is found on the X chromosome. Since males have only one X chromosome, they only need to inherit a single copy of the gene to have symptoms. Females would need to inherit two copies, which is far less likely.

Preparing for your exams

Introduction

- A Level students will sit three exam papers, each covering content from both years of A Level learning. The third paper will consist of synoptic questions that may draw on two or more different topics.
- A Level students will also have their competence in key practical skills assessed by their teacher in order to gain the Science Practical Endorsement. The endorsement will not contribute to the overall grade but the result (pass or not classified) will be recorded on the certificate.

The table below gives details of the exam papers.

A Level exam papers

Paper	Paper 1: Biological processes	Paper 2: Biological diversity	Paper 3: Unified biology	Science Practical Endorsement
Topics covered	Modules 1,2,3 and 5	Modules 1,2,4 and 6	Modules 1–6	Assessed by teacher throughout course. Does not count towards A Level grade but result (pass or not classified) will be reported on A Level certificate. It is likely that you will need to maintain a separate record of practical activities carried out during the course.
% of the A Level qualification	37%	37%	26%	
Length of exam	2 hours 15 minutes	2 hours 15 minutes	1 hour 30 minutes	
Marks available	100 marks	100 marks	70 marks	
Question types	15 marks multiple-choice followed by 85 marks structured response and extended writing including two level of response questions	15 marks multiple-choice followed by 85 marks structured response and extended writing including two level of response questions	Structured response and extended writing including two level of response questions	
Experimental methods	Yes	Yes	Yes	
Mathematics	A minimum of 10% of the marks across all three papers will be awarded for mathematics at GCSE higher tier level or above			

Exam strategy

Arrive equipped

Make sure that you have all of the correct equipment needed for your exam. In a transparent bag or pencil case, you should have at least:

- pen (black, ink or ball-point pen)
- pencil (HB)
- ruler (ideally 30 cm)
- rubber (make sure it's clean and doesn't smudge the pencil marks or rip the paper)
- calculator (scientific)

Ensure your answers can be read

Your handwriting does not have to be perfect but the examiner must be able to read it! When you're in a hurry it's easy to write key words that are difficult to decipher.

Plan your time

Note how many marks are available on the paper and how many minutes you have to complete it. This will give you an idea of how long to spend on each question. Be sure to leave some time at the end of the exam for checking answers. A rough guide of a minute a mark is a good start, but short answers and multiple choice questions may be quicker. Longer answers might require more time.

Understand the question

Always read the question carefully and spend a few moments working out what you are being asked to do. The command word used will give you an indication of what is required in your answer.

Be scientific and accurate, even when writing longer answers. Use the technical terms you have been taught.

Always show your working for any calculations. Marks may be available for individual steps, not just for the final answer. Also, even if you make a calculation error, you may be awarded marks for applying the correct technique.

Plan your answer

Questions marked with an * are the level of response questions. The examiners will be looking for a line of reasoning in your response. Here, marks will be awarded for your ability to logically structure your answer showing how the points that you make are related or follow on from each other where appropriate. Read the question fully and carefully (at least twice!) before beginning your answer.

Make the most of graphs and diagrams

Diagrams and sketch graphs can earn marks – often more easily and quickly than written explanations – but they will only earn marks if they are carefully drawn and fully annotated.

- If you are asked to read a graph, pay attention to the labels and numbers on the x- and y-axes. Remember that each axis is a number line.
- If asked to draw or sketch a graph, always ensure you use a sensible scale and label both axes with quantities and units. If plotting a graph, use a pencil and draw small crosses (X) or dots with a circle around them ($\odot$) for the points.
- Diagrams must always be neat, clear and fully labelled or annotated.

Check your answers

For open-response and extended writing questions, check the number of marks that are available. If three marks are available, have you made three distinct points? It can be helpful to number or bullet-point your response..

For calculations, read through each stage of your working. Substituting your final answer into the original question can be a simple way of checking that the final answer is correct. Another simple strategy is to consider whether the answer seems sensible. Pay particular attention to using the correct units.

Sample answers with comments

Question type: multiple choice

Processes i–v are steps that occur during respiration

i) *Glycolysis*
ii) *Link reaction*
iii) *Krebs cycle*
iv) *Oxidative phosphorylation*
v) *Fermentation*

Which row identifies the precise location at which each process takes place?

	Cytoplasm	Mitochondrial matrix	Inner mitochondrial membrane
A	v	i, ii, iii	iv
B		i, ii, iii, v	iv
C	i, v	ii, iii	iv
D	i, v	iv	ii, iii

Your answer:

Question analysis

- Multiple choice questions look easy until you try to answer them. Very often they require some working out and thinking.
- In multiple choice questions you are given the correct answer along with three or four incorrect answers (called distractors). You need to select the correct answer and write the appropriate letter in the box provided.
- If you change your mind, put a line through the box and write your new answer next to it.

> Think carefully about what each process involves. Remember that the mitochondria are involved only in aerobic respiration. You are allowed to write on the exam paper and it may be a good idea to write the location of each process on the paper next to the names of the processes – this will help to avoid being confused by the numbering system.

> Multiple choice questions always have one mark and the answer is given! For this reason students often make the mistake of thinking that they are the easiest questions on the paper. Unfortunately, this is not the case. These questions often require several points to be considered and an error in one of them will lead to the wrong answer being selected. The three incorrect answers supplied (distractors) will feature the answers that students arrive at if they make typical or common errors. The trick is to answer the question before you look at any of the answers.

Sample student answer

A

> Many students learn the biochemical pathways without considering the important aspect of where each occurs. Many students believe that all the pathways of aerobic respiration occur in the mitochondria – hence answers A and B.

Verdict

This is a weak answer because:

- This candidate has probably recalled that respiration occurs in the mitochondria – so the candidate thinks that glycolysis, link reaction and Krebs cycle occur in the matrix while the oxidative phosphorylation is on the inner membrane. This leaves fermentation for the cytoplasm.
- Fermentation actually uses the products of glycolysis so it makes sense that glycolysis also occurs in the cytoplasm and its products then enter fermentation or move into the mitochondrion.

If you have any time left at the end of the paper go back and check your answer to each part of a multiple choice question so that a slip like this does not cost you a mark.

Question type: extended writing level of response

Describe how the collecting ducts in the mammalian kidney contribute to the process of osmoregulation and how this process is achieved in someone on dialysis. [9]

Question analysis

- With any question worth four or more marks, think about your answer and the points that you need to make before you write anything down. It may be worth writing a few notes to help organise your thoughts – these could be written at the back of the examination paper.

- Keep your answer concise, and the information you write relevant to the question. You will not gain marks for writing biology that is not relevant to the question (even if correct) but it will cost you time.

- Remember that you can use bullet points or annotated diagrams in your answer.

- There will be at least two questions in each examination paper which will test your ability to organise your response with a clear and logical structure.

- These questions will be allocated either 6 or 9 marks.

- Your response will be assessed for the level of organisation and whether the information presented is relevant.

  - For level 1 (1–2 marks or 1–3 marks) the information is basic and presented in an unstructured way.

  - For level 2 (3–4 marks or 4–6 marks) there is a line of reasoning and some structure is evident. The information is for the most part relevant and is supported by some evidence.

  - For level 3 (5–6 marks or 7–9 marks) there is a well-developed line of reasoning and a logical structure. The information is relevant and substantiated.

- It helps with longer extended writing questions to think about the number of marks available and how they might be distributed.

Nine marks are available so nine points need to be made. The command word is 'describe' so there is no need to explain; however, a high level of detail will be expected.

You are given some hints in the question about what to write about – the terms 'collecting ducts', 'osmoregulation' and 'dialysis' are important.

You need to organise your answer so that it provides information about the collecting ducts, it describes osmoregulation and describes how osmoregulation is achieved through dialysis. Your response should follow a logical line of reasoning.

Sample student answer

The collecting ducts carry urine from the tubules in the cortex of the kidney down towards the ureter. As it passes down through the medulla, water is reabsorbed from the urine to make it more concentrated. The walls of the collecting ducts are permeable to water so that it can be absorbed. ADH makes the wall more permeable by inserting many aquaporins into the wall of the collecting duct so that more water can be reabsorbed and the urine gets more concentrated.

During dialysis the blood is taken out of the body and mixed with dialysis fluid. The dialysis fluid is carefully prepared so that it has all the right levels of sugars, salts and water so that the blood that returns to the body has all these levels corrected.

You need to write a section about the collecting ducts and how they allow water to be reabsorbed into the blood. This should be followed by a section about how the amount of water reabsorbed can be increased or decreased as and when required. Finally, you should provide information about dialysis.

Verdict

This is an average to weak answer because:

- The student clearly has a reasonable level of knowledge.
- The student has kept to relevant detail.
- The student has organised the information with a clear line of thought.
- The detail about inserting aquaporins is not quite accurate – these are inserted into the cell surface membranes of the cells in the wall of the collecting ducts.
- The candidate has not specified when the permeability of the collecting ducts should be increased or decreased, i.e. when is ADH released?
- The candidate has described blood mixing with dialysis fluid rather than being separated by a selectively permeable membrane. The role of osmosis is important for osmoregulation – the water potential of the dialysis fluid is very carefully regulated to ensure that the blood returning to the body has a suitable water potential.

Question type: Practical question

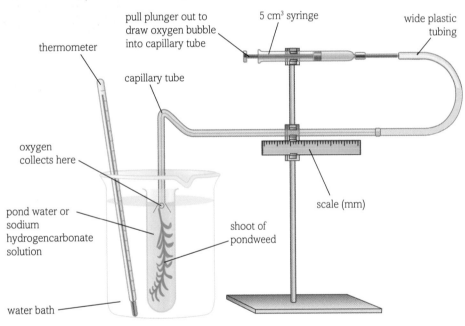

The diagram shows a photosynthometer, a device used to measure the rate of photosynthesis.

Describe how the apparatus could be used to investigate the effect of changing the light intensity on the rate of photosynthesis.

[4]

> Notice that the question says 'describe how the apparatus could be used'. The command word 'describe' requires a suitable level of detail. You should also notice that you are asked how to investigate the effect of changing the light intensity – don't forget to describe how the intensity would be changed.

Question analysis

- There will be questions in your exams which assess your understanding of practical skills and draw on your experience of the core practicals. For these questions, think about:
  - how apparatus is set up
  - the method of how the apparatus is to be used
  - how readings are to be taken

- how to make the readings reliable
- how to control any variables
- possible limitations and improvements.

● It helps with extended writing questions to think about the number of marks available and how they might be distributed.

● It is vital to plan out your answer before you write it down. There is always space given on an exam paper to do this so just jot down the points that you want to make before you answer the question in the space provided. This will help to ensure that your answer is coherent and logical and that you don't end up contradicting yourself.

Sample student answer

Shine light on the plant and allow the plant to acclimatise to the light intensity. Start the timer and collect the oxygen produced. Collect the gas for two minutes.
The gas moves along the capillary tube and the rate of movement can be used as a measure of the rate of photosynthesis. Repeat this measurement two times and calculate a mean value. Move the light to another position so that the intensity changes. Collect the gas for a further two minutes and measure the new rate of photosynthesis. Repeat for five different positions of the light. Suitable distances to set the light from the plant might be 5, 10, 15, 20 and 25 cm.

A suitable level of detail means stating how long you should allow for acclimatisation, how long to collect oxygen for and what distances the light should be set from the plant. You should also describe clearly how the rate of photosynthesis can be measured from the scale provided. It is also important to remember that the readings at each light intensity should be repeated to assess the reliability of the data.

Verdict

This is an average to good answer because:

● The student has stated the need for acclimatisation but has not quoted a length of time.

● The student has stated how long to allow photosynthesis to continue (2 minutes).

● The student has described repeating the readings to assess reliability.

● The student has suggested moving the light to alter the light intensity for five different levels and has even stated clearly what these distances should be.

● However, the student has misinterpreted how the apparatus is used. While the plant is photosynthesising the gas collects in the flared end of the capillary tube above the plant. The bubble collected can then be drawn into the capillary by pulling out the plunger of the syringe. The length of the bubble measured along the scale provided is a measure of the volume of oxygen collected in two minutes.

Question type: Structured question

Samphire Hoe is a Nature Reserve created at the foot of the Dover cliffs by using spoil from the channel tunnel. The soil dug out from under the sea was very low in organic matter and the minerals needed by plants. Samphire Hoe now contains a community of unusual diversity which includes a number of rare species.

(a) (i) *Mineral ions such as nitrates and phosphates are taken into plant roots by active transport. What component of a cell surface membrane enables active uptake?* [1]

   (ii) *State two organelles found in the cells of the root that contribute to the uptake of ions.* [2]

(b) *Outline how the energy used for active uptake is released from starch in the cells of the root.* [5]

(c) *Samphire Hoe is managed carefully. One aspect of that management is allowing a number of sheep and cattle to graze on the plants. Explain what effect the grazing will have on the biodiversity.* [3]

Question analysis

- The important thing to remember is that some questions will test different areas of the specification together.
- Biology is full of links between areas of the specification and all the areas tested will be linked in some way.
- Treat each part of the question separately and consider carefully the underlying biology.
- The question is introduced in the guise of a nature reserve. The question may lead on to aspects of conservation of biodiversity but parts a) – c) are not testing that part of the specification.

Sample student answer

(a) (i) proteins (ii) mitochondria and ribosomes

(b) The starch must be hydrolysed to glucose by enzyme action. The glucose then enters glycolysis and is converted to pyruvate. The pyruvate enters Krebs cycle in the mitochondrion and the energy is converted to ATP. This involves the release of hydrogens which pass along the electron transfer chains in the mitochondrial membranes causing chemiosmosis.

(c) The sheep will eat certain plants preventing their growth – so the biodiversity will be lower.

Verdict

This is an average answer because:

- The student has recalled the AS knowledge for part a)
- In part b) the student has remembered to include the breakdown of starch to glucose and has used a suitable term – hydrolysis.
- The student has realised that release of energy is through respiration and has given a fair outline of the process.
- However, towards the end of the description the student has become a little confused. Hydrogen atoms are split to a proton and an electron – it is this electron that passes down the electron transfer chain. Also the term chemiosmosis is not used clearly. The process of chemiosmosis involves the movement of hydrogen ions to create a concentration gradient which is then used to combine a phosphate group with ADP to make ATP.
- In part c) the student has not realised that grazing removes competition from dominant species, allowing smaller or less vigorous plants to grow in greater numbers – this increases biodiversity.

Question type: Calculation

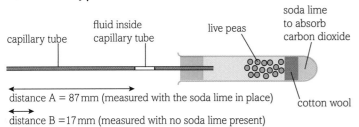

distance A = 87 mm (measured with the soda lime in place)

distance B = 17 mm (measured with no soda lime present)

The diagram shows a simple respirometer, an apparatus used to measure the rate of respiration.

As the peas respire they use oxygen and release carbon dioxide. The fluid in the capillary moves towards the peas. The distance moved along the capillary is a measure of the volume of gas absorbed.

The apparatus can be used to calculate the RQ value:

RQ = volume of carbon dioxide released / volume of oxygen absorbed.

Calculate the RQ value for these peas. [2]

Remember that the movement of the fluid is caused by absorption of gases.

You must interpret the diagram to work out what figure is used to measure the oxygen consumption and how the volume of carbon dioxide released must be calculated.

When the soda lime is in place the change in volume is caused by uptake of oxygen by the peas.

When the soda lime has been removed the change in volume is caused by a combination of oxygen uptake and release of carbon dioxide.

Question analysis

- You are provided with the formula to calculate RQ.
- The information in the diagram must be interpreted to work out what numbers must be used to calculate the RQ.
- In this case the actual maths component is relatively simple – but interpreting what figures to use may be more difficult.

Sample student answer

RQ = 17 / 87 = 0.2

The student has used the figures supplied without attempting to work out what the figures mean.

Verdict

This is a weak answer because:

- The movement of the fluid measured as distance B is caused by absorption of oxygen and release of carbon dioxide.
- The movement measured as distance A is caused by the absorption of oxygen.
- The release of carbon dioxide will push the fluid away from the peas. When the soda lime is in place the carbon dioxide is absorbed and this movement does not take place.
- When the soda lime is removed the fluid moves more slowly as oxygen is absorbed and carbon dioxide is released. Oxygen absorption moves the fluid towards the peas and carbon dioxide release moves the fluid away from the peas.
- The movement caused by the release of carbon dioxide can be calculated by subtracting distance B from distance A. Distance moved by fluid caused by carbon dioxide release = 87 – 17 mm = 70 mm.
- Therefore the correct calculation of RQ is given by: RQ = (87 – 17) / 87 = 0.8.

Glossary

abiotic components: components of an ecosystem that are non-living.

abiotic factors: non-living components of an ecosystem that affect other living organisms.

accelerans nerve: branch of the sympathetic nervous system that speeds up the heart rate.

acetylcholine: chemical that acts as a transmitter by diffusing across the synaptic cleft.

acetylcholinesterase: an enzyme in the synaptic cleft that breaks down acetylcholine. After it has triggered an action potential in the post-synaptic neurone, acetylcholine must be broken down otherwise it would remain in the synaptic cleft and would continue to open sodium ion channels in the post-synaptic membrane, causing action potentials.

acinus (pl. acini): small group of exocrine cells in a sac-like structure; in the pancreas these surround a tubule and secrete into the tubule.

actin and myosin: proteins involved in muscular contraction.

action potential: a brief reversal of the potential across the membrane of a neurone causing a peak of +40 mV compared to the resting potential of -60 mV.

adenyl cyclase: an intracellular enzyme that converts ATP to cyclic AMP (cAMP).

ADP: adenosine diphosphate.

adrenal cortex: the outer layer of the adrenal gland.

adrenal gland: one of a pair of glands lying above the kidneys, which release adrenaline and a number of other hormones known as corticoids (or corticosteroids) such as aldosterone.

adrenaline: a hormone released from the medulla of the adrenal glands, which stimulates the body to prepare for fight or flight.

adrenal medulla: the inner layer of the adrenal gland.

adrenocorticotropic hormone (ACTH): a hormone released by the pituitary, which stimulates the adrenal cortex to produce corticosteroid hormones.

adsorption: adhesion/binding to a surface.

aerobic: in the presence of oxygen.

agar: a polysaccharide of galactose obtained from seaweed, which is used to thicken the medium into a gel.

agarose: a type of sugar that can be incorporated into a type of agar gel.

alkaloids: organic nitrogen-containing bases that have important physiological effects on animals; includes nicotine, quinine, strychnine and morphine.

allele: a version of a gene.

allele frequency: proportion of a particular allele within the gene pool.

allopatric speciation: formation of two different species from one original species, due to geographical isolation.

alpha (α) cells: cells found in the islets of Langerhans that secrete glucagon.

ammonification: the production of ammonia by bacterial action in the decay of nitrogenous organic matter.

AMP: adenosine monophosphate.

anaerobic: in the absence of oxygen.

aneuploidy: abnormal chromosome number.

antagonist: something that works against another in opposite pairs, such as muscles arranged in opposing pairs, so one contracts as the other elongates.

antidiuretic hormone (ADH): a hormone, made in the hypothalamus, stored and released from the pituitary gland, that controls the permeability of the collecting duct walls in the kidneys.

apical dominance: inhibition of lateral buds further down the shoot by chemicals produced by the apical bud at the tip of a plant shoot.

apoptosis: programmed cell death.

artificial selection: selective breeding of organisms to produce desired phenotypes.

ascending limb: part of the loop of Henle that carries fluid back into the distal tubule in the cortex.

aseptic technique: sterile techniques used in culturing and manipulating microorganisms.

ATP: adenosine triphosphate.

autoclave: a pressurised device (pressure cooker) used to sterilise equipment by heating to 120 °C.

autoimmune response: a response in which the body's immune system attacks and destroys some of its own cells.

autonomic nervous system: part of the nervous system responsible for controlling the involuntary motor activities of the body.

autoradiograms: photographs made when photographic film is exposed to molecules labelled with radioactivity.

autosomal linkage: gene loci present on the same autosome (non-sex chromosome) that are often inherited together.

autosomes: chromosomes not concerned with sex determination.

autotrophic nutrition: type of nutrition where organic molecules are synthesised from inorganic molecules such as carbon dioxide and water; photosynthesis is one type of autotrophic nutrition where sunlight energy is converted to chemical energy that is used to synthesise large organic molecules from small inorganic molecules; chemosynthesis uses energy derived from chemical reactions (e.g. nitrifying bacteria in soil).

auxins: plant hormones responsible for regulating plant growth.

belt transect: continuous belt or series of quadrats used to estimate the distribution of organisms across a certain area such as a meadow or seashore.

beta (β) cells: cells found in the islets of Langerhans that secrete the hormone insulin.

bioinformatics: branch of biology that deals with storing, displaying and using large quantities of often complex data.

biomass transfer: transfer of biomass from one trophic level to another.

biotechnology: the use of living organisms or parts of living organisms in industrial processes. This could be to produce food, drugs or other products.

biotic components: components of an ecosystem that are living.

biotic factors: environmental factors associated with living organisms in an ecosystem that affect each other, e.g. predation, disease, competition, food availability.

blunt ends: cut ends of DNA where there is no staggered cut and no exposed unpaired bases.

bulb: an underground swollen stem with stored food and a bud; used for vegetative propagation.

Calvin cycle: metabolic pathway of the light-independent stage of photosynthesis, occurring (in eukaryotic cells) in the stroma of chloroplasts where carbon dioxide is fixed, with the products of the light-dependent stage, to make organic compounds.

cardiac muscle: muscle found in the heart walls.

cardiovascular centre: a part of the medulla oblongata in the brain that controls heart rate and other aspects of circulation.

carotid sinus: a small swelling in the carotid artery with stretch receptors in its walls.

carrying capacity: the maximum population size that can be maintained over a period of time in a particular habitat.

casein: the protein in milk.

cell signalling: the way in which cells communicate with each other.

cell surface membrane (plasma membrane): the membrane that surrounds every cell, forming the selectively permeable boundary between the cell and its environment.

central nervous system (CNS): the central part of the nervous system composed of the brain and spinal cord.

cerebellum: region of the brain coordinating balance and fine control of movement.

cerebral cortex: the outer layer of the cerebrum.

cerebrum: region of the brain dealing with the higher functions such as conscious thought; it is divided into two cerebral hemispheres.

chemiosmosis: flow of protons, down their concentration/electrochemical gradient, across a cell membrane, through a channel associated with ATP synthase, resulting in the formation of ATP.

chemoautotrophic bacteria: bacteria which derive energy from oxidation of certain inorganic compounds.

chemoreceptors: sensory receptors that detect changes in the concentration of chemicals, for example in the blood; they transduce a chemical signal into an electrical signal.

chemotropism: a directional growth response, in plants, to chemicals.

chi-squared test: statistical test designed to find out if the difference between observed and expected data is significant or due to chance.

cholinergic synapse: a synapse that uses acetylcholine as its neurotransmitter.

climax community: the final stable community that exists after the process of succession has occurred.

clones: genetically identical organisms or cells.

closed culture: a culture for growing microorganisms which has no exchange of nutrients or gases with the external environment.

coagulate: clump together.

codominant/codominance: where both alleles present in the genotype of a heterozygous individual contribute to the individual's phenotype.

coenzymes: organic non-protein molecules that act as cofactors, which aid enzymes catalysing biochemical reactions.

community: populations of different species living in the same place and time, that can interact with each other.

compensation point: where the rate of photosynthesis, in an organism, equals the rate of its respiration. There is no net loss or gain of carbohydrate.

conservation: maintenance of biodiversity, including diversity between species, genetic diversity within species, and maintenance of a variety of habitats and ecosystems.

conserved: has remained in all descendent species throughout evolutionary history.

consumers: any of the heterotrophic organisms in a food chain or web; primary consumers feed on plants, and are eaten by secondary consumers. These in turn are eaten by tertiary consumers.

continuous variation: variation that produces phenotypic variation where the quantitative traits vary by very small amounts between one group and the next.

coppicing: cutting a tree trunk close to the ground to encourage new growth.

corms: underground stem used for vegetative propagation.

corneal reflex: a protective blinking response resulting from physical stimulation of the cornea.

corpus callosum: nerve tract that connects the cerebral hemispheres of the brain.

corticotropin-releasing hormone (CRH): also known as corticotropin-releasing factor, is a hormone that causes release of adrenocorticotropic hormone from the pituitary gland.

cotyledon: the seed leaf of a plant embryo; acts as a food store for the developing plant during germination; in some dicotyledonous plants the cotyledons appear above ground after germination and act as the first leaves.

cranial reflex: a reflex mediated by neurones that pass into the brain.

creatine phosphate: an organic compound in muscle that acts as a store of phosphates and can supply phosphates to make ATP rapidly within muscle cells.

cristae: inner highly-folded mitochondrial membrane.

curd: the solid formed by coagulating milk proteins.

cyclic AMP (cAMP): a secondary messenger released inside cells to activate a response.

cyclic changes: rhythmic changes, such as tides, day length, and fluctuations in predator and prey species.

dark band: in striated muscle – also called the A band, which shows the length of the myosin filaments; there is some overlap with the action filaments within this band.

death (decline) phase: population size is decreasing as cells are dying faster than new ones are formed.

decarboxylation: removal of a carboxyl group from a substrate molecule.

decomposers: living organisms that feed on waste or dead organic matter.

deflected succession: when succession is stopped or interfered with, such as by grazing or when a lawn is mowed.

dehydrogenation: removal of hydrogen atoms from a substrate molecule.

depolarisation: where the inside of a cell becomes less negatively charged compared with the outside.

descending limb: part of the loop of Henle that carries fluid from the proximal tubule into the medulla.

diabetes mellitus: a condition in which blood glucose concentrations cannot be controlled effectively.

dihybrid: involving two gene loci.

directional changes: non-cyclic changes, in one direction only, such as coastal erosion.

directional selection: a type of natural selection that occurs when an environmental change favours a new phenotype and so results in a change in the population mean.

discontinuous variation: genetic variation producing discrete phenotypes – two or more non-overlapping categories.

DNA ligase: enzyme that catalyses the joining of sugar and phosphate groups within DNA.

DNA sequencing: a technique that allows genes to be isolated and read.

dry mass: the amount of biomass remaining after removal of all water.

ecosystem: a community of animals, plants and bacteria interrelated with the physical and chemical environment.

ectotherm: an organism that relies on external sources of heat to maintain its body temperature.

effector: a cell, tissue or organ that brings about a response.

electrofusion: application of an electric field to a recipient cell to make it more receptive to the vector carrying a novel gene.

electron carriers: molecules that can accept one or more electrons and then donate those electrons to another carrier. Proteins embedded in thylakoid membranes of chloroplasts and within cristae of mitochondria are electron carriers, and form an electron transport chain or system. Ferredoxin, NAD and NADP are also electron carriers.

electron transport chains: chain of iron-containing proteins that can undergo oxidation and reduction, accepting and donating electrons, thus passing the electrons along the chain. Electron transport chains are involved in the production of ATP.

electrophoresis: process used to separate proteins or DNA fragments of different sizes.

electroporation: method for introducing a vector with a novel gene into a cell; a pulse of electricity makes the recipient cell membrane more porous.

embryo twinning: splitting an embryo to create two genetically identical embryos.

endocrine glands: ductless glands from which hormones are released directly into the blood.

endocrine system: a communication system involving hormones as signalling molecules.

endotherm: an organism that uses heat from its metabolic reactions to maintain body temperature.

energy: property of objects, able to be transferred but not created nor destroyed; the capacity to do work.

entrapment: a technique in biotechnology that keeps the enzymes separate from the reaction mixture.

enucleation: removal of the cell nucleus.

envelope: double membrane.

epistasis: interaction of non-linked gene loci where one masks the expression of the other. From the Greek, *ephistanai*, meaning 'stoppage'.

ethanal: formerly acetaldehyde, CH_3CHO.

ethanol: ethyl alcohol C_2H_5OH.

excitatory post-synaptic potential (EPSP): a small post-synaptic potential that makes the neurone more likely to fire an action potential.

excretion: the removal of metabolic waste from the body.

exergonic: biochemical reaction that releases energy.

exocrine glands: glands that secrete substances into a duct.

exon: the coding, or expressed, region of DNA; a nucleotide sequence within a gene that remains in the final mRNA transcribed from that gene.

explant: tissue that is transferred from a plant and placed in a culture medium. It is usually a small part of the meristem.

facultative anaerobe: organism that will respire aerobically in the presence of oxygen, but can respire anaerobically in the absence of oxygen.

fermentation: the breakdown of organic molecules in the absence of oxygen; anaerobic respiration.

fermenter: a vessel used to grow populations of microorganisms.

first messengers: hormones that act as signalling molecules outside the cell that bind to the cell surface membrane and initiate an effect inside the cell.

founder effect: when a small sample of an original population establishes in a new area; its gene pool is not as diverse as that of the parent population.

frameshift: alteration to the base sequence of a length of DNA due to an insertion or deletion of a nucleotide base.

gas chromatography: a technique whereby a sample is vaporised in the presence of a gaseous solvent and passed into a long tube that is lined by an absorption agent. Each substance dissolves differently in the gas and stays in the gas phase for a unique, specific time, called the retention time. Eventually the substance comes out of the gas and is absorbed onto the agent lining the tube. This is then analysed to create a chromatogram.

gene locus: position of a gene on a chromosome/pair of homologous chromosomes.

gene pool: all the alleles and genes within a breeding population.

generator potential: the change in potential across a receptor membrane.

genetic bottleneck: a sharp reduction in size of a population due to environmental catastrophes such as earthquakes, floods, disease or human activities such as habitat destruction, overhunting or genocide, which reduces genetic diversity. As the population expands again it is less genetically diverse than before.

genetic modification: manipulating the genome of an organism.

genotype: genetic makeup of an organism.

geotropism: a directional growth response, made by plants, to gravity.

germ line gene therapy: gene therapy by inserting functional alleles into gametes or zygotes.

gibberellins: plant hormones that are responsible for control of stem elongation and seed germination.

glomerular filtration rate (GFR): the rate at which fluid enters the nephrons.

glucagon: a hormone that causes the breakdown of glycogen to glucose producing an increase in blood glucose concentration.

gluconeogenesis: process where amino acids and fats are converted into additional glucose.

glycerate-3-phosphate (GP): an intermediate compound in the Calvin cycle.

glycogenesis: the conversion of glucose to glycogen for storage.

glycogenolysis: glycogen converted to glucose by phosphorylase A.

glycolysis: first stage of respiration; a metabolic pathway that converts glucose to pyruvate.

granum (pl. grana): inner part of chloroplasts made of stacks of thylakoid membranes, where the light-dependent stage of photosynthesis takes place.

gross primary productivity: the rate at which plants convert light energy into chemical energy through photosynthesis.

growth medium: a liquid or jelly containing nutrients required for growth of microorganisms.

habitat: the place where an organism lives.

habituated: reduced neural signalling following repeated stimulation. This is due to reduced release of pre-synaptic neurotransmitter vesicles.

hazard: potential cause of harm or damage.

heterozygotes: individuals that have inherited different alleles at one or more gene loci and do not breed true for that/those characteristics.

hexose bisphosphate: 6-carbon sugar with two phosphate groups attached at C1 and C6; an intermediate compound in the glycolysis pathway.

heterozygous: not true-breeding; having different alleles at a particular gene locus on a pair of homologous chromosomes.

homeobox sequences: sequence of 180 base pairs (excluding introns) found within genes that are involved in regulating patterns of anatomical development in animals, fungi and plants.

homeodomain sequence: amino acid sequence, of 60 amino acids, encoded by the homeobox sequence. This sequence assumes a particular shape and can bind to DNA, acting as a transcription factor and regulating gene expression of genes involved in development.

homeostasis: maintaining a constant internal environment, despite changes in external and internal factors.

homologous chromosomes: matched in size and containing the same genes at the same gene loci. One of the homologous pair is of maternal and the other is of paternal origin.

homozygote: individual homozygous at a particular gene locus/at particular gene loci.

homozygous: true-breeding; having identical alleles at a particular gene locus on a pair of homologous chromosomes.

hormonal system: a communication system that uses molecules transported in the blood.

hormones: molecules (protein or steroids) that are released by endocrine glands directly into the blood. They act as messengers, carrying a signal from the endocrine gland to a specific target organ or tissue.

Hox **genes:** subset of homeobox genes, found only in animals; involved in formation of anatomical features in correct locations of body plan.

hybrid vigour: increased heterozygosity giving increased vigour of fertility, growth and survival; also called heterosis.

hyperglycaemia: where blood glucose concentrations remain high for longer than normal periods, which can lead to organ damage.

hypoglycaemia: abnormally low levels of glucose in the blood; it results in inadequate delivery of glucose to the body tissues and is particularly damaging to the brain.

hypostatic: the gene whose expression is masked by an epistatic gene.

hypothalamus: the part of the brain that coordinates homeostatic responses.

H zone: in skeletal muscle cells/sarcomeres – a region in the middle of the A band where there are only myosin filaments and no overlap with actin filaments.

immobilised enzyme: an enzyme that is held in place and not free to diffuse through the solution.

inbreeding depression: decreased vigour in terms of growth, survival and fertility, after generations of inbreeding.

incubation: maintenance of controlled environmental conditions to promote growth or development of microbial or tissue cultures, e.g. at a suitable temperature.

inhibitory post-synaptic potential: synaptic potential making a post-synaptic neurone less likely to generate an action potential; neurotransmitters may bind to the post-synaptic membrane and close ion channels.

inoculation: transferring a few microorganisms to a new growth medium.

insulin: the hormone, released from the pancreas, that causes cells to take in glucose from the blood, thus causing blood glucose levels to go down.

intercalated discs: specialised cell surface membranes fused to produce gap junctions that allow diffusion of ions between cells.

intermembrane space: space between inner and outer membranes of mitochondria (or chloroplasts).

interspecific competition: competition between individuals of different species.

intraspecific competition: competition between individuals of the same species.

intron: non-coding region of DNA; sequence of nucleotides within a gene that does not remain in the mRNA transcribed from that gene.

involuntary muscle: smooth muscle that contracts without conscious control.

isomerase enzymes: enzymes that catalyse the rearrangement of atoms within a molecule.

kappa-casein: one type of milk protein that keeps the other types in soluble form.

knee-jerk reflex: a reflex action that straightens the leg when the tendon below the knee cap is tapped.

k-**strategists:** species whose population size is determined by the carrying capacity.

lactate: a product of anaerobic respiration in mammals.

lag phase: the early growth phase where growth is very slow due to the organism adapting to the environment and genes being switched on to induce enzymes; time between adding microorganisms to a culture and the rapid exponential log phase of growth.

light band: in skeletal muscle fibres – also called the I band, either side of the Z region and consisting of only actin filaments, where there is no overlap with myosin filaments.

light intensity: level of light.

limiting factor: the factor whose magnitude slows down the rate of a natural process.

linked genes: genes on the same chromosome.

liposomes: small spherical vesicle of lipid bilayer, can be used in gene therapy to introduce a functional allele into a cell.

local currents: currents in the cytoplasm of the neurone, caused by depolarisation due to sodium ions flooding back into the neurone.

log phase: the most rapid phase during which growth is exponential.

medulla oblongata: region of the brain that controls many physiological processes such as breathing rate and blood flow.

meristems: parts of the plant with unspecialised cells that can divide and develop into any kind of cell.

metabolic waste: a substance that is produced in excess by the metabolic processes in the cells; it may become toxic.

methylation: addition of methyl (CH_3) groups to a molecule.

micropropagation: growing large numbers of new plants from meristem tissue taken from a sample plant.

missense mutation: point mutation leading to a different amino acid being incorporated into the polypeptide.

mitochondrial matrix: fluid-filled inner part of mitochondria where the reactions of the Krebs cycle take place.

monoclonal antibodies: antibodies made from one type of cell, are all specific to one complementary antigen molecule.

monogenic: determined by a single gene.

motor neurones: neurones that carry an action potential from the CNS to the effector (muscle or gland).

motor system: the motor nerves conducting action potentials to the effectors.

mRNA: messenger RNA.

multiple alleles: characteristic for which there are three or more alleles in the population's gene pool.

muscle spindle: a stretch receptor in muscles that respond to changes in length.

mutagen: physical or chemical agent that increases the chance of a mutation above the spontaneous rate of 1 in 10^8 base pairs.

myofibril: cylindrical structure extending along the muscle fibre, containing protein filaments (actin and myosin).

myelinated neurone: a neurone with an individual layer of myelin around the axon.

myelin sheath: an insulating fatty layer around a neurone that consists of several layers of membrane and thin cytoplasm from a Schwann cell.

myogenic: muscle that generates its own excitation for contraction.

NADP: nicotinamide adenine dinucleotide phosphate; coenzyme and electron and hydrogen carrier.

nanotechnology: manipulation of matter on a molecular or atomic scale.

nastic response: a non-directional response, usually by plants, to a stimulus.

negative feedback: the mechanism that reverses a change, bringing the system back to the optimum.

nephron: functional unit of the kidney.

net primary productivity: the proportion of energy from the Sun available to enter the food chain.

neuromuscular junction: the structure at which a nerve meets the muscle, similar in action to a synapse.

neuronal system: an interconnected network of neurones that signal to each other across synapse junctions.

neurotransmitter: a chemical used as a signalling molecule between two neurones in a synapse.

niche: the role of an organism within its habitat.

node: a part of the stem from which leaves grow.

nodes of Ranvier: gaps in the myelin sheath at intervals of 1–3 mm.

non-myelinated neurone: a neurone with no individual layer of myelin around its axon.

nonsense mutation: point mutation leading to a truncated protein/ polypeptide due to a base triplet coding for an amino acid changing and becoming a stop triplet.

operon: a group of genes that function as a single transcription unit; first identified in prokaryote cells.

optical reflex: a reflex action resulting from stimulation of the retina – e.g. blinking due to bright light.

ornithine cycle: a series of biochemical reactions that convert ammonia to urea.

osmoreceptor: a sensory receptor that detects changes in water potential.

oxidation: removal of hydrogen atom or electrons from substrate molecules.

oxidative phosphorylation: the formation of ATP using energy released in the electron transport chain (of mitochondria) and in the presence of oxygen. It is the last stage in aerobic respiration.

Pacinian corpuscle: a pressure sensor found in the skin.

parasympathetic system: part of the autonomic system that regulates physiological functions when the body is at rest.

peripheral nervous system (PNS): the sensory and motor nerves connecting the sensory receptors and effectors to the CNS.

Petri dish: a shallow plastic or glass dish used to contain agar gel for culturing microorganisms.

phenotype: visible characteristic of an organism.

pheromone: any chemical substance released by one living thing, which influences the behaviour or physiology of another living thing.

phosphorylation: addition of one or more phosphate groups to a molecule.

photolysis: the enzyme-catalysed splitting of water in the presence of light.

photoperiod: the duration of an organism's daily exposure to light.

photophosphorylation: the generation of ATP from ADP and inorganic phosphate, in the presence of light.

photosynthetic pigment: pigment that absorbs specific wavelengths of light and traps the energy associated with the light; such pigments include chlorophyll a and b, carotene and xanthophyll.

photosynthometer: apparatus for collecting and measuring a volume of oxygen produced by a plant in unit time.

photosystem: system of photosynthetic pigments found in thylakoids of chloroplasts; each photosystem contains about 300 molecules of chlorophyll that trap photons and pass their energy to a primary pigment reaction centre, a molecule of chlorophyll a, during the light-dependent stage of photosynthesis.

phototropism: a directional growth response, made by plants, to light.

P_i: inorganic phosphate.

pioneer species: the species that begin the process of succession, often colonising an area as the first living things.

pituitary gland: endocrine gland at the base of the brain, below but attached to the hypothalamus; the anterior lobe secretes many hormones; the posterior lobe stores and releases hormones made in the hypothalamus.

plasma membrane (cell surface membrane): the membrane that surrounds every cell, forming the selectively permeable boundary between the cell and its environment.

plasmids: small loops of DNA in prokaryotic cells.

polarised: cell membrane more negatively charged inside compared with the outside.

polygenic: characteristic determined by two or more gene loci.

polymerase chain reaction (PCR): a biomedical technology in molecular biology that can amplify a short length of DNA to thousands of millions of copies.

polymorphism: having more than one form – may refer to variations within the DNA sequences of a particular gene/segment of DNA.

polyploidy: having more than two sets of chromosomes.

pons: a part of the brain stem that connects the cerebellum and cerebrum.

population: members of a species living in the same place and at the same time, that can interbreed.

positive feedback: the mechanism that increases a change, taking the system further away from the optimum.

potassium ion channels: channels in a cell membrane specific to potassium ions.

preservation: maintenance of habitats and ecosystems in their present condition, minimising human impact.

pre-synaptic bulb (or pre-synaptic knob): the swelling at the end of the pre-synaptic neurone.

primary consumers: in food chains and webs – herbivores, that feed on plants, and are eaten by carnivores.

primers: short (10–20 base pairs) single-stranded sequences of DNA needed for sequencing reactions and for polymerase chain reactions.

producers: organisms such as plants that produce biomass by converting energy from the Sun into chemical energy in biological molecules.

productivity: the rate of production of new biomass by producers.

promoter region: part of a DNA molecule to which RNA polymerase can bind and initiate transcription of a gene.

pyramid of biomass: diagrammatic representation of biomass at different trophic levels of a food chain; the area of each bar in the pyramid is proportional to the dry mass of all individuals at that trophic level.

pyramid of numbers: diagrammatic representation of the numbers of organisms at each trophic level in a food chain; the area of each bar in the pyramid is proportional to the number of individuals, at that trophic level.

pyruvate: 3-carbon compound; final product of glycolysis. It is the carboxylate anion of pyruvic acid ($CH_3COCOOH$), formula CH_3COCOO^-.

quadrat: a sampling plot, usually one square metre, used to study and analyse species present in a habitat.

recombinant DNA: a composite DNA molecule created *in vitro* by joining foreign DNA with a vector molecule such as a plasmid.

recombinant gametes: gametes containing new combinations of alleles due to crossing over at prophase 1 of meiosis.

reduction: addition of hydrogen atoms or electrons to substrate molecules.

reflex action: a response that does not involve any processing by the brain.

reflex arc: a direct pathway that does not involve any thought processes in the higher parts of the brain.

relay neurones: neurones within the central nervous system that allow sensory neurones to communicate with motor neurones.

releasing factors: hormones made in the hypothalamus, such as corticotropin-releasing factor and thyrotropin-releasing factor, which stimulate the release of another hormone from the pituitary gland.

renal dialysis: the most common treatment for kidney failure; a mechanism used to artificially regulate the concentrations of solutes in the blood.

rennet: a substance extracted from the stomachs of calves that contains the enzyme rennin (chymosin).

rennin (chymosin): a coagulating enzyme found in rennet, that can curdle milk.

respiratory substrate: an organic substance that can be oxidised by respiration, releasing energy to make molecules of ATP.

respirometer: apparatus used to measure the rate of respiration of living organisms, by measuring the rate for exchange of oxygen and carbon dioxide.

response: a change in an organism resulting from a stimulus.

resting potential: the potential difference across the membrane while the neurone is at rest.

restriction enzymes: endonuclease enzymes that cleave DNA molecules at specific recognition sites.

ribulose bisphosphate (RuBP): a five-carbon compound present in chloroplasts; a carbon dioxide acceptor.

risk: hazard × exposure to the hazard.

rhizomes: horizontal underground stem that sends out roots and shoots from its nodes; used for vegetative propagation.

***r*-strategists:** species where the population size increases so quickly that it can exceed the carrying capacity of the habitat before limiting factors start to affect population size.

RuBisCO: enzyme that catalyses the acceptance of carbon dioxide by ribulose bisphosphate.

runners or stolons: horizontal stems lying on the ground that can form roots and new plants at points; are used for vegetative propagation.

saltatory conduction: method of rapid impulse transmission in myelinated neurones where the action potential jumps from one node to the next.

saprotrophs: organisms, such as bacteria and fungi, that feed on dead and decaying matter.

sarcolemma: the membrane surrounding muscle fibres.

sarcomere: contractile unit of a skeletal muscle cell/fibre between two Z membranes.

sarcoplasm: specialised cytoplasm in muscle fibres.

sarcoplasmic reticulum: specialised endoplasmic reticulum in muscle fibres; stores calcium ions.

second messenger: a signalling molecule inside a cell that stimulates a change in the activity of the cell.

secondary consumers: in food chains or webs – carnivores, that eat primary consumers (herbivores).

sensory neurones: neurones that carry an action potential from the sensory receptor to the CNS.

sensory receptors: cells/sensory nerve endings that respond to a stimulus in the internal or external environment of an organism and can create action potentials.

sensory system: the sensory nerves conducting action potentials into the CNS.

serial dilution: a sequence of dilutions used to reduce the concentration of a solution or suspension.

sex chromosomes: chromosomes involved in sex determination.

sex-linked: gene present on (one of) the sex chromosomes.

silent mutation: point mutation where one base within a triplet is changed, but the triplet still codes for the same amino acid.

skeletal (striated) muscle: muscle under voluntary nerve control, attached to bones.

smooth muscle: involuntary muscle found in the walls of the digestive system and blood vessels.

sodium ion channels: protein channels in cell membranes that, when open, allow movement of sodium ions into or out of the cell.

sodium/potassium pumps: protein carriers in cell membranes that actively transport sodium ions out of the cell and potassium ions into the cell.

somatic cell gene therapy: gene therapy by inserting functional alleles into body cells.

somatic cell nuclear transfer (SCNT): a technique that involves transferring the nucleus from a somatic cell to an egg cell.

somatic nervous system: the motor neurones under conscious control.

spatial summation: summation (adding up) due to action potentials arriving from several different pre-synaptic neurones – can lead to an action potential in the post-synaptic neurone.

speciation: the splitting of a genetically similar population into two or more populations that undergo genetic differentiation and eventually reproductive isolation, leading to the evolution of two or more new species.

spinal reflex: a reflex action mediated by neurones that pass into the spinal cord.

stabilising selection: natural selection leading to constancy within a population. Intermediate phenotypes are favoured and extreme phenotypes selected against. Alleles for extreme phenotypes may be removed from the population. Stabilising selection reduces genetic variation within the population.

stationary phase: phase during which there is no population growth and the number of new cells formed is equal to the number of cells which are dying. This is when secondary metabolites are formed.

stem cells: unspecialised cells that have the potential to keep dividing by mitosis and develop into any type of cell.

sticky ends: exposed unpaired nucleotide bases on the ends of cut DNA that has been cleaved by a staggered cut.

stimulus: a change in the environment that brings about a response.

stroma: fluid-filled matrix of chloroplasts surrounding the thylakoids, where the light-independent stage of photosynthesis takes place.

substrate-level phosphorylation: production of ATP from ADP and P_i during glycolysis and the Krebs cycle.

succession: progressive change in a community of organisms over time.

suckers: new stems arising from roots of plants.

summation: occurs when the effects of several excitatory post-synaptic potentials (EPSPs) are added together.

sympathetic system: part of the autonomic nervous system that prepares the body for activity.

sympatric speciation: formation of two different species from one original species, due to reproductive isolation, while the populations inhabit the same geographical location.

symport: transport protein within cell membranes that transports two ions or molecules in the same direction.

synapse: a nerve junction.

synaptic cleft: a small gap between two neurones, approximately 20 nm wide.

synthetic biology: science of making useful biological devices and systems.

tannins: phenolic compounds, located in cell vacuoles or in surface wax on plants.

target cells: cells that possess a specific receptor on their plasma (cell surface) membrane. The shape of the receptor is complementary to the shape of the hormone (or other signalling) molecule. Many similar cells together form a target tissue.

temporal summation: summation (adding up) of several action potentials in the same pre-synaptic neurone – can lead to an action potential in the post-synaptic neurone.

tertiary consumers: animals that eat secondary consumers.

thigmonasty: a non-directional response, in plants, to the stimulation of contact.

thigmotropism: a directional response, in plants, to the stimulation of contact.

thylakoid: flattened membrane-bound sac found inside chloroplasts; contains photosynthetic pigments/photosystems and is the site of the light-dependent stage of photosynthesis.

thyroid-stimulating hormone (TSH): a hormone that stimulates the thyroid glands to release thyroxine.

thyrotropin-releasing hormone (TRH/TRF): also known as thyrotropin-releasing factor, is a hormone that causes release of thyroid stimulating hormone from the pituitary gland.

tissue culture: growing new tissues, organs or plants from certain tissues cut from a sample plant.

tissue typing: donor and recipient tissues can be typed prior to transplantation, matching tissues reduce the risk of rejection of the transplant.

totipotent: a stem cell that can differentiate into any type of cell found in the organism.

transcription: formation of mRNA using a length of DNA as a template.

transcription factor: protein or short non-coding RNA that can combine with a specific site on a length of DNA and inhibit or activate transcription of the gene.

transducer: a cell that converts one form of energy into another – in the case of sensory receptors, to an electrical impulse.

transect (line transect): a line cutting across a habitat, for sampling and measuring changes in species present.

translation: assembly of a protein from amino acids, following the code on a piece of mRNA.

transpiration: loss of water vapour from leaves.

triose phosphate (TP): a three-carbon compound, and the product of the Calvin cycle; can be used to make other larger organic molecules. Serves as an intermediate in several metabolic pathways, including the glycolysis pathway.

trophic level: the level at which an organism feeds in a food chain.

tropic hormones: hormones such as adreno-corticotropic hormone and thyroid stimulating hormone, which stimulate endocrine glands to release a hormone.

tropism: a directional growth response, made by plants, in which the direction of the response is determined by the direction of the external stimulus.

tropomyosin: in skeletal muscle fibres – a protein strand wrapped around the actin strands and involved in regulation of muscle contraction.

troponin: in skeletal muscle fibres – a globular protein with calcium binding sites, attached to tropomyosin.

tuber: underground stem used for vegetative propagation.

ultrafiltration: in the glomerulus of nephrons – filtration of the blood at a molecular level under high pressure.

unpredictable/erratic changes: extreme random changes, such as those caused by storms.

vagus nerve: a branch of the parasympathetic system stimulation of which reduces heart rate.

vector: in gene technology, anything that can carry/insert DNA into a host organism; examples of such vectors include plasmids, viruses and certain bacteria.

vegetative propagation: reproduction from vegetative parts of a plant – usually an over-wintering organ.

voltage-gated ion channels: ion channels on the cell surface membrane that open when the membrane reaches a certain stage of depolarisation.

water stress: the condition a plant will experience when water supply becomes limiting.

wet mass: entire biomass/freshmass which is variable as water content of tissue can change.

Z line: in skeletal muscle fibre – Z lines form the borders of each sarcomere; actin filaments are anchored to the Z line.

zona fasciculata: the middle layer of the cortex of the adrenal gland; secretes cortisol and corticosterone plus small amounts of androgens and oestrogens.

zona glomerulosa: the outermost layer of the adrenal gland; secretes aldosterone.

zona reticularis: innermost layer of the adrenal cortex; secretes precursor molecules, adrenal androgens and oestrogens, that are used to make sex hormones.

Index